Interdisciplinary Biotechnological Advances

This series is an authoritative resource that explains recent advances and emerging directions in biotechnology, reflecting the forefront of research clearly and reliably, without excessive hype. Each volume is written by authors with excellent reputations and acknowledged expertise in the topic under discussion. The volumes span the entire field from an interdisciplinary perspective, covering everything from biotechnology principles and methods to applications in areas including genetic engineering, transgenic plants and animals, environmental problems, genomics, proteomics, diagnosis of disease, gene therapy, and biomedicine. The significance of these applications for the achievement of UN Sustainable Development Goals is highlighted. The series will be highly relevant for Master's and PhD students in Biotechnology, Nanochemistry, Biochemical Engineering, and Microbiology, medical students, academic and industrial researchers, agricultural scientists, farmers, clinicians, industry personnel, and entrepreneurs.

Mustafa Nadhim Owaid • Yee Shin Tan
Chia Wei Phan
Editors

Mushroom Production

Springer

Editors
Mustafa Nadhim Owaid
Department of Environment, College of Applied Sciences-Hit
University of Anbar
Heet, Anbar, Iraq

Department of Heet Education, General Directorate of Education in Anbar
Ministry of Education
Heet, Anbar, Iraq

Chia Wei Phan
Department of Pharmaceutical Life Sciences
University of Malaya
Kuala Lumpur, Kuala Lumpur, Malaysia

Yee Shin Tan
Institute of Biological Sciences
University of Malaya
Kuala Lumpur, Kuala Lumpur, Malaysia

ISSN 2730-7069 ISSN 2730-7077 (electronic)
Interdisciplinary Biotechnological Advances
ISBN 978-981-95-7792-7 ISBN 978-981-95-7793-4 (eBook)
https://doi.org/10.1007/978-981-95-7793-4

This Springer imprint is published by the registered company Springer Nature Singapore Pte Ltd.
The registered company address is: 152 Beach Road, #21-01/04 Gateway East, Singapore 189721, Singapore

Foreword

Mushroom biotechnology has rapidly evolved into a multidisciplinary field intersecting microbiology, agricultural engineering, bioprocessing, and environmental sustainability. The chapters collected in *Mushroom Production* illustrate how modern technologies—robotic modular systems, bioreactors, and bio-conversion of lignocellulosic wastes—are transforming the cultivation of edible and medicinal mushrooms.

From *Pleurotus ostreatus* to *Agaricus bisporus* and *Lentinula edodes*, the species covered in this volume represent both traditional and frontier models of fungal biotechnology. The topics extend beyond cultivation into *therapeutic properties, genetic improvement, and biotechnological recycling*—providing readers with an integrated framework of sustainable production.

This book stands as a timely contribution to the scientific community, advancing global understanding of how mushrooms can address challenges of food security, waste management, and human health.

Co-Editors, University of Malaya
Kuala Lumpur, Malaysia

Yee Shin Tan,
Chia Wei Phan

Preface

The book *Mushroom Production* brings together contemporary insights, technological innovations, and cross-disciplinary applications in the field of mushroom biotechnology and sustainable cultivation systems. Mushrooms, long valued as a nutritional delicacy, have now emerged as a vital component of the global bioeconomy—serving as renewable biological tools for waste valorization, environmental remediation, nutraceutical development, and rural socioeconomic enhancement.

This volume aims to serve as both a scientific reference and a practical guide for *researchers, graduate students, and entrepreneurs* who aspire to explore and advance mushroom production from biological, technological, and industrial perspectives. The 14 chapters contained herein are authored by distinguished specialists in *biotechnology, microbiology, bioprocess engineering, and agricultural sciences*, each contributing a focused discussion on recent breakthroughs and methodologies in mushroom research and production.

Collectively, the chapters cover a diverse array of subjects: from *submerged cultivation biotechnology* and *bioreactor-based production systems* to *solid-state fermentation processes, pathogen control strategies*, and *genetic improvement for sustainable yields*. In addition, several chapters examine the *therapeutic and immunological properties* of edible and medicinal mushrooms, their role in *bio-circular economy models*, and innovative approaches for *waste-to-value conversion* using lignocellulosic residues and cereal by-products.

This book is envisioned as a bridge between *academic research and industrial application*, addressing the growing global interest in mushrooms as renewable resources and sustainable bioproducts. It provides comprehensive coverage that connects molecular biology and bioprocess technology with environmental sustainability and economic feasibility—making it an essential reference for both scholars and practitioners.

The preparation of this book has been an inspiring journey of international collaboration. I extend my deepest appreciation to all contributors, reviewers, and scientific partners who made this work possible. Special recognition is given to the *Iraqi Mushroom and Truffle Platform (IQMT)*, which continues to promote knowledge exchange and collaboration among mushroom researchers and enthusiasts both locally and globally, fostering innovation within Iraq and beyond.

I sincerely hope that this volume will inspire future research, entrepreneurial ventures, and educational development in the fascinating world of mushroom biotechnology.

Heet, Anbar, Iraq Mustafa Nadhim Owaid

Acknowledgments

The editors gratefully acknowledge the valuable support and guidance received from **Professor Dr. Jayanta Kumar Patra**, *Series Editor of "Interdisciplinary Biotechnological Advances" (Dongguk University, Republic of Korea)*, whose vision and commitment to advancing interdisciplinary research in biotechnology have made this volume possible within the Springer book series.

We also extend our sincere appreciation to all the contributing authors for their insightful chapters and dedicated efforts in shaping the book *Mushroom Production*. Their expertise in mycology, microbial biotechnology, and sustainable agriculture has significantly enriched the scientific depth and multidisciplinary scope of this publication.

Special thanks are due to Assist. Prof. **Dr. Yee Shin Tan** and Assist. Prof. **Dr. Chia Wei Phan** of the *University of Malaya, Malaysia*, for their valuable assistance as co-editors throughout the editorial and peer-review process. Their contributions in coordinating authors, refining manuscripts, and ensuring scientific accuracy have been instrumental in achieving the high standards of this book.

The editors are also grateful to the **Department of Environment, College of Applied Sciences-Hit, University of Anbar (Iraq)** and the **University of Malaya (Malaysia)** for providing institutional support, research facilities, and academic collaboration that greatly facilitated the preparation of this volume.

In addition, the editors wish to express their deep appreciation to the **Iraqi Mushroom and Truffle Platform (IQMT)** for its continuous cooperation and active engagement with local and international researchers and enthusiasts in the field of mushroom science. The platform's efforts in promoting knowledge exchange and fostering collaboration have played a significant role in advancing mushroom research and production in Iraq and beyond.

Finally, we acknowledge the editorial and production teams at **Springer Nature** for their professionalism, meticulous attention to detail, and continuous support throughout the publication process. This collective effort exemplifies the spirit of scientific collaboration that drives innovation in interdisciplinary biotechnological research.

Contents

Recent Biotechnology of Mushroom Growth Through Submerged Cultivation

Sakthivel Muthu, Mythileeswari Lakshmikanthan, Kathiravan Krishnan, and Lakshmanan Govindan

Abstract Submerged cultivation techniques in mushroom biotechnology have advanced significantly, yielding diverse bioactive compounds. Managed by specialized bioreactors, this method optimizes fungal growth, producing valuable molecules like proteins, enzymes, and polysaccharides. Temperature, agitation, pH, dissolved oxygen, and inoculum quality must be precisely controlled to maximize biomass and bioactive compound yields. Stirred-tank and airlift bioreactors are commonly used across different scales for mushroom cultivation. Research aims to enhance yields and streamline production processes, offering promising avenues for nutrition, pharmaceuticals, and industry. Optimizing cultivation conditions significantly impacts fungal growth and bioactive compound synthesis. Various carbon sources like glucose and xylose affect cellulase production differently in *Pleurotus ostreatus* strains. Nitrogen sources, organic and inorganic, play pivotal roles in cellulolytic enzyme biosynthesis. Temperature regulation is crucial due to species-specific growth optima. Dissolved oxygen levels profoundly influence aerobic cell growth, requiring adequate oxygen transfer. pH adjustments maintain favorable growth conditions, impacting mycelial growth and metabolite synthesis. Proper agitation balances oxygen transfers while minimizing shear stress, optimizing biomass production. Inoculum size and age significantly influence mycelial yield and bioactive compound production. Mushroom mycelia show potential in various industrial sectors due to their bioactive compounds, despite regulatory challenges.

S. Muthu (✉)
Department of Dermatology, Saveetha Medical College and Hospital (SMCH), Saveetha Institute of Medical and Technical Sciences (SIMATS), Saveetha University, Chennai, Tamil Nadu, India

M. Lakshmikanthan · K. Krishnan
Department of Biotechnology, University of Madras, Chennai, Tamil Nadu, India

L. Govindan (✉)
Sri Lakshmi Narayana Institute of Medical Sciences (affiliated to Bharath Institute of Higher Education and Research), Puducherry, India

M. N. Owaid et al. (eds.), *Mushroom Production*, Interdisciplinary Biotechnological Advances, https://doi.org/10.1007/978-981-95-7793-4_1

Keywords Submerged cultivation · Bioreactors · Biomass production · Bioactive compounds · Mushroom biotechnology · Nutritional applications

1 Introduction

Mushrooms have long been revered worldwide for their multifaceted utility in nutraceuticals, pharmaceuticals, and gastronomy. As members of the Basidiomycota phylum, alongside truffles from the Ascomycota phylum, they constitute the esteemed realm of "higher fungi." Within an estimated 12,000 mushroom species, a mere 2000 are deemed edible, with approximately 35 species undergoing commercial cultivation (Fazenda et al. 2008; Sandargo et al. 2019). Among these, *Agaricus bisporus* holds the title of the most globally cultivated, closely followed by *Lentinus edodes* and various *Pleurotus species* (Fazenda et al. 2008; Elisashvili 2012; Corrêa et al. 2016). Noteworthy for their nutritional richness, mushrooms boast a robust profile of vitamins, minerals, and proteins, making them a coveted component of a balanced diet. Of particular interest is their protein composition, which encompasses all essential amino acids, including those renowned for imparting umami taste sensations, positioning mushrooms as a prized source of vegan protein. Carbohydrates found in mushrooms, such as β-glucans, α-glucans, mannans, xylans, and galactans, resist pancreatic enzyme digestion, suggesting their potential as prebiotics (Kalač 2009; Singdevsachan et al. 2016). Recent investigations have unveiled the presence of polyunsaturated fatty acids in edible mushrooms, complemented by their low-fat and low-calorie attributes. Beyond nutrition, mushrooms harbor an array of bioactive compounds within their fruiting bodies, encompassing polysaccharides, polyketides, phenolic compounds, triterpenoids, steroids, proteins, nucleotides, fatty acids, and lactones. These compounds demonstrate a wide range of biological effects, including anti-inflammatory, antioxidant, antitumor, immunomodulatory, and antimicrobial properties (Tsai et al. 2006; Zhang et al. 2010; Gallego et al. 2020). However, conventional fruiting body cultivation methods are beset by inefficiencies such as time intensity, resource demand, and variable product quality (Corrêa et al. 2016; Khan et al. 2018). Solid-substrate cultivation on agricultural residues faces its own set of hurdles, including pH fluctuations, heat management, mass transfer limitations, water activity constraints, and substrate nonuniformity (Pérez-Chávez et al. 2019). Considering these obstacles, submerged liquid cultivation of mycelia has emerged as a promising alternative. This method offers finer control over environmental parameters and expedites cultivation processes. Submerged cultivation involves nurturing mushrooms in a liquid milieu with augmented oxygen provision through agitation, enabling precise regulation of physical, chemical, and biological factors (Junior Letti et al. 2017). Although submerged cultivation presents its own set of challenges, recent progress underscores its potential for yielding bioactive compounds tailored for diverse applications in nutrition, medicine, and pharmaceuticals. Scaling up submerged cultivation from laboratory flasks to industrial bioreactor systems is imperative for fine-tuning parameters such as pH, oxygen concentration, gas composition, stirring rate, and temperature (Fazenda et al. 2008; Bentil et al. 2018). Bioreactors confer numerous advantages including reduced contamination risks and improved biomass quality, rendering them ideal for conducting clinical investigations on health-promoting effects of mushrooms (Petre and Petre 2015). Various bioreactor configurations, spanning air-lift and stirred tank designs, accommodate different cultivation strategies such as batch, fed-batch, and repeated fed-batch cultures (Kim et al. 2007; Tang et al. 2007). Despite these advancements, challenges persist, including escalating culture density leading to biofilm formation and occlusion of sampling apparatuses. Ongoing research endeavors aim to surmount these obstacles, enhancing cultivation efficiency and product consistency. In essence, submerged cultivation of mushrooms heralds a promising frontier preparation of bioactive substances, with ramifications spanning the realms of food science, pharmaceuticals, and nutraceuticals (Turło 2014; Camacho-Morales and Sánchez 2015).

Mushroom utilization has expanded beyond traditional uses, spurring research into their structure, biochemical composition, and cultivation techniques. This review highlights recent advancements in submerged cultivation for bioactive metabolite production over the past 5 years, focusing on factors affecting biomass and compound yields, especially in large-scale bioreactors. It examines mycelia applications in the functional food and dietary supplement sectors, and investigates strategies to boost the production of bioactive compounds, aiming to update the current state and future prospects in medicinal mushroom cultivation for new pharmaceutical products.

2 Advancements in Submerged Cultivation Techniques for Edible Mushrooms

Advancements in mushroom cultivation have transcended traditional methods, embracing submerged cultivation (SCM), where fungal growth occurs in a liquid medium rather than in its natural terrestrial habitat. This innovative technique involves creating an artificial environment conducive to optimal fungal development while mitigating contamination risks (Petre and Petre 2015; Dudekula et al. 2020). However, mushrooms, inherently adapted to solid substrates, encounter challenges in liquid environments due to their reliance on aerobic respiration and oxygen availability. Overcoming this obstacle requires precise control of dissolved oxygen levels, facilitated by specialized equipment to sustain fungal metabolic processes. Despite initial challenges in oxygenation, the mycelium adapts, exhibiting circular movement within the medium driven by centrifugal forces (Gowthaman et al. 2001; Fazenda et al. 2010). The inception of submerged cultivation, dating back to the 1950s, revolutionized mushroom production, ensuring enhanced safety and reproducibility through bioreactors (Fazenda et al. 2008). This method has spurred extensive research, particularly in bioprospecting for bioactive compounds.

Once valued primarily for culinary purposes, mushrooms are now recognized as rich sources of biologically active molecules (BAMs), housing compounds with medicinal and pharmaceutical potential. Species such as *Pleurotus*, *Agaricus*, *Lentinus*, *Cordyceps*, *Morchella*, and *Tuber* have garnered significant attention. *Pleurotus* species, notably *Pleurotus ostreatus* and *Pleurotus eryngii*, have attracted interest for their medicinal properties attributed to polysaccharides, with submerged cultivation experiments elucidating optimal growth conditions. Similarly, submerged cultivation studies on *Agaricus* species, including *Agaricus bisporus* and *Agaricus blazei*, focus on factors influencing polysaccharide production (Valverde et al. 2015; Bhambri et al. 2022). *Lentinus edodes*, or shiitake mushroom, is valued for its cholesterol-lowering lentinan polysaccharide, with submerged cultivation optimizing biomass production and polysaccharide content. Cordyceps species, particularly *Cordyceps militaris*, have gained attention for their immunomodulatory and antitumor properties, with submerged cultivation crucial for maximizing their bioactive potential. Morchella species, known for their distinct honeycomb structure, are explored in submerged cultivation to enhance bioactive compound production (Bisen et al. 2010; Sheng et al. 2021). Tuber species, prized as "underground gold," present unique challenges, yet submerged cultivation endeavors aim to extract bioactive compounds with both nutritional and pharmaceutical value (Tang et al. 2015; Bakratsas et al. 2021).

3 Functional Metabolites Generated Through Mushroom-Submerged Culture

Mushroom biomass production in liquid medium represents a significant advancement in fungal biotechnology, aiming to match the bioactive compound yields achieved through traditional fruiting body cultivation. This method offers notable benefits, including shorter cultivation periods and increased reproducibility, making it ideal for large-scale mushroom production. Within liquid media, mycelia produce a variety of functional bioactives, including proteins, enzymes, polysaccharides, phenolic compounds, terpenoids, ergosterols, fatty acids, and ceramides (Elisashvili 2012; Bakratsas et al. 2021). This segment will categorize research efforts focused on producing these bioactive compounds based on their characteristics, while also examining methods that influence yield production to identify the most effective strains, cultivation techniques, and extraction methods. Traditionally valued for their culinary qualities, mushrooms are now recognized for their nutritional richness and as potent sources of biologically active molecules (BAMs). These substances, present in fungal fruiting structures, hyphal networks, and fermentation broths, offer diverse medicinal and pharmaceutical benefits with minimal toxicity (Bains et al. 2021; Kumar et al. 2021). Recent findings have revealed the capacity of various basidiomycetes to thrive as mycelial biomass in submerged cultures, with a particular focus on polysaccharide and ganoderic acid production (Wagner et al.

2003; Rosales-López et al. 2022). However, investigations into the submerged mushroom cultures exhibit minimal antioxidant and hemagglutinating activities (Elisashvili 2012). The growing potential of mushrooms to produce active compounds through submerged cultures not only holds promise for therapeutic applications but also opens new avenues in nutritional science. Submerged cultivation enables controlled mycelial biomass proliferation, ensuring consistent oxygen and nutrient supply crucial for optimal metabolic function (Arshadi et al. 2023). This method has shown efficacy across various mushroom species, each possessing its own unique reservoir of bioactive compounds.

3.1 Protein

In today's market, the scarcity of protein sources, particularly for vegan diets, has spurred interest in alternative options such as microorganisms like microalgae and mycelium fungi, renowned for their high protein content (Ahlborn et al. 2019). Notably, certain mushroom species, notably *Pleurotus*, are rich in protein, comprising 10.5–49% of their dry weight (Mumpuni et al. 2017). Studies indicate that cultivating these mushrooms on substrates like corn stover can significantly enhance protein yields, reaching concentrations between 40% and 49% (Chahal 1989; Bakratsas et al. 2021). Moreover, utilizing agro-industrial waste like paddy straw has been effective in increasing protein content in *Pleurotus* species, promoting sustainable waste management practices. Additionally, exploring cultivation methods employing natural waste materials, such as peat extract medium, has shown potential in augmenting protein production (Manu-Tawiah and Martin 1987). Phytohormones like indole-3-acetic acid (IAA) have been explored for their role in promoting fungal development and enhancing protein levels. Amino acid profiling of *Pleurotus* mycelium reveals a rich composition of essential amino acids, especially leucine, aspartic acid, phenylalanine, and lysine. This underscores the nutritional importance of *Pleurotus* mycelial proteins and underscores the need to understand the factors influencing their production for optimization (Mukhopadhyay et al. 2005). Similarly, submerged cultivation of other fungi like *Agaricus* spp. and *Lentinus edodes* has demonstrated potential in protein extraction, with reported concentrations ranging from 10% to 53.5% depending on the medium composition (Lobanok et al. 2003). Notably, *Agaricus bisporus* cultivated in ram horn hydrolysate exhibited a protein concentration of 47%, surpassing that of its fruiting bodies. Amino acid analysis further confirms the nutritionally dense of these proteins. *Cordyceps militaris* and *Morchella* spp. have also been acknowledged for their high essential amino acid content, underscoring the nutritional richness of mycelial proteins across various fungal species (Bakratsas et al. 2021). Understanding the underlying factors governing mycelial protein production is essential for maximizing their potential in addressing the growing demand for sustainable protein sources.

3.2 Enzymes

Biotechnology heavily relies on a diverse array of enzymes to catalyze specific biochemical processes. Fungi, thriving on various plant residues, are prolific producers of these essential enzymes. For instance, lignocellulosic biomass, rich in soluble carbohydrates, serves as a substrate for synthesizing ligninolytic and hydrolytic enzymes like lignin peroxidase, manganese peroxidase, laccase, cellulases, hemicellulases, and xylanases. Research highlights the remarkable potential of submerged cultivation techniques in enhancing enzyme production (Bettin et al. 2019). For example, cultivating *Tuber maculatum* with carboxymethyl cellulose significantly boosts cellulase activity, with peak yields of 1.70 U/ml achieved under optimal conditions (Bedade et al. 2017). Similarly, submerged cultivation of *Pleurotus dryinus* yields five times higher cellulase activity compared to solid-state methods, indicating the superiority of submerged techniques in maximizing hydrolytic enzyme yields (Elisashvili et al. 2008). Moreover, submerged cultivation proves advantageous for xylanase production, with *Tuber maculatum* mycelium reaching a maximum output of 13.15 U/ml under optimized conditions (Bedade et al. 2017). Additionally, laccase, crucial for various applications, is abundantly produced by genera like *Pleurotus* and *Agaricus*. Submerged cultivation of *Agaricus blazei* on sugarcane molasses has been optimized to reach laccase levels as high as 9635 U/ml, with further enhancements achieved through the addition of aromatic compounds and $CuSO_4$ to the growth medium (Do Valle et al. 2014). Environmental applications of mushroom-derived laccases, particularly in phenol removal from wastewater, have gained significant attention. Studies on *Pleurotus sajor-caju* demonstrate increased phenol removal efficiency under submerged cultivation conditions (Bettin et al. 2019). These findings underscore the efficacy of submerged cultivation in enzyme production and its potential in biotechnological applications, suggesting a future focus on enzyme production through submerged cultivation methodologies in macro fungi research.

3.3 Lectin

Lectins, which are proteins or glycoproteins originating from nonimmune sources, possess the ability to agglutinate by precipitating glycoconjugates, displaying a specific affinity for particular sugars that can inhibit their cell-binding activity (Xu et al. 2011b). This specificity allows lectins to selectively bind to carbohydrates, aiding in the elucidation of polysaccharide structures in solution. They engage with saccharides through diverse mechanisms like hydrogen bonding, metal coordination, van der Waals forces, and hydrophobic interactions, facilitating the examination of cell surface structures and the isolation of carbohydrate polymers. The utilization of lectin microarray technology offers profound insights into the glucan molecule backbones' structure (Devi et al. 2013; El Enshasy and Hatti-Kaul 2013).

Lectins sourced from edible mushrooms exhibit various biological activities, including immunomodulation, anti-proliferation, antitumor, and inhibition of HIV-reverse transcriptase. They play pivotal roles in processes like cancer cell metastasis and lymphocyte transportation from the bloodstream to lymphoid organs (Zhang et al. 2010). Oligomerization often enhances lectins' binding affinity, enabling multivalent interactions with similar carbohydrates. In vitro studies have demonstrated the antiproliferative and antitumor effects of lectins from mushrooms such as *Tricholoma mongolicum* and *Pleurotus ostreatus* on different cell lines. The cytotoxicity of lectins is affected by preincubation with specific sugars, emphasizing the significance of the binding site in their activity (Wang et al. 1995, 2000). For instance, the lectin derived from *Grifola frondosa* fruiting bodies displays cytotoxic effects against HeLa cells; however, preincubation with haptenic sugar abolishes this cytotoxicity (Kawagishi et al. 1990). Ongoing exploration of mushroom lectins holds substantial promise for the pharmaceutical and nutraceutical industries, offering valuable insights into their diverse biological activities and therapeutic applications.

3.4 Polysaccharides

Recent research in mushroom cultivation has primarily focused on the production of carbohydrates, particularly polysaccharides, owing to their prebiotic effects and their role in promoting the growth of lactic acid bacteria in the colon (Singdevsachan et al. 2016). These polysaccharides and oligosaccharides derived from mushrooms are recognized as functional foods and are classified into intracellular polysaccharides (IPS), such as glycogen, pullulans, β-glucans, and trehalose, and extracellular polysaccharides (EPS), which hold considerable economic value in industries like food, biotechnology, and pharmaceuticals (Diamantopoulou et al. 2014; Wu et al. 2012). Cold aqueous extracts from submerged cultivation of *Pleurotus* spp. have exhibited higher carbohydrate content, reaching up to 70.5%, compared to hot aqueous extracts from fruiting body cultivation, which only achieved 33%. Investigations into the impact of agitation and cultivation duration on polysaccharide content revealed that *Pleurotus pulmonarius* cultured in glucose-based submerged cultures showed higher glucose monosaccharide content in IPS. Mannitol was predominant at 12 days, constituting 20.5% of total IPS, while static cultures at 20 days displayed IPS with 88.5% glucose (Wu et al. 2012; Diamantopoulou et al. 2014). *Cordyceps tuberculata* cultivation identified 150 rpm as the optimal rotation speed for maximizing IPS production (264 mg/l) (Bakratsas et al. 2021). EPS production is influenced by carbon and nitrogen sources, light penetration in culture, and nitrogen concentration, with organic sources such as tryptone proving to be optimal for EPS production in *Pleurotus eryngii* (Assis et al. 2013; Jing et al. 2013). These polysaccharides and β-glucans from various mushrooms exhibit medicinal properties, including antioxidant and antitumor activities, which are dependent on their molecular weight (Jing et al. 2013). β-glucans from mushrooms like *Lentinus edodes* and *Ganoderma lucidum* have been widely acknowledged for their significant health benefits and clinical applications (Manzi and Pizzoferrato 2000; El Enshasy and Hatti-Kaul 2013).

3.5 Lipids

Mushrooms produce a diverse array of valuable bioactive compounds, including lipids such as phospholipids, triglycerides, sterols, oils, terpenoids, and fatty acids. Triglycerides, the predominant neutral lipids in fungi, are typically composed of 16–18 carbon chains. Common polar lipids include diacylglyceryltrimethylhomoserine and phosphatidylcholine (Gebreyohannes and Sbhatu 2023). These lipid compounds have significant applications in pharmaceuticals and nutrition, with the fatty acid composition influencing the taste and aroma of mushrooms (Das et al. 2010). Studies on *Pleurotus ostreatus* reveal that the total fat content in the mycelia is comparable to that in the fruiting bodies, with a higher proportion of saturated fatty acids in the mycelia (Hadar and Cohen-Arazi 1986). Submerged cultivation of *Pleurotus sajor-caju* results in a higher total fat content (10.2%) compared to fruiting body cultivation (Confortin et al. 2008). Furthermore, agitation during the cultivation of *Pleurotus pulmonarius* increases lipid production: agitated cultures produced 3.7 g/L of lipids over 24 days, compared to 1.03 g/L in static cultures. Similarly, *Morchella esculenta* produced 2.3 g/L of lipids in agitated cultures versus 1.97 g/L in static cultures (Diamantopoulou et al. 2014). Research indicates that lipid production peaks in the early growth phases and degrades in later stages (Diamantopoulou et al. 2014). Fatty acids derived from mushrooms are particularly valuable. In *Pleurotus pulmonarius*, unsaturated fatty acids predominate in the mycelium biomass, with linoleic acid comprising 33–68% of total fatty acids, depending on the carbon source, with the highest percentage observed when glucose is used. Among saturated fatty acids, palmitic acid is most prevalent, reaching 17.3% of total fatty acids with xylose as the carbon source (Smiderle et al. 2012). In *Morchella crassipes* cultivated in waste sulfite liquors, linolenic acid is the dominant fatty acid in both magnesium- and ammonia-supplemented mediums, with other unsaturated fatty acids such as palmitoleic, oleic, and linoleic acids also detected. Saturated fatty acids included myristic, pentadecanoic, palmitic, heptadecanoic, and stearic acids (LeDuy et al. 1974). Linoleic, stearic, oleic, and palmitic acids were identified in *Pleurotus ostreatus* cultivated in a batch stirred tank bioreactor, indicating that industrial-scale cultivation does not affect the production of essential fatty acids (Papaspyridi et al. 2012). Lipid fractions in mushrooms enhance organoleptic properties, with unsaturated and essential fatty acids boosting nutritional value. The taste and aroma are influenced by the autooxidation products of unsaturated fatty acids such as palmitoleic, oleic, linoleic, and linolenic acids (Sun et al. 2020).

3.6 Carotenoid Pigments

Carotenoids, pigments ranging from light yellow to deep red, are synthesized by both plants and microorganisms, with animals and humans acquiring them through their diets, primarily in the stable trans isomer form. These pigments can exist as cis and trans isomers due to isomerization, facilitated by their conjugated double bonds which are responsible for their characteristic coloration. Carotenoids' reactivity is influenced by their structural properties and functional groups, and they are broadly classified into carotenes and xanthophylls. Carotenes, such as β-carotene, α-carotene, and γ-carotene, serve as provitamin A carotenoids, while xanthophylls, including lutein, cryptoxanthin, and canthaxanthin, act as antioxidants (Ribeiro et al. 2011; Vacharathit 2014). Although approximately 600 carotenoids have been identified, the majority possess a 40-carbon skeleton, while some have a 30-carbon structure. Carotenoid biosynthesis occurs via the mevalonate and isoprene pathways (Vacharathit 2014). Due to their notable properties such as anticancer, antioxidant, and anti-HIV activities, as well as their role in enhancing photoreception, carotenoids find applications in various industries including cosmetics, nutraceuticals, and pharmaceuticals. Interestingly, despite being pigments, carotenoids do not consistently contribute to the coloration of mushroom fruiting bodies. For instance, while β-carotene was found in fresh samples of *Cantharellus cibarius*, other orange-colored mushrooms like *P. nameko* and *S. bovinus* did not contain carotenes. *Cordyceps militaris* stands out for its high carotenoid content. Research indicates that exposure to light at 6000 lx inhibits carotenoid accumulation, whereas heat treatment at 25 °C for 20 days enhances carotenoid levels, albeit with variations among species (Wu et al. 2016; Jiaojiao et al. 2018).

4 The Role of Bioreactors in Submerged Mushroom Cultivation

Scaling up mushroom biotechnology for industrial applications necessitates the use of bioreactor vessels to optimize cultivation and growth processes. Achieving high yields, comparable to or surpassing those of small-scale operations, is essential for the nutritional and pharmaceutical uses of mushroom-derived bioactive compounds (Petre and Petre 2015). Successful large-scale cultivation relies on precise control of culture parameters, such as temperature, agitation, pH, dissolved oxygen, substrate, and metabolite concentrations, to produce high-quality biomass and bioactive compounds (Tang et al. 2007). This section will discuss the types of bioreactors used in submerged mushroom cultivation and the factors influencing their efficiency on a large scale. Common bioreactor types include stirred-tank (Fig. 1) and airlift bioreactors (Fig. 2), such as bubble columns. Cultivation methods can be batch, fed-batch, or continuous (Fig. 3) (Tang et al. 2007; Petre and Petre 2015). Stirred-tank bioreactors, featuring an agitator or impeller (axial or radial), facilitate heat and

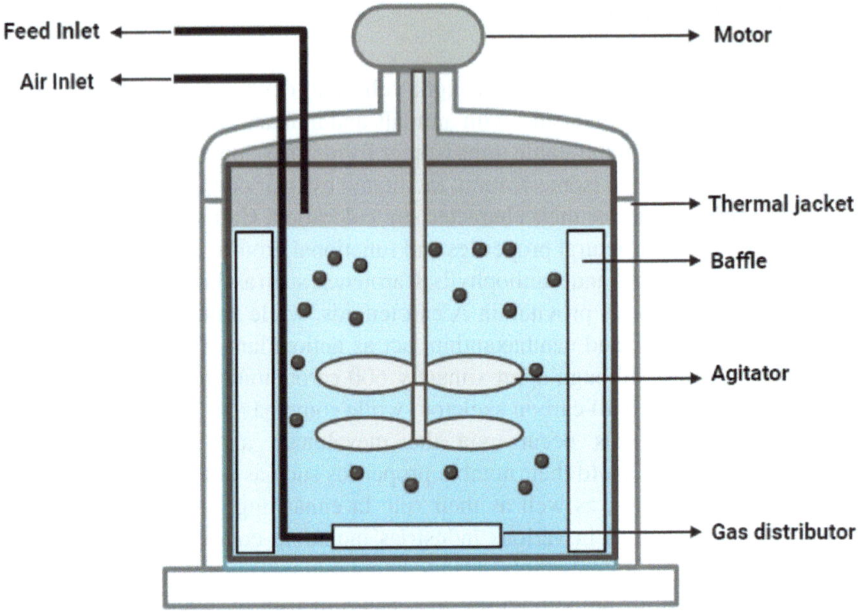

Fig. 1 Fundamental design of a stirred tank bioreactor for mushroom cultivation

mass transfer, aeration, and mixing (Fig. 1). These bioreactors range from a few liters to industrial-scale capacities up to 400 m^3. Key factors for optimal performance include impeller size, off-bottom clearance, baffle width, the ratio of liquid height to tank diameter, and the use of multiple impellers (Wang and Zhong 2007). The homogenization provided by stirred-tank bioreactors is advantageous for mushroom cultivation, as agitation control promotes the formation of mycelium pellets (Wang and Zhong 2007; Petre and Petre 2015). Numerous studies have utilized stirred-tank bioreactors from 2-L bench-scale to 15-L pilot-scale. For example, a significant increase in EPS production was observed in a 15-L stirred-tank bioreactor for *Pleurotus ostreatus* (El-Enshasy et al. 2010). Increased total protein concentration and earlier laccase production were noted in a 3.5-L stirred-tank bioreactor for *Pleurotus sajor-caju* compared to flask cultivation (Bettin et al. 2011). In 100-L fermenters, *Pleurotus eryngii* achieved a biomass concentration of 12.54 g/L with high concentrations of vitamin D2 and micronutrients within 10 days (Singh et al. 2020). Automated sampling systems in stirred-tank bioreactors demonstrated high yields of *Pleurotus pulmonarius* mycelia, promising significant energy and labor savings for industrial production (Abdullah et al. 2013). Airlift bioreactors and bubble columns are the second most common bioreactor types for mushroom cultivation. Airlift bioreactors differ from bubble columns by having an internal or external loop, enhancing effectiveness through gentler shear force transfer, increased circular liquid flow, and improved mass and heat transfer (Wang and Zhong 2007). Preferred for filamentous fermentation, airlift bioreactors feature a conical bottom

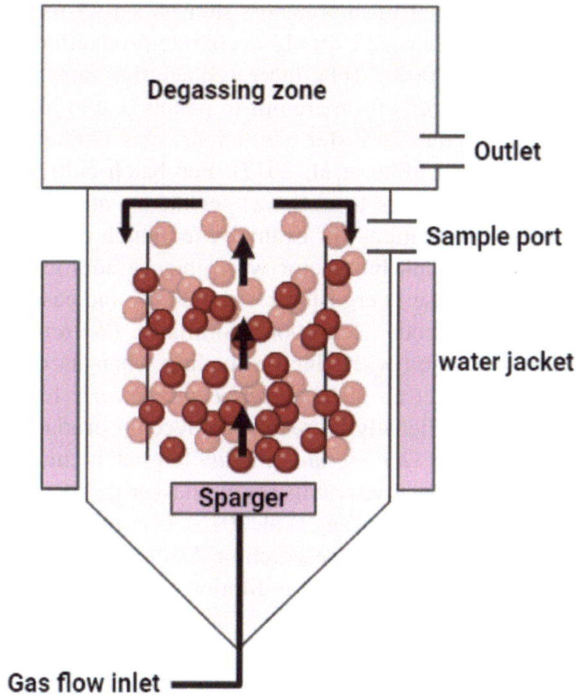

Fig. 2 Fundamental design of an airlift bioreactor for mushroom cultivation

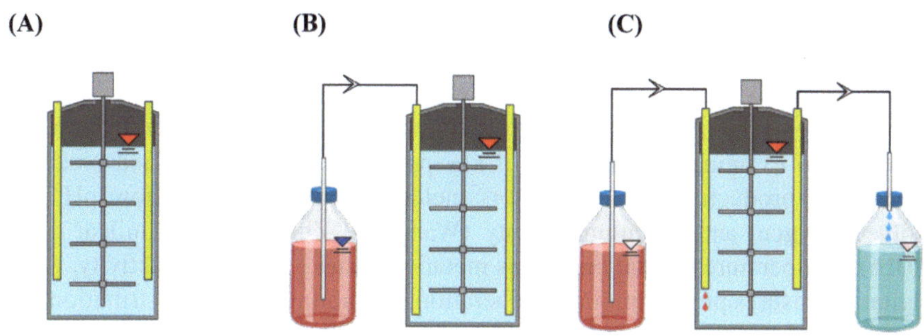

Fig. 3 Fundamental design of batch (**a**), fed batch (**b**), and continuous culture bioreactor (**c**) for mushroom cultivation

with a sparger, a cylindrical middle section, a top degassing zone, and a water jacket for temperature control (Fig. 2) (Choi et al. 2007, 2011).

In bubble columns, cell flow is more randomized, leading to potential cell precipitation. Multistage bubble columns with perforated plates can reduce these issues by minimizing back-mixing and enhancing mass transfer (Tsuge et al. 2004). A comparative study by Kawagoe et al. found higher yields of L-malic acid in an airlift bioreactor compared to a bubble column for *Schizophyllum commune* cultivation due to reduced cell precipitation. Research on *Pleurotus flabellatus* in airlift bioreactors showed a 12.5-fold increase in biomass productivity and enhanced exopolysaccharide production compared to flask cultivation (Choi et al. 2007, 2011). Increased aeration rates in airlift bioreactors can transform fungal cells from clumps to pellets due to higher superficial gas velocities, volumetric mass transfer coefficients, gas holdup, and shear forces (Wang and Zhong 2007; Filho et al. 2017). Fed-batch cultivation is widely used in large-scale bioprocesses to optimize secondary metabolite production by preventing substrate limitation. For example, fed-batch cultivation of *Agaricus brasiliensis* in a stirred-tank bioreactor with glucose addition after 72 hours resulted in a 17.2% increase in ergosterol and a 26.9% increase in EPS production compared to batch methods (Zou 2006). Similarly, *Pleurotus ostreatus* showed enhanced EPS production with glucose feeding when its concentration fell below 2 g/L (El-Enshasy et al. 2010). *Cordyceps militaris* fed-batch cultivation with NH^{4+} feeding significantly increased cordycepin production to 346.1 mg/L (Mao and Zhong 2006). *Ganoderma lucidum* fed-batch cultivation in stirred-tank bioreactors yielded positive results for biomass, EPS, IPS, and GA production compared to flask cultivation (Tang et al. 2015). Continuous cultivation of mushroom mycelia is less studied, but research on *Lentinus edodes* showed a 14-fold increase in mycelium productivity at a dilution rate of 0.03 h^{-1} compared to batch culture (Fukushima et al. 1993).

5 Factors Affecting Mushroom-Submerged Cultivation in Bioreactors

To understand the critical role of bioreactor systems in microbial cultivation, it's essential to examine the influence of key parameters on biomass and bioactive compound production. Crucial factors include temperature, agitation, pH, dissolved oxygen, and inoculum quality (Wang and Zhong 2007; Amilcar et al. 2016). Temperature regulation affects metabolic rates and enzyme activity, while agitation ensures proper mixing and aeration, promoting nutrient distribution and gas exchange for uniform growth. Maintaining optimal pH is vital for cellular stability and nutrient bioavailability. Adequate dissolved oxygen levels are necessary to meet the respiratory needs of aerobic microbes. The quality and quantity of inoculum impact the initial growth rate and overall yield. By precisely adjusting and monitoring these parameters, bioreactor efficiency can be optimized, enhancing microbial biomass and bioactive compound yields. Such meticulous control is essential for consistent, high-quality production in industrial biotechnology.

5.1 Carbon

Selecting an optimal carbon source is crucial for enhancing cellulase production in *Pleurotus ostreatus*, particularly the PO 14 strain. This strain was cultivated on nutrient media with soluble sugars like maltose, glucose, and xylose at concentrations ranging from 3% to 15%, which did not inhibit cellulase production (Zhang and Cheung 2011). To study cellulose biodegradation by *P. ostreatus* in a batch system, various carbon sources in suspensions (5–10% dry matter) were tested, using pure cellulose (Schuhardt) as the control. Natural carbon sources from industrial cereal grain wastes yielded higher enzymatic activity compared to the control. These agricultural wastes were classified as solid substrates (80–90% dry matter), semisolid substrates (30–40% dry matter), and colloidal suspensions (5–10% dry matter). Growth on polysaccharides like starch was slower due to limited depolymerization, leading to faster growth on monosaccharide substrates (Pierre Beguin 1994). The final fungal biomass correlated with substrate particle size, influenced by the volume-to-surface ratio in submerged cultivation processes. The carbon source is essential for microbial growth and bioactive metabolite production. Although glucose is commonly used, various sugars have been compared to optimize fungal growth and product yield. For example, in submerged cultivation of *Humphrea coffeata*, lactose and glucose at 35 g/L produced the highest biomass concentrations (11.9 and 10.8 g/L, respectively), while sucrose resulted in significantly lower biomass (2.9 g/L) due to catabolic repression. *H. coffeata* exhibited distinct growth phases with lactose and glucose, but not with sucrose. Similarly, *Lyophyllum decastes* showed optimal mycelial growth and polysaccharide production with lactose, glucose, and fructose at 30 g/L. Glucose was particularly effective for extracellular polysaccharide (EPS) production, while fructose resulted in the lowest intracellular polysaccharide (IPS) production. Other fungi, such as *Antrodia cinnamomea* and *Ganoderma applanatum*, also demonstrated variability in biomass and EPS yields depending on the carbon source. *A. cinnamomea* grew best with xylose, and *G. applanatum* produced high biomass with fructose and maltose, but higher EPS with glucose. *Morchella esculenta* yielded the highest EPS with glucose and significant biomass with xylose. The type of carbon source influenced not only the amount of biomass and EPS but also their molecular weight and biological activity. Overall, while glucose often proved to be the best carbon source for both biomass and EPS production, specific preferences varied among different fungal species and strains.

5.2 Nitrogen

In experimental investigations into nitrogen sources for mushroom cultivation, it was discovered that certain organic compounds (e.g., yeast extracts, malt extracts, peptones, and triptones) do not directly provide accessible nitrogen molecules to the cells. However, they deliver essential vitamins and minerals crucial for mushroom development across various growth cycles. For many mushroom species, these compounds act as combined sources of carbon and nitrogen. Specifically, for *Pleurotus ostreatus* strain PO 14, the optimal nitrogen source for cellulase synthesis was identified by testing both organic (urea, yeast extract, peptone) and inorganic (ammonium nitrate, potassium nitrite) compounds. Maintaining a total nitrogen concentration of 0.5 g/L in the culture fluid highlighted the necessity of both organic compounds (e.g., urea, peptone, yeast, or meat extracts) and inorganic salts (e.g., ammonium nitrate) as preferred nitrogen sources for cellulolytic enzyme production (Tanaka and Matsuno 1985; Songulashvili et al. 2005). Experimental data showed that ammonium nitrate at 0.5% and 1% concentrations significantly stimulated cellulolytic enzyme biosynthesis in *Lentinula edodes* and *P. ostreatus* batch cultures, with maximum enzyme activities observed with a mix of beef extract, peptone, and inorganic salts like ammonium nitrate or potassium nitrate (Cohen et al. 2002). The nitrogen source is crucial for efficient mushroom growth and polysaccharide production, significantly influencing fungal enzyme synthesis in both primary and secondary metabolisms. Nitrogen can be supplied as ammonium or nitrate ions, or in organic forms such as amino acids or proteins. Commonly, yeast extract and peptone are added to the culture medium at concentrations of 1–5 g/L. Pokhrel and Ohga (2007) compared various nitrogen sources, finding that yeast extract provided the highest mycelial growth (7.03 g/L), as well as extracellular polysaccharides (EPS) and intracellular polysaccharides (IPS). Polypeptone also supported substantial mycelial growth. Although ammonium sulfate resulted in lower mycelial growth, it still produced significant EPS and IPS. Varying yeast extract concentrations revealed 1% as optimal for mycelial growth and IPS production, while 2% improved EPS production. For *Grifola frondosa*, soytone was the best organic nitrogen source for superoxide anion scavenging activity and reducing power. Among inorganic sources, ammonium chloride yielded the highest superoxide anion scavenging activity, followed by sodium nitrate and ammonium acetate. Amino acids like glutamic acid and arginine also enhanced these activities. *Antrodia cinnamomea* showed increased EPS production with calcium nitrate, while organic nitrogen sources generally supported better mycelial biomass. In contrast, inorganic sources were less effective for mycelial growth but improved EPS yield. For *Agaricus nevoi* and *Inonotus levis*, KNO3 and (NH4)2SO4 were effective for EPS accumulation. Yeast extract and peptone were optimal for *Morchella esculenta* biomass and EPS yield, whereas (NH4)2HPO4 was best for phenol and flavonoid synthesis in *Leucopaxillus giganteus*, correlating with improved antioxidant properties. Overall, organic nitrogen sources typically stimulate mycelial growth, while inorganic sources enhance polysaccharide production, indicating that optimal EPS production does not necessarily correlate with high mycelial growth.

5.3 Temperature

Microorganisms have specific optimal growth temperatures, with deviations impacting protein expression and metabolite synthesis adversely. Higher temperatures can cause thermal death, while lower temperatures slow cellular growth and product synthesis. This phenomenon aligns with the Arrhenius equation, which posits that cell growth rates increase with temperature up to an optimal point, beyond which growth rates decline due to thermal damage (Wang and Zhong 2007). In mushroom cultivation, optimal mycelial growth and exopolysaccharide (EPS) production for species like *Mycoleptodonoides aitchisonii* and *Antrodia cinnamomea* occur at 25 °C. Temperatures outside this range reduce growth and EPS yield (Choi et al. 2011). Notably, the optimal temperatures for biomass production and bioactive compound synthesis often differ. For example, *Flammulina velutipes* produces maximum biomass at 30 °C, but angiotensin-converting enzyme synthesis peaks at 20 °C (Kim et al. 2002). *Lentinus edodes* produces lectin optimally at a different temperature than its best growth temperature (Nikitina et al. 2007). Pleurotus species such as *P. ostreatus* and *P. cystidiosus* grow best at 28 °C, but *P. cystidiosus* experiences significant morphological changes and reduced viability at 16 °C and 36 °C (Hoa and Wang 2015). *Ganoderma applanatum*, in contrast, achieves maximum biomass and intracellular polysaccharide production at the low temperature of 10 °C (Lee et al. 2007). Maintaining optimal temperature conditions in bioreactors is essential for maximizing fungal growth and metabolite production. While most mushrooms thrive between 20 °C and 25 °C, there is no consistent correlation between biomass growth and bioactive metabolite synthesis (Elisashvili 2012). For example, *A. cinnamomea* achieves optimal biomass production at 25 °C and EPS production at 28 °C, with performance declining outside these temperatures (Lin and Sung 2006). Phellinus linteus reaches maximum biomass at 30 °C, with optimal EPS production between 20 °C and 35 °C (Hwang et al. 2003). *Grifola frondosa* achieves peak biomass and EPS production at 25 °C (Lee et al. 2004), while *Tremella fuciformis* shows reduced biomass production at higher temperatures (Cho et al. 2006). *Pleurotus cystidiosus* grows best at 28 °C, 32 °C, and 24 °C, respectively (Hoa and Wang 2015). Temperature significantly influences biosynthetic processes, impacting specific growth rates as described by the Arrhenius equation. For *Lentinus edodes*, enzyme activities vary with temperature, peaking at 23 °C under stationary conditions and at 28 °C with stirring. Stirring enhances enzymatic activity by 30–40% compared to stationary cultivation, underscoring the importance of both temperature control and mixing in optimizing microbial growth and metabolite production (Jiang 2010; Petre and Petre 2015).

5.4 Dissolved Oxygen

Dissolved oxygen (DO) is crucial for aerobic cell growth in bioreactors, along with temperature. Effective oxygen transfer in a well-mixed liquid system is essential for cell metabolism. Inadequate oxygen can limit cell growth, impair respiration, and reduce biomass and bioactive compound production (Wang and Zhong 2007; Elisashvili 2012). The volumetric mass transfer coefficient, K_La, is a key factor in oxygen transfer, influenced by aeration rate, fluid dynamics, media composition, agitation speed, cell type and morphology, and biocatalyst properties (Wang and Zhong 2007). Studies on *Ganoderma lucidum* indicated that increasing K_La from 16.4 to 96 h^{-1} resulted in larger mycelia and a 1.8-fold increase in ganoderic acid production, with the highest biomass yield at 78.2 h^{-1} (Elisashvili 2012). For *Cordyceps militaris*, increasing initial K_La from 11.5 to 113.8 h^{-1} boosted biomass from 13.59 to 20.12 g/L, with maximum cordycepin yield at K_La of 54.5 h^{-1} (Mao and Zhong 2004). Oxygen supply, as measured by dissolved oxygen in the medium, affects hyphal length, cell wall properties, clump roughness, and sample viscosity (Fazenda et al. 2010). Controlled DO levels can double biomass concentration compared to uncontrolled conditions (Fazenda et al. 2010). Successful scale-up from flasks to stirred-tank bioreactors can be achieved with DO levels between 20% and 35% air saturation (Tang et al. 2007). For *Ganoderma lucidum*, high air saturation (~25%) was optimal for intracellular polysaccharide (IPS) and ganoderic acid production, while extracellular polysaccharide (EPS) production favored lower air saturation (~10%) (Tang and Zhong 2003). A two-stage DO control strategy, optimizing levels for specific bioactive compounds, enhanced cordycepin productivity by 30% compared to conventional methods (Mao and Zhong 2006). Fungi, being strictly aerobic, require adequate aeration for optimal metabolism. Studies with varying surface-to-volume ratios and airflow rates showed the sensitivity of *Pleurotus ostreatus* strains to uniform air dispersion in nutrient media (Mikiashvili et al. 2006). In batch cultivation, increased aeration significantly boosted dried protein biomass from 15.9 to 23.8 g% d.m., compared to 2.1 to 4.0 g% d.m. without aeration. However, airflow rates above 5 m^3/h did not further enhance metabolic activity. The highest mycelial biomass was at aeration rates of 0.5–1.5 vvm, with fungal pellets forming early and increasing in size during submerged cultivation. In chemostat systems, protein biomass conversion yield increased to 14.5–15.9 g% d.m. from an initial 2.0–4.0 g% d.m. Optimal mycelial growth occurred at aeration rates of 0.7–1.2 vvm, peaking at 1.5 vvm and declining at 1.7 vvm.

5.5 pH

pH is a crucial factor influencing cell growth and metabolite synthesis in microorganisms, effective only within specific ranges. In bioreactor cultivation, pH adjustments are essential and are typically managed using acidic or alkaline solutions, or CO_2, as described by the equation: $CO_2(aq) + H_2O \leftrightarrow H^+ + HCO_3^-$ (Wang and

Zhong 2007). For most microorganisms, excluding acidophiles and alkalophiles, the optimal pH range is between 5 and 7 (Wang and Zhong 2007). Numerous studies on mushroom submerged cultivation have highlighted the impact of pH on mycelial growth. For instance, *Ganoderma lucidum* shows optimal biomass production at pH 6.5; ganoderic acid and IPS production at pH 5.5–6.5 and 5.5–7.0, respectively; and EPS production peaks at pH 3.5–4.5 (Elisashvili 2012). *Rigidoporus ulmarius* prefers a higher initial pH, as growth is hindered at pH 4.0 (Chao et al. 2011). Conversely, *Antrodia camphorata* thrives at pH 4.0, with maximum EPS production at pH 5.0 (Shu and Lung 2004). *Fomitopsis pinicola* has optimal growth and EPS production at pH 6.0, with reduced activity above pH 7 and below pH 5.5. *Mycoleptodonoides aitchisonii* achieves maximum biomass and EPS production at pH 6.0 and 6.5, respectively. A two-stage pH control strategy has been proposed for optimizing growth and bioactive compound production. For *Ganoderma lucidum* in an airlift bioreactor, shifting the pH from 3 to 6 during the exponential growth phase significantly increases EPS yields (Lee et al. 1999). This method also affects pellet morphology, producing typical circular pellets at pH 4.5 and less dense pellets when shifting from pH 3.0 to 6.0. For *Cordyceps sinensis*, pellet formation is optimal between pH 4.0 and 6.0, with higher pH values resulting in smaller, less abundant pellets due to changes in electrostatic forces on the mycelial surface (Liu and Wu 2012). Extensive research has examined pH effects on substrates for edible mushroom cultivation, particularly in the range of 3.0–7.5. For *Lentinula edodes* and *Pleurotus ostreatus*, maximum cellulase activity occurs at pH 5.5–6.0, with peak enzymatic activities between 2.80 and 3.70 U/mL. Maintaining pH at 5.5 during submerged cultivation of the PO 14 strain enhances enzyme activity by 20–23% compared to unadjusted pH. Variations in pH from 4 to 7 result in a 30–40% decrease in endoglucanase activity for the L.E.07 strain, compared to the standard pH of 7.0. In flask studies, pH significantly impacts biomass growth and EPS production in fungi. While pH control is not feasible in flasks, initial pH is critical for assessing its effects. Changes in pH influence cell morphology and nutrient uptake, varying with the microorganism and medium composition (Elisashvili 2012). For example, *Phellinus linteus* has optimal mycelial composition at pH 5 and EPS production at pH 4 (Hwang et al. 2003). *Antrodia cinnamomea* shows optimal biomass and EPS production at pH 5.5, with production declining sharply beyond this pH (Lin and Sung 2006). *Grifola frondosa*, known for its polysaccharides with antitumor properties, has optimal biomass and EPS production at pH 5.5, consistent with other mushroom species (Lee et al. 2004). The pH of the culture medium also affects monosaccharide production rates in EPS (Di Hsu et al. 2017).

5.6 Agitation

Agitation and aeration are critical for optimizing biomass production and the yield of bioactive compounds in fungal cultures. Proper agitation balances oxygen transfer with minimal shear stress, as excessive agitation can damage cells due to hydrodynamic forces (Wang and Zhong 2007; Elisashvili 2012). Innovations in impeller design, such as cell lift and centrifugal impellers, help mitigate these effects and protect delicate cells (Ahmed et al. 2014). Filamentous fungi cultures exhibit non-Newtonian behavior and high apparent viscosities, necessitating robust agitation. However, this can negatively impact growth, mycelial morphology, and bioactive compound yield (Wang and Zhong 2007). In *Aspergillus oryzae*, protein yield was unaffected by agitation intensity in a chemostat culture with 75% dissolved oxygen and increased in a fed-batch culture at 50% dissolved oxygen (Amanullah et al. 2002). Agitation also influences pellet formation; higher agitation enhances dissolved oxygen tension and hyphal branching, resulting in stronger pellets (Zhou et al. 2018). For instance, in *Cordyceps militaris*, low agitation (50 rpm) produced large, feathery pellets, while high agitation (300 rpm) resulted in smaller, smoother pellets, with maximum biomass yield at 300 rpm and optimal exobiopolymer production at 150 rpm (Park et al. 2002). In *Aspergillus niger*, maximum growth was observed at 200 rpm, with peak pellet size at 150 rpm; enzyme production was highest at 150 rpm (Ibrahim 2015). Similarly, optimal biomass production in *Ganoderma lucidum* occurred at 200 rpm and pH 4.5 (Agudelo-Escobar et al. 2017). Proper nutrient dispersion is critical, emphasizing the need for a stirring regime tailored to fungal cultures' specific needs (Glazebrook et al. 1992). Without controlled stirring, uneven nutrient distribution can lead to metabolic imbalances and suboptimal growth due to shear forces near impeller blades (Carlile 1995). Each fungal species requires a specific stirring regime to support morphogenesis and minimize cell damage. Agitation and aeration significantly impact product formation in fermenter tanks by affecting oxygen availability, biomass morphology, and culture medium viscosity. High agitation and aeration rates can enhance hyphal density and extracellular polysaccharide (EPS) production, while high aeration combined with low agitation can yield high biomass and EPS (Cui et al. 2016). For *Tremella fuciformis*, an agitation rate of 200 rpm maximized biomass and EPS production at 7.03 g/L and 2.00 g/L, respectively, with an optimal aeration rate of 2 vvm to maintain high dissolved oxygen levels. Conversely, high shear stress from increased agitation rates can be detrimental to biomass growth, as observed in *Agaricus campestris*, where 150 rpm was optimal (Martin 1983). Agitation affects morphology, growth, and product formation by generating shear forces that disrupt mycelial structures. Increased aeration reduces culture medium viscosity, enhancing oxygen uptake and nutrient consumption, with final cell morphology and broth rheology dependent on shear forces during fermentation (Márquez-Rocha et al. 1999).

5.7 *Inoculum*

The yield of mycelia and production of bioactive compounds are significantly affected by the size and age of the inoculum. For *Pleurotus tuber-regium*, an inoculum volume of 2 ml per 100 ml medium was optimal for biomass production, outperforming volumes ranging from 0.5 to 12 ml (Wu et al. 2003). In *Ganoderma lucidum*, increasing the inoculum concentration within a similar range enhanced mycelium yield and pellet formation, though pellet size decreased (Yang and Liau 1998; Wu et al. 2003). *Tricholoma mongolicum* showed optimal intracellular and extracellular polysaccharide production at inoculum concentrations of 10% and 12% v/v, respectively (Xiaotong Wu 2012). For *Armillaria luteo-virens Sacc.*, a 5% v/v inoculum concentration was ideal for extracellular polysaccharide production, whereas a 2% v/v inoculum delayed metabolism (Xu et al. 2011a). In *Cordyceps jiangxiensis* JXPJ 0109, a 10-day-old inoculum resulted in lower biomass and EPS production compared to a 4–8-day-old inoculum (Xiao et al. 2004). *Aspergillus niger* MK-15 had optimal pectinolytic enzyme synthesis with a 3-day-old inoculum at a 3% concentration. Inoculum size and age are critical in microbial metabolism, impacting biomass and metabolite production. For *Armillaria luteo-virens*, a 2% inoculum yielded the highest cell dry weight, while 0.5% and 5% inocula resulted in lower biomass, likely due to nutrient limitations (Xu et al. 2011a). *Agaricus campestris* achieved maximum biomass yield with a 4% v/v inoculum, with higher inoculum percentages leading to decreased biomass, possibly due to inhibitory substances transferred from the seed culture (Martin 1983). *Grifola frondosa* showed optimal biomass and EPS production at a 3% w/v inoculum (Lee et al. 2004).

6 Applications of Mushroom Mycelia in the Nutraceutical and Food Industries

The extensive variety of edible and medicinal mushrooms, coupled with their rich array of bioactive compounds, positions mushroom mycelium as a promising candidate for industrial applications in the food and nutraceutical sectors. Many mushrooms are already used as nutraceuticals, producing a range of biotechnological products with therapeutic effects available worldwide (Łysakowska et al. 2023). For instance, polysaccharide krestin (PSK) from *Trametes versicolor* is marketed in Japan as a mycopharmaceutical for cancer treatment (Turło 2014; Sandargo et al. 2019). Another notable example is pleuromutilin, derived from *Pleurotus passeckerianus*, with its analog retapamulin marketed by GlaxoSmithKline as Altargo (Sandargo et al. 2019). Compounds like erinacines and hericenones from *Hericium erinaceus* have been shown to stimulate nerve growth factor, promoting nerve cell growth. Preclinical studies suggest that diets enriched with *Hericium erinaceus* mycelia containing erinacines can improve conditions such as Alzheimer's disease, Parkinson's disease, ischemic stroke, and depression (Li et al. 2018; Rathore et al. 2019). Additionally, the health supplement Organic ReiShi-Gen, combining *Lentinus edodes* and *Ganoderma lucidum*, is recognized for its immune-boosting properties (Rathore et al. 2019). In the realm of functional foods, substituting 5% of wheat flour with mushroom mycelia in bread has been shown to maintain rheological, textural, and sensory qualities while preserving high levels of ergothioneine and γ-aminobutyric acid post-baking (Ulziijargal et al. 2013). Mycelial products are also being explored as meat analogs; for example, mycelia from *Agaricus bisporus* offer superior umami characteristics, texture, and sensory attributes compared to soy protein analogs (Kim and Yun 2005). Despite promising research outcomes, stringent regulations by the European Food Safety Authority (EFSA) pose significant barriers to market entry in Europe. Challenges include the regulation of RNA content, toxic metabolites, microbial contamination, GMO use, and the need for substrate safety checks (Lähteenmäki-Uutela et al. 2021). Mushrooms are also prized for their umami flavor, attributable to monosodium glutamate content, making them ideal for enhancing the taste of various dishes. Key volatile compounds responsible for this flavor include oct-1-en-3-ol, octan-3-ol, octan-3-one, and oct-1-en-3-one, along with unsaturated fatty acids and amino acid derivatives (Rathore et al. 2019). Submerged fermentation of mushrooms in bioreactors can optimize the production of biomass rich in these flavor compounds, presenting a lucrative opportunity for industrial-scale flavor extraction. Extensive research explores the pharmaceutical and nutraceutical potential of mushroom biomass and its bioactive metabolites. Mushrooms offer a natural alternative to synthetic drugs, which often have adverse side effects. For instance, sclerotium powder from *Lignosus rhinocerus* has shown inhibitory effects on HIV-1 protease and reverse transcriptase, with water extracts demonstrating efficacy without cytotoxicity (Sillapachaiyaporn and Chuchawankul 2020). Additionally, powder from *Taiwanofungus camphoratus* has shown promise in reducing tumor size in liver cancer patients (Lee et al. 2019). The medicinal mushroom *Cantharellus cibarius* possesses antioxidant and anti-inflammatory properties, with effective iron-chelating capacity for treating anemia. Similarly, polysaccharides from *Pleurotus ostreatus* have demonstrated antitumor activity against gastric cancer in both in vitro and in vivo studies, significantly reducing tumor size and cell migration in treated mice. However, extensive in vivo and in vitro studies are essential to validate the clinical and nutraceutical applications of these bioactive constituents (Khalili et al. 2015).

7 Conclusions and Future Prospects

Mushroom cultivation, particularly through submerged methods, offers significant research and industrial potential by allowing precise control of growth conditions and supporting various edible strains. Utilizing bioreactors like stirred-tank and air-lift systems produces high-value, safe biomass for nutraceuticals, pharmaceuticals, food supplements, and vegan proteins. Optimizing temperature, dissolved oxygen, pH, agitation, and inoculum size is crucial for maximizing yield and ensuring safety. In the EU, strict regulations limit mushroom-derived products, despite their popularity in Asia and the USA. Compounds such as sterols, stanols, chitosan, and beta-glucans have EFSA approval. Future research should address bioreactor issues like agitation, viscosity, and biofilm formation, and explore LED lighting to boost bioactive compounds. Genomic and proteomic studies will help enhance mycelial bioactive content to match fruiting bodies. To expand the market for submerged cultivated mushroom products, focus on enhancing umami-related amino acids, standardizing amino acid profiling, and ensuring safety per FDA and EFSA standards. Prioritize research into bioactive peptides with health benefits, educate consumers on medicinal uses of mushroom compounds, and investigate novel fungal metabolites for medical applications.

References

Abdullah N, Ismail R, Mohd N et al (2013) Scientia Horticulturae Production of liquid spawn of an edible grey oyster mushroom, Pleurotus pulmonarius (Fr.) Quél by submerged fermentation and sporophore yield on rubber wood sawdust. Sci Hortic (Amsterdam) 161:65–69. https://doi.org/10.1016/j.scienta.2013.06.026

Agudelo-Escobar LM, Gutiérrez-López Y, Urrego-Restrepo S (2017) Effects of aeration, agitation and pH on the production of mycelial biomass and exopolysaccharide from the filamentous fungus Ganoderma lucidum1. Dyna 84:72–79. https://doi.org/10.15446/dyna.v84n200.57126

Ahlborn J, Stephan A, Meckel T et al (2019) Upcycling of food industry side streams by basidiomycetes for production of a vegan protein source. Int J Recycl Org Waste Agric 8:447–455. https://doi.org/10.1007/s40093-019-00317-4

Ahmed S, Bellary I, Samad A (2014) GT2014-25217 improvement of efficiency by design optimization of a centrifugal pump leading edge. 1–11

Amanullah A, Christensen LH, Hansen K et al (2002) Dependence of morphology on agitation intensity in fed-batch cultures of Aspergillus oryzae and its implications for recombinant protein production. Biotechnol Bioeng 77:815–826. https://doi.org/10.1002/bit.10181

Amilcar J, Insuasti P, Patricia C et al (2016) Design of a bioreactor for the bioconversion of lignocellulosic residues into protein for human feeding. https://doi.org/10.9790/1959-0506037279

Arshadi N, Nouri H, Moghimi H (2023) Increasing the production of the bioactive compounds in medicinal mushrooms: an omics perspective. Microb Cell Factories 22:1–34. https://doi.org/10.1186/s12934-022-02013-x

Assis IS, Chaves MB, Silveira MLL et al (2013) Production of bioactive compounds with antitumor activity against sarcoma 180 by pleurotus sajor-caju. J Med Food 16:1004–1012. https://doi.org/10.1089/jmf.2012.0267

Bains A, Chawla P, Kaur S et al (2021) Bioactives from mushroom: health attributes and food industry applications. Materials (Basel) 14. https://doi.org/10.3390/ma14247640

Bakratsas G, Polydera A, Katapodis P, Stamatis H (2021) Recent trends in submerged cultivation of mushrooms and their application as a source of nutraceuticals and food additives. Futur Foods 4:100086. https://doi.org/10.1016/j.fufo.2021.100086

Bedade D, Berezina O, Singhal R et al (2017) Extracellular xylanase production from a new xylanase producer Tuber maculatum mycelium under submerged fermentation and its characterization. Biocatal Agric Biotechnol 11:288–293. https://doi.org/10.1016/j.bcab.2017.07.008

Bentil JA, Thygesen A, Mensah M et al (2018) Cellulase production by white-rot basidiomycetous fungi: solid-state versus submerged cultivation. Appl Microbiol Biotechnol 102:5827–5839. https://doi.org/10.1007/s00253-018-9072-8

Bettin F, Cousseau F, Martins K et al (2019) Phenol removal by laccases and other phenol oxidases of Pleurotus sajor-caju PS-2001 in submerged cultivations and aqueous mixtures. J Environ Manage 236:581–590. https://doi.org/10.1016/j.jenvman.2019.02.011

Bettin F, Osório L, Montanari Q et al (2011) Growth kinetics , production , and characterization of extracellular laccases from. Process Biochem 46:758–764. https://doi.org/10.1016/j.procbio.2010.12.002

Bhambri A, Srivastava M, Mahale VG et al (2022) Mushrooms as potential sources of active metabolites and medicines. Front Microbiol 13. https://doi.org/10.3389/fmicb.2022.837266

Bisen PS, Baghel RK, Sanodiya BS et al (2010) Lentinus edodes: a macrofungus with pharmacological activities. Curr Med Chem 17:2419–2430. https://doi.org/10.2174/092986710791698495

Camacho-Morales RL, Sánchez JE (2015) Biotechnological use of fungi for the degradation of recalcitrant agro-pesticides. Mushroom Biotechnol Dev Appl:203–214. https://doi.org/10.1016/B978-0-12-802794-3.00012-6

Carlile MJ (1995) The success of the hypha and mycelium. In: Grow Fungus, pp 3–19. https://doi.org/10.1007/978-0-585-27576-5_1

Chahal DS (1989) Production of protein-rich mycelial biomass of a mushroom, Pleurotus sajor-caju, on corn stover. J Ferment Bioeng 68:334–338. https://doi.org/10.1016/0922-338X(89)90008-1

Chao CH, Wu HJ, Lu MK (2011) Promotion of fungal growth and underlying physiochemical changes of polysaccharides in Rigidoporus ulmarius, an edible Basidiomycete mushroom. Carbohydr Polym 85:609–614. https://doi.org/10.1016/j.carbpol.2011.03.023

Cho EJ, Oh JY, Chang HY, Yun JW (2006) Production of exopolysaccharides by submerged mycelial culture of a mushroom Tremella fuciformis. J Biotechnol 127:129–140. https://doi.org/10.1016/j.jbiotec.2006.06.013

Choi D, Maeng J, Ding J, Cha W (2007) Exopolysaccharide production and mycelial growth in an air-lift bioreactor using Fomitopsis pinicola. J Microbiol Biotechnol 17:18051607

Choi DB, Lee JH, Kim YS et al (2011) A study of mycelial growth and exopolysaccharide production from a submerged culture of Mycoleptodonoides aitchisonii in an air-lift bioreactor. Korean J Chem Eng 28:1427–1432. https://doi.org/10.1007/s11814-011-0109-2

Cohen R, Persky L, Hadar Y (2002) Biotechnological applications and potential of wood-degrading mushrooms of the genus Pleurotus. Appl Microbiol Biotechnol 58:582–594. https://doi.org/10.1007/s00253-002-0930-y

Confortin FG, Marchetto R, Bettin F et al (2008) Production of Pleurotus sajor-caju strain PS-2001 biomass in submerged culture. J Ind Microbiol Biotechnol 35:1149–1155. https://doi.org/10.1007/s10295-008-0394-x

Corrêa RCG, Brugnari T, Bracht A et al (2016) Biotechnological, nutritional and therapeutic uses of Pleurotus spp. (Oyster mushroom) related with its chemical composition: a review on the past decade findings. Trends Food Sci Technol 50:103–117. https://doi.org/10.1016/j.tifs.2016.01.012

Cui FJ, Chen XX, Liu WM et al (2016) Control of Grifola frondosa morphology by agitation and aeration for improving mycelia and exo-polymer production. Appl Biochem Biotechnol 179:459–473. https://doi.org/10.1007/s12010-016-2006-y

Das SK, Masuda M, Sakurai A, Sakakibara M (2010) Medicinal uses of the mushroom Cordyceps militaris: current state and prospects. Fitoterapia 81:961–968. https://doi.org/10.1016/j.fitote.2010.07.010

Devi KSP, Roy B, Patra P et al (2013) Characterization and lectin microarray of an immunomodulatory heteroglucan from Pleurotus ostreatus mycelia. Carbohydr Polym 94:857–865. https://doi.org/10.1016/j.carbpol.2013.02.017

Diamantopoulou P, Papanikolaou S, Komaitis M et al (2014) Patterns of major metabolites biosynthesis by different mushroom fungi grown on glucose-based submerged cultures. Bioprocess Biosyst Eng 37:1385–1400. https://doi.org/10.1007/s00449-013-1112-2

Do Valle JS, Vandenberghe LP d S, Santana TT et al (2014) Optimization of Agaricus blazei laccase production by submerged cultivation with sugarcane molasses. Afr J Microbiol Res 8:939–946. https://doi.org/10.5897/ajmr2013.6508

Dudekula UT, Doriya K, Devarai SK (2020) A critical review on submerged production of mushroom and their bioactive metabolites. 3 Biotech 10:1–12. https://doi.org/10.1007/s13205-020-02333-y

El Enshasy HA, Hatti-Kaul R (2013) Mushroom immunomodulators: unique molecules with unlimited applications. Trends Biotechnol 31:668–677. https://doi.org/10.1016/j.tibtech.2013.09.003

Elisashvili V (2012) Submerged cultivation of medicinal mushrooms: bioprocesses and products (review). Int J Med Mushrooms 14:211–239. https://doi.org/10.1615/IntJMedMushr.v14.i3.10
Elisashvili V, Penninckx M, Kachlishvili E et al (2008) Lentinus edodes and Pleurotus species lignocellulolytic enzymes activity in submerged and solid-state fermentation of lignocellulosic wastes of different composition. Bioresour Technol 99:457–462. https://doi.org/10.1016/j.biortech.2007.01.011
Fazenda ML, Harvey LM, McNeil B (2010) Effects of dissolved oxygen on fungal morphology and process rheology during fed-batch processing of Ganoderma lucidum. J Microbiol Biotechnol 20:844–851. https://doi.org/10.4014/jmb.0911.11020
Fazenda ML, Seviour R, McNeil B, Harvey LM (2008) Submerged culture fermentation of "higher fungi": the macrofungi. Adv Appl Microbiol 63:33–103. https://doi.org/10.1016/S0065-2164(07)00002-0
Filho PFS, Zamani A, Taherzadeh MJ (2017) Production of edible fungi from potato protein liquor (PPL) in airlift bioreactor. Fermentation. https://doi.org/10.3390/fermentation3010012
Fukushima Y, Okada K, Kawai G, Motai DH (1993) Efficient production of mycelium of Lentinus edodes by a continuous culture and the effect of lignin on growth. J Ferment Bioeng 76:45–48
Gallego M, Arnal M, Talens P et al (2020) Effect of gelatin coating enriched with antioxidant tomato by-products on the quality of pork meat. Polymers (Basel) 12:1–17. https://doi.org/10.3390/POLYM12051032
Gebreyohannes G, Sbhatu DB (2023) Wild mushrooms: a hidden treasure of novel bioactive compounds. Int J Anal Chem 2023. https://doi.org/10.1155/2023/6694961
Glazebrook MA, Vining LC, White RL (1992) Growth morphology of Streptomyces akiyoshiensis in submerged culture: influence of pH, inoculum, and nutrients. Can J Microbiol 38:98–103. https://doi.org/10.1139/m92-016
Gowthaman MK, Krishna C, Moo-Young M (2001) Fungal solid state fermentation - an overview. Appl Mycol Biotechnol 1:305–352. https://doi.org/10.1016/S1874-5334(01)80014-9
El-Enshasy H, Daba A, El-Demellawy M, Ibrahim A, El Sayed S, El-Badry I (2010) Bioprocess development for large scale production of anticancer exo-polysaccharide by Pleurotus ostreatus in submerged culture. J Appl Sci 10:2523–2529
Hadar Y, Cohen-Arazi E (1986) Chemical composition of the edible mushroom Pleurotus ostreatus produced by fermentation. Appl Environ Microbiol 51:1352–1354. https://doi.org/10.1128/aem.51.6.1352-1354.1986
Hoa HT, Wang CL (2015) The effects of temperature and nutritional conditions on mycelium growth of two oyster mushrooms (Pleurotus ostreatus and Pleurotus cystidiosus). Mycobiology 43:14–23. https://doi.org/10.5941/MYCO.2015.43.1.14
Di Hsu K, Wu SP, Lin SP et al (2017) Enhanced active extracellular polysaccharide production from Ganoderma formosanum using computational modeling. J Food Drug Anal 25:804–811. https://doi.org/10.1016/j.jfda.2016.12.006
Hwang HJ, Kim SW, Choi JW, Yun JW (2003) Production and characterization of exopolysaccharides from submerged culture of Phellinus linteus KCTC 6190. Enzym Microb Technol 33:309–319. https://doi.org/10.1016/S0141-0229(03)00131-5
Ibrahim D (2015) Effect of agitation speed on the morphology of Aspergillus Niger HFD5A-1 hyphae and its pectinase production in submerged fermentation. World J Biol Chem 6:265. https://doi.org/10.4331/wjbc.v6.i3.265
Jiang L (2010) Optimization of fermentation conditions for pullulan production by Aureobasidium pullulan using response surface methodology. Carbohydr Polym 79:414–417. https://doi.org/10.1016/j.carbpol.2009.08.027
Jiaojiao Z, Fen W, Kuanbo L et al (2018) Heat and light stresses affect metabolite production in the fruit body of the medicinal mushroom Cordyceps militaris. Appl Microbiol Biotechnol 102:4523
Jing X, Mao D, Geng L, Xu C (2013) Medium optimization, molecular characterization, and bioactivity of exopolysaccharides from Pleurotus eryngii. Arch Microbiol 195:749–757. https://doi.org/10.1007/s00203-013-0927-1

Junior Letti LA, Destéfanis Vítola FM, Vinícius de Melo Pereira G et al (2017) Solid-state fermentation for the production of mushrooms. In: Current developments in biotechnology and bioengineering

Kalač P (2009) Chemical composition and nutritional value of European species of wild growing mushrooms: a review. Food Chem 113:9–16. https://doi.org/10.1016/j.foodchem.2008.07.077

Kawagishi H, Nomura A, Mizuno T et al (1990) Isolation and characterization of a lectin from Grifola frondosa fruiting bodies. Biochimica et Biophysica Acta (BBA) 1034:247–252. https://doi.org/10.1016/0304-4165(90)90045-X

Khalili M, Ebrahimzadeh MA, Kosaryan M et al (2015) Iron chelation and liver disease healing activity of edible mushroom (Cantharellus cibarius), in vitro and in vivo assays. RSC Adv 5:4804–4810

Khan AA, Gani A, Khanday FA, Masoodi FA (2018) Biological and pharmaceutical activities of mushroom β-glucan discussed as a potential functional food ingredient. Bioact Carbohydrates Diet Fibre 16:1–13. https://doi.org/10.1016/j.bcdf.2017.12.002

Kim HO, Yun JW (2005) A comparative study on the production of exopolysaccharides between two entomopathogenic fungi Cordyceps militaris and Cordyceps sinensis in submerged mycelial cultures. J Appl Microbiol 99:728–738. https://doi.org/10.1111/j.1365-2672.2005.02682.x

Kim JM, Ra KS, Noh DO, Suh HJ (2002) Optimization of submerged culture conditions for the production of angiotensin converting enzyme inhibitor from Flammulina velutipes. J Ind Microbiol Biotechnol 29:292–295. https://doi.org/10.1038/sj.jim.7000306

Kim SW, Hwang HJ, Lee BC, Yun JW (2007) Submerged production and characterization of Grifola frondosa polysaccharides - a new application to cosmeceuticals. Food Technol Biotechnol 45:295–305

Kumar K, Mehra R, Guiné RPF et al (2021) Edible mushrooms: a comprehensive review on bioactive compounds with health benefits and processing aspects. Foods 10:1–22. https://doi.org/10.3390/foods10122996

Lähteenmäki-Uutela A, Rahikainen M, Lonkila A, Yang B (2021) Alternative proteins and EU food law. Food Control 130:108336. https://doi.org/10.1016/j.foodcont.2021.108336

LeDuy A, Kosaric N, Zajic JE (1974) Morel mushroom mycelium growth in waste sulfite liquors as source of protein and flavouring. Can Inst Food Sci Technol J 7:44–50. https://doi.org/10.1016/s0315-5463(74)73845-7

Lee BC, Bae JT, Pyo HB et al (2004) Submerged culture conditions for the production of mycelial biomass and exopolysaccharides by the edible Basidiomycete Grifola frondosa. Enzyme Microb Technol 35:369–376. https://doi.org/10.1016/j.enzmictec.2003.12.015

Lee IS, Kirschner R, Chen ST (2019) Medicinal mushroom Taiwanofungus camphoratus: a potential cure for cancer. Funct Foods Health Dis 9:508–520. https://doi.org/10.31989/ffhd.v9i8.576

Lee KM, Lee SY, Lee HY (1999) Bistage control of pH for improving exopolysaccharide production from mycelia of Ganodema lucidum in an air-lift fermentor. J Biosci Bioeng 88:646–650. https://doi.org/10.1016/S1389-1723(00)87094-2

Lee WY, Park Y, Ahn JK et al (2007) Factors influencing the production of endopolysaccharide and exopolysaccharide from Ganoderma applanatum. Enzyme Microb Technol 40:249–254. https://doi.org/10.1016/j.enzmictec.2006.04.009

Li IC, Lee LY, Tzeng TT et al (2018, 2018) Neurohealth properties of Hericium erinaceus mycelia enriched with erinacines. Behav Neurol:10.1155/2018/5802634

Lin ES, Sung SC (2006) Cultivating conditions influence exopolysaccharide production by the edible Basidiomycete Antrodia cinnamomea in submerged culture. Int J Food Microbiol 108:182–187. https://doi.org/10.1016/j.ijfoodmicro.2005.11.010

Liu YS, Wu JY (2012) Effects of Tween 80 and pH on mycelial pellets and exopolysaccharide production in liquid culture of a medicinal fungus. J Ind Microbiol Biotechnol 39:623–628. https://doi.org/10.1007/s10295-011-1066-9

Lobanok AG, Babitskaia VG, Plenina LV et al (2003) Composition and biological activity of submerged mycelium of the xylotrophic basidiomycete Lentinus edodes. Prikl Biokhim Mikrobiol 39:69–73

Łysakowska P, Sobota A, Wirkijowska A (2023) Medicinal mushrooms: their bioactive components, nutritional value and application in functional food production—a review. Molecules 28. https://doi.org/10.3390/molecules28145393
Manu-Tawiah W, Martin AM (1987) Chemical composition of Pleurotus ostreatus mycelial biomass. Food Microbiol 4:303–310. https://doi.org/10.1016/S0740-0020(87)80004-7
Manzi P, Pizzoferrato L (2000) Beta-glucans in edible mushrooms. Food Chem 68:315–318. https://doi.org/10.1016/S0308-8146(99)00197-1
Mao XB, Zhong JJ (2004) Hyperproduction of cordycepin by two-stage dissolved oxygen control in submerged cultivation of medicinal mushroom Cordyceps militaris in bioreactors. Biotechnol Prog 20:1408–1413. https://doi.org/10.1021/bp049765r
Mao XB, Zhong JJ (2006) Significant effect of NH4+ on cordycepin production by submerged cultivation of medicinal mushroom Cordyceps militaris. Enzym Microb Technol 38:343–350. https://doi.org/10.1016/j.enzmictec.2004.10.010
Márquez-Rocha FJ, Guillén NGK, Sánchez VJE, Vázquez-Duhalt R (1999) Growth characteristics of Pleurotus ostreatus in bioreactors. Biotechnol Tech 13:29–32. https://doi.org/10.1023/A:1008861432337
Martin AM (1983) Submerged production of Agaricus campestris mycelium in peat extracts. J Food Sci 48:206–207
Mikiashvili N, Wasser SP, Nevo E, Elisashvili V (2006) Effects of carbon and nitrogen sources on Pleurotus ostreatus ligninolytic enzyme activity. World J Microbiol Biotechnol 22:999–1002. https://doi.org/10.1007/s11274-006-9132-6
Mukhopadhyay R, Chatterjee S, Chatterjee BP, Guha AK (2005) Enhancement of biomass production of edible mushroom Pleurotus sajor-caju grown in whey by plant growth hormones. Process Biochem 40:1241–1244. https://doi.org/10.1016/j.procbio.2004.05.006
Mumpuni A, Ekowati N, Purnomowati P, Purwati ES (2017) Growth and protein content establishment of Pleurotus ostreatus on liquid and solid medium. Biosaintifika 9:572. https://doi.org/10.15294/biosaintifika.v9i3.11660
Nikitina VE, Tsivileva OM, Pankratov AN, Bychkov NA (2007) Lentinula edodes biotechnology - from lentinan to lectins. Food Technol Biotechnol 45:230–237
Papaspyridi LM, Aligiannis N, Topakas E et al (2012) Submerged fermentation of the edible mushroom Pleurotus ostreatus in a batch stirred tank bioreactor as a promising alternative for the effective production of bioactive metabolites. Molecules 17:2714–2724. https://doi.org/10.3390/molecules17032714
Park JP, Kim YM, Kim SW et al (2002) Effect of agitation intensity on the exo-biopolymer production and mycelial morphology in Cordyceps militaris. Lett Appl Microbiol:433–438
Pérez-Chávez AM, Mayer L, Albertó E (2019) Mushroom cultivation and biogas production: a sustainable reuse of organic resources. Energy Sustain Dev 50:50–60. https://doi.org/10.1016/j.esd.2019.03.002
Petre M, Petre V (2015) Biotechnology of mushroom growth through submerged cultivation. Elsevier Inc.
Pierre Beguin JPA (1994) The biological degradation of cellulose. FEMS Microbiol Rev 13:25–58
Pokhrel CP, Ohga S (2007) Submerged culture conditions for mycelial yield and polysaccharides production by Lyophyllum decastes. Food Chem 105:641–646. https://doi.org/10.1016/j.foodchem.2007.04.033
Rathore H, Prasad S, Kapri M et al (2019) Medicinal importance of mushroom mycelium: mechanisms and applications. J Funct Foods 56:182–193. https://doi.org/10.1016/j.jff.2019.03.016
Ribeiro B, De Pinho PG, Andrade PB et al (2011) Do bioactive carotenoids contribute to the color of edible mushrooms? Open Chem Biomed Methods J 4:14–18
Rosales-López C, Vargas-López A, Monge-Artavia M, Rojas-Chaves M (2022) Evaluation of conditions to improve biomass production by submerged culture of Ganoderma sp. Microorganisms 10. https://doi.org/10.3390/microorganisms10071404
Sandargo B, Chepkirui C, Cheng T et al (2019) Biological and chemical diversity go hand in hand: basidiomycota as source of new pharmaceuticals and agrochemicals. Biotechnol Adv 37:107344. https://doi.org/10.1016/j.biotechadv.2019.01.011

Sheng K, Wang C, Chen B et al (2021) Recent advances in polysaccharides from Lentinus edodes (Berk.): isolation, structures and bioactivities. Food Chem 358:129883. https://doi.org/10.1016/j.foodchem.2021.129883
Shu CH, Lung MY (2004) Effect of pH on the production and molecular weight distribution of exopolysaccharide by Antrodia camphorata in batch cultures. Process Biochem 39:931–937. https://doi.org/10.1016/S0032-9592(03)00220-6
Sillapachaiyaporn C, Chuchawankul S (2020) HIV-1 protease and reverse transcriptase inhibition by tiger milk mushroom (Lignosus rhinocerus) sclerotium extracts: in vitro and in silico studies. J Tradit Complement Med 10:396–404. https://doi.org/10.1016/j.jtcme.2019.08.002
Singdevsachan SK, Auroshree P, Mishra J et al (2016) Mushroom polysaccharides as potential prebiotics with their antitumor and immunomodulating properties: a review. Bioact Carbohydr Diet Fibre 7:1–14. https://doi.org/10.1016/j.bcdf.2015.11.001
Singh U, Gautam A, Singha TK et al (2020) Mass production of Pleurotus eryngii mycelia under submerged culture conditions with improved minerals and vitamin D2. LWT Food Sci Technol:109665. https://doi.org/10.1016/j.lwt.2020.109665
Smiderle FR, Olsen LM, Ruthes AC et al (2012) Exopolysaccharides, proteins and lipids in Pleurotus pulmonarius submerged culture using different carbon sources. Carbohydr Polym 87:368–376. https://doi.org/10.1016/j.carbpol.2011.07.063
Songulashvili GG, Elisashvili V, Penninckx M et al (2005) Bioconversion of plant raw materials in value-added products by Lentinus edodes (Berk.) singer and Pleurotus spp. Int J Med Mushrooms 7:467–468. https://doi.org/10.1615/intjmedmushr.v7.i3.990
Sun L b, Zhang Z y, Xin G et al (2020) Advances in umami taste and aroma of edible mushrooms. Trends Food Sci Technol 96:176–187. https://doi.org/10.1016/j.tifs.2019.12.018
Tanaka M, Matsuno R (1985) Conversion of lignocellulosic materials to single-cell protein (SCP): recent developments and problems. Enzym Microb Technol 7:197–206. https://doi.org/10.1016/S0141-0229(85)80002-8
Tang YJ, Liu RS, Li HM (2015) Current progress on truffle submerged fermentation: a promising alternative to its fruiting bodies. Appl Microbiol Biotechnol 99:2041–2053. https://doi.org/10.1007/s00253-015-6379-6
Tang YJ, Zhong JJ (2003) Role of oxygen supply in submerged fermentation of Ganoderma lucidum for production of Ganoderma polysaccharide and ganoderic acid. Enzym Microb Technol 32:478–484. https://doi.org/10.1016/S0141-0229(02)00338-1
Tang YJ, Zhu LW, Li HM, Li DS (2007) Submerged culture of mushrooms in bioreactors - challenges, current state-of-the-art, and future prospects. Food Technol Biotechnol 45:221–229
Tsai SY, Huang SJ, Mau JL (2006) Antioxidant properties of hot water extracts from Agrocybe cylindracea. Food Chem 98:670–677. https://doi.org/10.1016/j.foodchem.2005.07.003
Tsuge H, Otatsume S, Kobayashi K et al (2004) Liquid circulation and mass transfer in an external-loop airlift reactor with partitioning plates. J Chem Eng Jpn 37:941–946
Turło J (2014) The biotechnology of higher fungi - current state and perspectives. Folia Biol Oecol 10:49–65. https://doi.org/10.2478/fobio-2014-0010
Ulziijargal E, Yang JH, Lin LY et al (2013) Quality of bread supplemented with mushroom mycelia. Food Chem 138:70–76. https://doi.org/10.1016/j.foodchem.2012.10.051
Vacharathit V (2014) Production of provitamina carotenoids in mushrooms and uses thereof. 2
Valverde ME, Hernández-Pérez T, Paredes-López O (2015) Edible mushrooms: improving human health and promoting quality life. Int J Microbiol 2015. https://doi.org/10.1155/2015/376387
Wagner R, Mitchell DA, Sassaki GL et al (2003) Current techniques for the cultivation of Ganoderma lucidum for the production of biomass, Ganoderic acid and polysaccharides. Food Technol Biotechnol 41:371–382
Wang H, Gao J, Ng TB (2000) A new lectin with highly potent antihepatoma and antisarcoma activities from the oyster mushroom Pleurotus ostreatus. Biochem Biophys Res Commun 275:810–816. https://doi.org/10.1006/bbrc.2000.3373
Wang HX, Ng TB, Liu WK et al (1995) Isolation and characterization of two distinct lectins with antiproliferative activity from the cultured mycelium of the edible mushroom Tricholoma

mongolicum. Int J Pept Protein Res 46:508–513. https://doi.org/10.1111/j.1399-3011.1995.tb01606.x

Wang SJ, Zhong JJ (2007) Chapter 6. Bioreactor engineering. In: Bioprocessing for value-added products from renewable resources, pp 131–161. https://doi.org/10.1016/B978-044452114-9/50007-4

Wu J, Chen H, Chen M et al (2012) Quantitative analysis of LED effects on exopolysaccharide and mycelial growth productions in Pleurotus eryngii cultures. Soc Chem Ind. https://doi.org/10.1002/jctb.04038

Wu JZ, Cheung PCK, Wong KH, Huang NL (2003) Studies on submerged fermentation of Pleurotus tuber-regium (Fr.) Singer - part 1: physical and chemical factors affecting the rate of mycelial growth and bioconversion efficiency. Food Chem 81:389–393. https://doi.org/10.1016/S0308-8146(02)00457-0

Wu Y, Choi M, Li J et al (2016) Mushroom cosmetics: the present and futures. Cosmetics 3:22. https://doi.org/10.3390/cosmetics3030022

Xiao JH, Chen DX, Liu JW et al (2004) Optimization of submerged culture requirements for the production of mycelial growth and exopolysaccharide by Cordyceps jiangxiensis JXPJ 0109. J Appl Microbiol 96:1105–1116. https://doi.org/10.1111/j.1365-2672.2004.02235.x

Wu X (2012) Factors affecting extracellular and intracellular polysaccharide production in submerged cultivation of Tricholoma Mongolicum. Afr J Microbiol Res 6. https://doi.org/10.5897/ajmr-11-632

Xu DQ, Fu ML, Chen QH, Liu J (2011a) Effect of submerged culture conditions on exopolysaccharides production by Armillaria luteo-virens Sacc QH and kinetic modeling. Bioprocess Biosyst Eng 34:103–111. https://doi.org/10.1007/s00449-010-0451-5

Xu X, Yan H, Chen J, Zhang X (2011b) Bioactive proteins from mushrooms. Biotechnol Adv 29:667–674. https://doi.org/10.1016/j.biotechadv.2011.05.003

Yang FC, Liau CB (1998) Effects of cultivating conditions on the mycelial growth of Ganoderma lucidum in submerged flask cultures. Bioprocess Eng 19:233–236. https://doi.org/10.1007/s004490050512

Zhang BB, Cheung PCK (2011) A mechanistic study of the enhancing effect of Tween 80 on the mycelial growth and exopolysaccharide production by Pleurotus tuber-regium. Bioresour Technol 102:8323–8326. https://doi.org/10.1016/j.biortech.2011.06.021

Zhang G, Sun J, Wang H, Ng TB (2010) First isolation and characterization of a novel lectin with potent antitumor activity from a Russula mushroom. Phytomedicine 17:775–781. https://doi.org/10.1016/j.phymed.2010.02.001

Zhou Y, Han LR, He HW et al (2018) Effects of agitation, aeration and temperature on production of a novel glycoprotein gp-1 by streptomyces kanasenisi zx01 and scale-up based on volumetric oxygen transfer coefficient. Molecules 23:1–14. https://doi.org/10.3390/molecules23010125

Zou X (2006) Fed-batch fermentation for hyperproduction of polysaccharide and ergosterol by medicinal mushroom Agaricus brasiliensis. Process Biochem 41:970–974. https://doi.org/10.1016/j.procbio.2005.10.030

Cultivation of Medicinal Mushroom in Bioreactors

Mythileeswari Lakshmikanthan, Sakthivel Muthu, Kathiravan Krishnan, Selvakumari Jeyaperumal, and Lakshmanan Govindan

Abstract The cultivation of medicinal mushrooms holds significant promise for biotechnological applications, driven by their rich reservoir of bioactive compounds with therapeutic potential. This study investigates the submerged cultivation of *Cordyceps sinensis*, *Grifola frondosa*, and solid-state cultivation of *Ganoderma lucidum* and *Hericium erinaceus*, focusing on optimizing growth parameters for enhanced biomass and metabolite production. For *C. sinensis*, carbon and nitrogen sources significantly influenced mycelial growth, with peptone and beef extract demonstrating superior performance. Temperature and pH were identified as crucial factors for optimal growth and metabolite synthesis. In *G. frondosa*, pH ranging from 4.0 to 7.0 and temperature at 25 °C favored biomass and exopolysaccharide production. Aeration and agitation rates influenced fermentation performance and morphological characteristics. *G. lucidum* cultivation on sawdust substrate exhibited notable biomass production, with temperature and pH optimization enhancing growth and exopolysaccharide accumulation. *Hericium erinaceus* cultivation in a stirred tank reactor showed promising results, with substrate composition and aeration strategies contributing to biomass yield. These findings underscore the impor-

M. Lakshmikanthan
Department of Research , Saveetha College of Nursing (SCON), Saveetha Institute of Medical and Technical Sciences (SIMATS), Thandalam, Chennai, India

S. Muthu
Natural Biomedicine Laboratory, Saveetha Medical College and Hospital (SMCH), Saveetha Institute of Medical and Technical Sciences (SIMATS), Thandalm, Chennai, India

K. Krishnan (✉)
Department of Biotechnology, University of Madras, Guindy Campus, Chennai, Tamil Nadu, India

S. Jeyaperumal
National Centre for Disease Control, Kozhikode, Thiruvananthapuram Field Unit, Iranimuttam, Kerala, India

L. Govindan (✉)
Central Research Laboratory, Sri Lakshmi Narayana Institute of Medical Sciences (affiliated to Bharath Institute of Higher Education and Research), Puducherry, India

M. N. Owaid et al. (eds.), *Mushroom Production*, Interdisciplinary Biotechnological Advances, https://doi.org/10.1007/978-981-95-7793-4_2

tance of precise control over cultivation parameters in maximizing the biotechnological potential of medicinal mushrooms.

Keywords Medicinal mushrooms · Bioreactor cultivation · Submerged cultivation · Solid-state cultivation · Growth optimization

1 Introduction

Medicinal mushrooms have an established history of use in traditional oriental medicine. Numerous mushrooms traditionally used for their medicinal properties come from genera such as *Tremella*, *Ganoderma*, *Flammulina*, *Lentinus*, *Grifola*, *Trametes*, and *Auricularia*. These mushrooms have been scientifically proven to offer substantial health benefits (Ajith and Janardhanan 2007).

Bioreactor cultivation of medicinal mushrooms represents a cutting-edge approach in the field of mycology, offering numerous advantages over traditional methods (Berovic et al. 2022). Bioreactors, which are vessels designed to provide a controlled environment for biological reactions, enable the precise regulation of growth conditions such as temperature, pH, agitation, nutrient feeding, and sterility, oxygen, and nutrient supply (Regonesi 2023). This technological advancement is crucial for the cultivation of medicinal mushrooms, which are often challenging to grow in natural settings due to their specific environmental requirements and susceptibility to contamination (Zou et al. 2024). The use of bioreactors allows for the consistent and scalable production of high-quality mycelium and fruiting bodies, which are valued for their therapeutic bioactive compounds like polysaccharides, triterpenoids, and ergosterols. These compounds have significant medicinal properties, including immunomodulatory, anti-inflammatory, neuroprotective, renoprotective effects, and antiproliferative. By optimizing the cultivation conditions within bioreactors, it is possible to enhance the yield and potency of these bioactive substances (Rathore et al. 2019).

Furthermore, bioreactor cultivation mitigates the limitations of seasonal and geographical variability, making it possible to produce medicinal mushrooms year-round in any location (Elisashvili 2012). This method also reduces the risk of contamination from competing microorganisms, which is a common challenge in traditional outdoor or semi-controlled cultivation systems. Advanced bioreactors can be equipped with sensors and automation technologies to monitor and adjust the growth parameters in real-time, ensuring optimal conditions throughout the cultivation process (Chong et al. 2023).

The scalability of bioreactor systems is another significant advantage, allowing for the mass production of medicinal mushrooms to meet increasing global demand (Berovic and Zhong 2023). Whether for pharmaceutical applications, nutraceutical products, or functional foods, bioreactor cultivation ensures a reliable supply of high-quality medicinal mushrooms. This modern approach not only enhances the efficiency and sustainability of mushroom production but also opens new possibilities for research and development in mycology and biotechnology (Bakratsas et al. 2021).

In summary, bioreactor cultivation of medicinal mushrooms offers a controlled, efficient, and scalable method to produce these valuable fungi, ensuring consistent quality and maximizing the production of their therapeutic compounds. This innovation marks a significant advancement in the field, supporting the growing needs of both the medical and commercial sectors.

2 Medicinal Mushrooms

Various types of medicinal mushrooms are cultured in bioreactors to harness their bioactive compounds for therapeutic purposes (Łysakowska et al. 2023). Some of the most commonly cultured medicinal mushrooms include the following:

Cordyceps sinensis: Valued for its anti-inflammatory, antitumor, and immunomodulatory activities, *Cordyceps* produces bioactive compounds such as cordycepin, adenosine, and polysaccharides (Das et al. 2021).

Lentinula edodes (shiitake): Shiitake mushrooms are recognized for their antiviral, antibacterial, and immune-enhancing properties. They contain lentinan, a polysaccharide with potential anticancer effects, and eritadenine, which may help lower cholesterol levels (Roszczyk et al. 2022).

Grifola frondosa (maitake): Known for its immune-boosting and anticancer properties, maitake mushrooms produce polysaccharides like beta-glucans, which are significant for their therapeutic benefits (Wu et al. 2021).

Ganoderma lucidum (reishi): Known for its immune-boosting properties and potential anticancer effects, reishi mushrooms are rich in triterpenoids, polysaccharides, and peptidoglycans (Oke et al. 2022).

Hericium erinaceus (lion's mane): This mushroom is reputed for its neuroprotective effects, potentially supporting brain health and cognitive function. It contains hericenones and erinacines, compounds that promote nerve growth factor (NGF) synthesis (Spelman et al. 2017).

Trametes versicolor (turkey tail): Turkey tail mushrooms are used for their immunomodulatory and anticancer properties, largely due to the presence of polysaccharide-K (PSK) and polysaccharopeptide (PSP) (Habtemariam 2020).

Inonotus obliquus (Chaga): Chaga mushrooms are valued for their antioxidant, anti-inflammatory, and anticancer properties. They produce compounds such as betulinic acid, polysaccharides, and melanin (Plehn et al. 2023).

Agaricus blazei: This mushroom is known for its immune-enhancing and antitumor activities, attributed to its polysaccharides, particularly beta-glucans (Firenzuoli et al. 2008).

Fig. 1 The different types of medicinal mushrooms

These medicinal mushrooms are cultured in bioreactors using either submerged fermentation (SmF) or solid-state fermentation (SSF) methods to optimize the production of their bioactive compounds. The controlled environment of bioreactors allows for precise manipulation of growth conditions, leading to consistent and high yields of medicinally valuable biomass (Fig. 1) (Costa et al. 2017; Dudekula et al. 2020).

3 Selection of Medicinal Mushroom Strains

3.1 Genetic Diversity and Strain Selection

Genetic diversity is crucial in selecting strains of medicinal mushrooms due to its impact on growth rate, yield, and the production of bioactive compounds (Arshadi et al. 2023). Different strains demonstrate distinct characteristics that can be advantageous for various applications. To evaluate genetic diversity and identify strains with optimal traits, such as elevated polysaccharide levels or strong antioxidant properties, scientists utilize molecular techniques. These methods enable the precise assessment of genetic variation and facilitate the selection of superior strains for cultivation and medicinal use (Venturella et al. 2021).

3.2 *Phenotypic Screening for Bioactivity*

Phenotypic screening is a pivotal method employed in the assessment of medicinal mushroom strains to gauge their therapeutic viability. This process entails the meticulous evaluation of discernible characteristics, encompassing the morphology of fruiting bodies, patterns of mycelial growth, and the production of metabolites. Strains exhibiting auspicious bioactivity, such as immunomodulatory capabilities, anti-inflammatory attributes, or antimicrobial potency, are accorded precedence for subsequent cultivation endeavors.

3.3 *Adaptation to Cultivation Conditions*

Strain selection in medicinal mushroom cultivation hinges significantly upon the adaptability of strains to specific environmental conditions. Mushrooms exhibiting compatibility with precise parameters such as temperature, humidity, aeration and substrate composition are prioritized for cultivation endeavors. Moreover, strains showcasing proficiency in swift colonization and fruiting within controlled environments are greatly coveted for enhancing production efficiency, particularly in bioreactor systems (Sher et al. 2010).

3.4 *Stability and Reproducibility*

Ensuring the stability and reproducibility of medicinal mushroom strains is paramount in the selection process, as it directly impacts the consistency of yields and product quality. Strains exhibiting phenotypic constancy across successive generations, without significant phenotypic variance, are highly favored. Achieving this requires comprehensive scrutiny and continual monitoring of strains throughout the cultivation cycle to evaluate their genetic stability and reproductive fidelity.

3.5 *Preservation and Maintenance*

Preservation techniques, such as cryopreservation or lyophilization, are employed to maintain the long-term viability of selected mushroom strains. This ensures a continuous supply of genetically uniform cultures for future cultivation. Additionally, comprehensive databases and culture collections are established to catalog and manage diverse mushroom strains for research and commercial purposes (García-García et al. 2014).

3.6 Cultivation of Medicinal Mushrooms

The cultivation of medicinal mushrooms has evolved significantly from traditional methods reliant on natural habitats. Historically, mushrooms were predominantly grown on wood logs, a practice dating back centuries (Bijalwan et al. 2020). However, due to their scarcity in nature and the increasing demand for these fungi, cultivation techniques have diversified and advanced. Contemporary methods now include cultivating mushrooms on artificial media, such as sawdust substrates in trays, beds, sterilized plastic bags, or bottles. These methods offer controlled environments conducive to optimal mushroom growth and yield. Additionally, techniques have been developed for producing fungal mycelia in bioreactors, utilizing both submerged cultivation in liquid media and solid-state substrates. These advancements in cultivation technology have enabled efficient and scalable production of medicinal mushrooms, meeting the rising demand while also ensuring consistent quality and purity (Berovic and Podgornik 2015).

3.7 Nutrient Requirements for Medicinal Mushrooms

Effective media preparation and optimization are crucial for the successful cultivation of medicinal mushrooms in bioreactors. Nutritional supplementation of the cultivation substrate is a critical agronomic practice in mushroom cultivation. The growth, yield, and quality parameters of mushrooms are significantly influenced by varying levels of supplementation. Incorporating various additives, including nitrogen sources, into substrates has been demonstrated to enhance the growth, yield, and quality of mushrooms (Onyango et al. 2011). These additives typically alter the decomposition rate and sequence of substrate components. The effectiveness of agricultural waste as a substrate is often significantly improved when supplemented with protein-rich materials (Frimpong-Manso et al. 2011). For instance, supplementing substrates with nutrients has been shown to increase the yields of *Pleurotus sajor-caju* mushrooms. Common supplements such as urea, ammonium sulfate, gram flour, soybean meal, mustard cake, cottonseed cake, and molasses are recommended to enhance oyster mushroom cultivation. Substrate supplementation is a well-established practice to boost the productivity of *Pleurotus* species. Adding supplements to mushroom substrates is particularly important for substrates with low protein content, as it promotes better growth and yield. Most agricultural wastes used as substrates are lignocellulosic materials, which are typically low in protein and inadequate for commercial mushroom cultivation. These substrates require various supplements or additives containing sufficient amounts of nitrogen, phosphorus, potassium, and vitamins to improve mushroom growth and yield (Alam et al. 2010). Supplementing the substrate is a common method to increase productivity, often measured by biological efficiency. The most common supplements are organic nitrogen sources, such as cereal bran, which are essential for the growth of the mycelial mass but can potentially affect the productivity and biological efficiency of the mushrooms (Assan et al. 2014).

4 Production of Biomass in Bioreactors

The production of medicinal mushroom biomass in bioreactors involves a sophisticated biotechnological process designed to optimize the growth and yield of valuable fungal species. Medicinal mushrooms, such as *Ganoderma lucidum* (reishi), *Cordyceps sinensis*, and *Lentinula edodes* (shiitake), are known for their bioactive compounds with therapeutic properties, including polysaccharides, triterpenoids, and ergosterol (Arshadi et al. 2023). Bioreactors provide a controlled environment for cultivating these mushrooms, allowing precise regulation of parameters such as temperature, pH, oxygen levels, and nutrient supply. This control is essential to maximize the production of the desired bioactive compounds (Dudekula et al. 2020). Typically, two main types of bioreactors are used: submerged fermentation (SmF) and solid-state fermentation (SSF) (Webb 2017).

In submerged fermentation, the mycelium of the mushroom is cultured in a liquid medium. This method is advantageous for producing a high yield of fungal biomass and extracting water-soluble metabolites. The liquid medium usually contains carbon and nitrogen sources, vitamins, and minerals, which support the growth and metabolic activity of the mycelium. Agitation and aeration are crucial in SmF to ensure homogenous nutrient distribution and sufficient oxygen supply (Fig. 2) (Kachrimanidou et al. 2023).

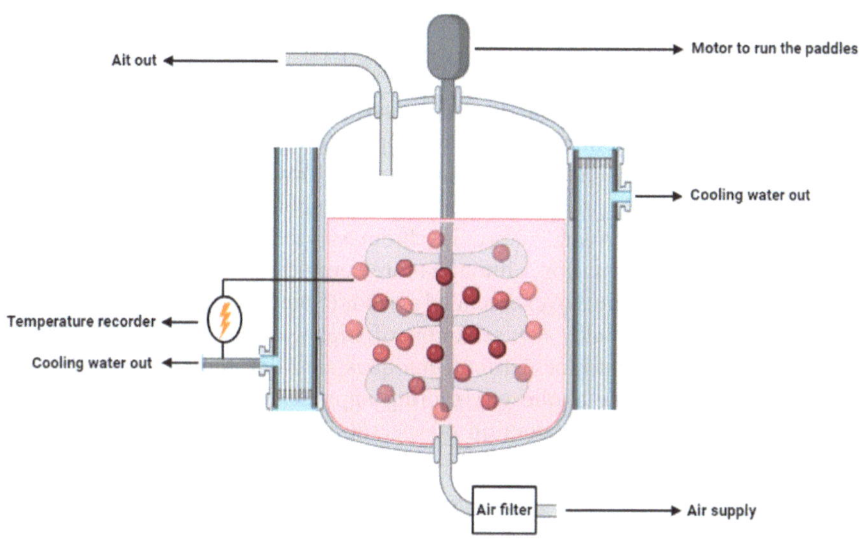

Fig. 2 The bioreactor for submerged cultivation

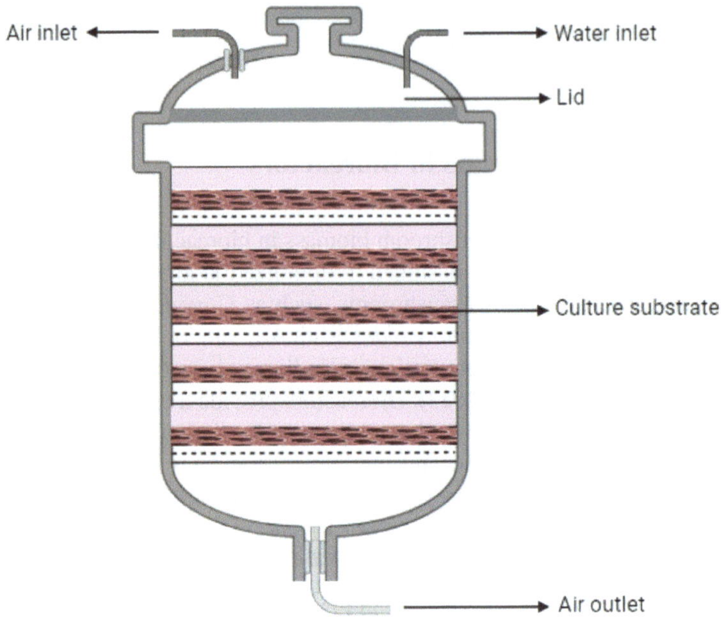

Fig. 3 The bioreactor for solid-state cultivation

Solid-state fermentation, on the other hand, involves the cultivation of mushrooms on a solid substrate, such as agricultural residues (e.g., rice bran and wheat straw). SSF is often preferred for the production of mushrooms that naturally grow on solid materials, as it closely mimics their natural environment. This method is particularly effective for producing complex secondary metabolites and enzymes (Fig. 3) (Lübeck and Lübeck 2022).

Scaling up the production of medicinal mushroom biomass in bioreactors requires careful consideration of bioprocess parameters and reactor design. Batch, fed-batch, and continuous modes of operation can be employed, each with specific advantages. Batch fermentation is straightforward and suitable for initial studies and small-scale production. Fed-batch fermentation allows for the incremental addition of nutrients, improving biomass yield and metabolite production. Continuous fermentation, although more complex, provides a steady supply of biomass and metabolites over extended periods (Spier et al. 2011). Optimization of the bioprocess includes the selection of high-yielding mushroom strains, development of efficient inoculation techniques, and fine-tuning of culture conditions. Genetic and metabolic engineering approaches are increasingly being explored to enhance the production of specific bioactive compounds (Hussain et al. 2022).

5 Submerged Cultivation of *Cordyceps sinensis*

5.1 Effects of Carbon and Nitrogen Source

The effect of various carbon and nitrogen sources on the mycelial growth of *Cordyceps sinensis* was investigated. For carbon sources, different types of monosaccharides (D-arabinose, D-xylose, D-fructose, D-galactose, D-glucose), disaccharides (cellobiose, lactose, maltose, sucrose), polysaccharides (cellulose, dextrin, malt extract, pectin, soluble starch), an organic acid (citric acid), and alditols (D-mannitol, D-sorbitol) were tested. Each carbon source was added to the basal medium at a concentration of 12 g of pure carbon per liter, calculated based on the carbon content of the molecule (Kim and Yun 2005). For the nitrogen source experiments, sucrose was used as the carbon source at 12 g/L, as determined from the carbon source tests. Various nitrogen sources, including inorganic nitrogen compounds (potassium nitrate, sodium nitrate, ammonium nitrate, ammonium sulfate, ammonium chloride, urea), amino acids (glycine, L-aspartic acid, L-glutamine, L-lysine, L-methionine, L-tryptophan), and complex organic nitrogen sources (beef extract, casein, peptone, yeast extract), were added individually to the basal medium at a concentration of 0.96 g of pure nitrogen per liter, calculated based on the nitrogen content of the molecule. Among the 16 nitrogen sources examined, peptone was the most effective for the mycelial growth of *C. sinensis*, with beef extract also proving to be a good nitrogen source. There was no significant difference in mycelial yield between media containing peptone and beef extract. Both sources resulted in significantly higher mycelial yields compared to other nitrogen sources. Nitrate sources (ammonium, potassium, and sodium nitrates) supported better mycelial growth than ammonium salts (ammonium chloride and ammonium sulfate), although the differences were not statistically significant. Single amino acids did not serve as good nitrogen sources for growth, with aspartic acid supporting the best mycelial growth among the amino acids tested. Media containing glycine, L-lysine, L-methionine, and L-tryptophan resulted in only weak growth. Ammonium salts, glycine, L-lysine, L-methionine, L-tryptophan, and urea were not utilized effectively by the fungus and had a negative effect on growth, as evidenced by lower mycelial dry weight compared to the nitrogen-free control (Dong and Yao 2005; Mao et al. 2005; Yoon et al. 2008).

5.2 Effects of Temperature and pH

The cultivation of *Cordyceps sinensis* in submerged fermentation is significantly influenced by both temperature and pH, which are critical parameters in optimizing growth and metabolite production. The mycelial growth and bioactive compound synthesis of *C. sinensis* are highly sensitive to temperature variations. Typically, the optimal temperature range for its cultivation lies between 20 °C and 25 °C, facilitating robust mycelial biomass accumulation. Deviations from this range can lead to suboptimal growth rates and reduced yields of key metabolites such as cordycepin and adenosine (Mag et al. 2015). Similarly, the pH of the growth medium plays a pivotal role in the metabolic activities of *C. sinensis*. An optimal pH range, generally between 6.0 and 7.0, is crucial for maintaining enzymatic activities and nutrient availability. pH levels outside this range can adversely affect cellular processes, leading to diminished growth and lower production of valuable secondary metabolites. Therefore, precise control of temperature and pH is essential in submerged fermentation to maximize the biotechnological potential (Mag et al. 2015; Park et al. 2001).

5.3 Effects of Aeration and Agitation

Aeration and agitation are critical parameters influencing the cultivation of *Cordyceps sinensis* in submerged fermentation. Optimal aeration (1000-mL shake flask:aeration rate 1 L/min) ensures sufficient oxygen transfer rates, which are vital for the metabolic activities and biomass accumulation of this fungus. Proper oxygenation facilitates aerobic respiration, enhancing the production of bioactive compounds such as cordycepin and adenosine. Concurrently, agitation speed 130 rpm plays a significant role in maintaining homogeneity within the culture medium, preventing cell sedimentation and promoting uniform nutrient distribution. Adequate agitation also helps in the dispersal of mycelial clumps, thereby increasing the surface area for nutrient uptake and gas exchange. However, excessive agitation may lead to shear stress, potentially damaging the delicate mycelial structures and adversely affecting growth and metabolite production (Park et al. 2002; Yan et al. 2020).

6 Submerged Cultivation of *Grifola frondosa*

6.1 Inoculum Preparation

G. frondosa was initially cultured on potato dextrose agar (PDA) plates. Subsequently, ten approximately 5 × 5 mm sections of the mycelial culture were excised using a sterilized, custom-designed cutter and transferred to the seed medium. Flask fermentation was carried out in a 500-mL flask containing 150 mL of medium, inoculated with 10% (v/v) of the seed culture (Cui et al. 2016).

6.2 pH and Temperature

The pH of the medium plays a crucial role in the growth of fungal biomass and the production of exopolysaccharides (EPS). Although precise pH control is challenging during flask studies, the initial pH is essential for assessing its effects on biomass accumulation and EPS synthesis. pH fluctuations, due to metabolic activity, influence cell morphology and nutrient uptake, underscoring the relationship between pH, the chosen microorganism, and the nutrient medium composition (Elisashvili 2012). Dudekula et al. (2020) found that *Grifola frondosa*, which produces antitumor polysaccharides, reaches optimal biomass and EPS production at pH 4.0–7.0. This finding aligns with observations in other mushroom species. The effects of varying temperatures (20–30 °C) on the growth of *Grifola frondosa* were investigated, and the results demonstrated that maximum biomass and exopolysaccharide (EPS) production were achieved at 25 °C.

6.3 Agitation and Aeration Rate

A series of submerged cultivations were performed in a 25-L stirred-tank bioreactor with a working volume of 16 L. The seed culture, prepared in a 500-mL flask, was inoculated at 10% (v/v). A range of agitation rates, from 60 to 150 rpm at a fixed aeration rate of 0.5 vvm, and various aeration rates, from 0.25 to 1.0 vvm at a constant agitation speed of 90 rpm, were employed to investigate their effects on the fermentation performance and morphological characteristics of *Grifola frondosa*. All fermentations were conducted at 28 °C for 7 days, with continuous monitoring of temperature, pH, dissolved oxygen, agitation speed, and aeration rate. Samples were periodically collected to determine residual glucose levels, mycelial morphology, mycelial biomass, and exo-biopolymer concentration (Cui et al. 2016).

7 Solid-State Cultivation of *Ganoderma lucidum*

7.1 Effects of Substrate

Solid-state cultivation of *Ganoderma lucidum* offers numerous advantages, including low investment costs, ease of operation, and minimal waste production. However, it also presents challenges such as extended cultivation periods and difficulties in separating and isolating the biomass and pharmaceutically active compounds from the solid substrate. Research on solid-state cultivation of *G. lucidum* in bioreactors is limited (Kachrimanidou et al. 2023). Gregori et al. (2015) followed a method for growing *G. lucidum* on a solid substrate using a horizontal stirred bioreactor, utilizing sawdust as the substrate. This process resulted in significant biomass production, including pharmaceutically valuable products. In their studies (2000, 2002), the substrate composition included beech sawdust, olive oil, $(NH_4)2SO_4$, KH_2PO_4, $CaCl_2 \cdot 2H_2O$, $MgSO_4 \cdot 7H_2O$, $FeSO_4 \cdot 7H_2O$, and distilled water. During cultivation, the exopolysaccharide (EPS) content in the solids increased rapidly over the first 7 days, stabilized until day 21, and then declined, indicating polysaccharide degradation in the later stages. This decrease in polysaccharide content corresponded with a rapid drop in the water mass fraction from 70–80% in the first 21 days to 20% by day 35.

7.2 Effects of Temperature, pH, Aeration, and Agitation

Song et al. (2007) explored the use of whey permeate as an alternative additive for *G. lucidum* mycelia cultivation in solid-state conditions, finding optimal growth at pH 4.4 and 25.0 ± 0.5 °C (10 days). Their results suggested that using whey permeate could be a cost-effective solution for treating cheese whey permeates, though proper mixing with other solid media was necessary to adjust the carbon/nitrogen ratio. In an infrequent study, cultivation in a horizontal stirred tank bioreactor (STB) with a working volume of 30 L was described. Conditions were maintained at 30 °C with an airflow of 2 L/min and an agitation rate of 150 rpm for 2 min every other day during the first 7 days, and daily agitation during later stages (Kachrimanidou et al. 2023).

7.3 Effects of Inoculum

The inoculum comprised five 1-cm^2 sections of a 7-day-old culture grown on potato dextrose agar (PDA). Following the inoculation of 100 mL of potato dextrose broth in 250 mL Erlenmeyer flasks at pH 5.8, the vegetative inoculum biomass was incubated on a rotary shaker at 100 rpm for 14 days at 28–30 °C. Subsequently, the culture was transferred to bioreactors and maintained under the same conditions for an additional 14 days (Gregori et al. 2015).

8 Cultivation of *Hericium erinaceus*

8.1 Effects of Temperature, Aeration, Agitation, and Inoculums

Cultivation experiments were conducted in a 15-L horizontal stirred tank reactor (HSTR) of custom construction and design. The cultivation conditions were maintained at 24 °C with an airflow rate of 5 L/min. Periodic mixing was performed at

80 rpm for 2 min, every other day during the first 7 days, and daily during the latter part of the cultivation period. A humidification system utilizing sterile distilled water was connected to the bioreactor to prevent substrate desiccation.

The cultivation substrate comprised beech sawdust, paddy millet, and hulled millet, with a paddy millet to hulled millet ratio of 1:1 and a substrate moisture content of 50%. In contrast to glass jar systems, aeration of the solid matrix was implemented in the solid-state reactor. Sawdust, constituting 10% of the substrate, was incorporated to enhance porosity and oxygen permeability. Based on substrate optimization results from glass jar experiments, the substrate was supplemented with 0.56% NaCl and 3.4% casein peptone (dry weight/dry weight).

The inoculum was prepared from a mixture of beech sawdust and paddy millet in a 4:1 ratio, incubated and colonized by *Hericium erinaceus* mycelia, and accounted for 7% of the total substrate weight. The cultivation of fungal biomass in the HSTR was carried out over an 8-week period (Andrej Gregori 2015).

9 Conclusion

The cultivation of medicinal mushrooms such as *Cordyceps sinensis*, *Lentinula edodes*, *Grifola frondosa*, *Ganoderma lucidum*, *Hericium erinaceus*, *Trametes versicolor*, *Inonotus obliquus*, and *Agaricus blazei* has progressed through biotechnological methods. Utilizing submerged (SmF) and solid-state fermentation (SSF) in bioreactors allows precise control of growth conditions and consistent bioactive compound yields. Crucial aspects include careful strain selection, genetic diversity assessment, phenotypic screening, and adaptability to cultivation conditions, with stability and reproducibility maintained through meticulous monitoring and preservation. Optimizing nutrient requirements and parameters such as temperature, pH, aeration, and agitation enhances mycelial growth and metabolite production. Efficient inoculation techniques and bioprocess optimization are key for scaling up production, ensuring a steady supply of high-quality medicinal mushrooms to meet growing therapeutic demands.

References

Ajith TA, Janardhanan KK (2007) Indian medicinal mushrooms as a source of antioxidant and antitumor agents. J Clin Biochem Nutr 40(3):157–162. https://doi.org/10.3164/jcbn.40.157

Alam N, Amin R, Khair A, Lee TS (2010) Influence of different supplements on the commercial cultivation of milky white mushroom. Mycobiology 38(3):184. https://doi.org/10.4489/myco.2010.38.3.184

Andrej Gregori BG (2015) Solid state cultivation of Hericium erinaceus biomass and Erinacine: a production. J Bioprocess Biotech 05(03). https://doi.org/10.4172/2155-9821.1000210

Arshadi N, Nouri H, Moghimi H (2023) Increasing the production of the bioactive compounds in medicinal mushrooms: an omics perspective. Microb Cell Factories 22(1):1–34. https://doi.org/10.1186/s12934-022-02013-x

Assan N, Assan N, Mpofu T (2014) The influence of substrate on mushroom productivity-a review. Sci J Crop Sci 3(7):86–91. https://doi.org/10.14196/sjcs.v3i7.1502

Bakratsas G, Polydera A, Katapodis P, Stamatis H (2021) Recent trends in submerged cultivation of mushrooms and their application as a source of nutraceuticals and food additives. Future Foods 4(September):100086. https://doi.org/10.1016/j.fufo.2021.100086

Berovic M, Podgornik BB (2015) Cultivation of medicinal fungi in bioreactors. In: Mushroom biotechnology: developments and applications. Elsevier Inc. https://doi.org/10.1016/B978-0-12-802794-3.00009-6

Berovic M, Zhong JJ (2023) Advances in production of medicinal mushrooms biomass in solid state and submerged bioreactors. Adv Biochem Eng Biotechnol 184:125–161. https://doi.org/10.1007/10_2022_208

Berovic M, Podgornik BB, Gregori A (2022) Cultivation technologies for production of medicinal mushroom biomass: review. Int J Med Mushrooms 24(2):1–22. https://doi.org/10.1615/IntJMedMushrooms.2021042445

Bijalwan A, Singh A, Chaudhary S, Tyagi A, Thakur MP, Thakur TK (2020) Insights of medicinal mushroom (Ganoderma lucidum): prospects and potential in India. Biodivers Int J 4(5):202–209. https://doi.org/10.15406/bij.2020.04.00186

Chong JL, Chew KW, Peter AP, Ting HY, Show PL (2023) Internet of things (IoT)-based environmental monitoring and control system for home-based mushroom cultivation. Biosensors 13(1). https://doi.org/10.3390/bios13010098

Costa JAV, Treichel H, Kumar V, Pandey A (2017) Advances in solid-state fermentation. In: Current developments in biotechnology and bioengineering: current advances in solid-state fermentation. Elsevier B.V. https://doi.org/10.1016/B978-0-444-63990-5.00001-3

Cui FJ, Chen XX, Liu WM, Sun WJ, Huo S, Yang Y (2016) Control of Grifola frondosa morphology by agitation and aeration for improving mycelia and Exo-polymer production. Appl Biochem Biotechnol 179(3):459–473. https://doi.org/10.1007/s12010-016-2006-y

Das G, Shin HS, Leyva-Gómez G, Prado-Audelo MLD, Cortes H, Singh YD, Panda MK, Mishra AP, Nigam M, Saklani S, Chaturi PK, Martorell M, Cruz-Martins N, Sharma V, Garg N, Sharma R, Patra JK (2021) Cordyceps spp.: a review on its immune-stimulatory and other biological potentials. Front Pharmacol 11(February):1–31. https://doi.org/10.3389/fphar.2020.602364

Dong CH, Yao YJ (2005) Nutritional requirements of mycelial growth of Cordyceps sinensis in submerged culture. J Appl Microbiol 99(3):483–492. https://doi.org/10.1111/j.1365-2672.2005.02640.x

Dudekula UT, Doriya K, Devarai SK (2020) A critical review on submerged production of mushroom and their bioactive metabolites. 3 Biotech 10(8):1–12. https://doi.org/10.1007/s13205-020-02333-y

Elisashvili V (2012) Submerged cultivation of medicinal mushrooms: bioprocesses and products (review). Int J Med Mushrooms 14(3):211–239. https://doi.org/10.1615/IntJMedMushr.v14.i3.10

Firenzuoli F, Gori L, Lombardo G (2008) The medicinal mushroom Agaricus blazei murrill: review of literature and pharmaco-toxicological problems. Evid Based Complement Alternat Med 5(1):3–15. https://doi.org/10.1093/ecam/nem007

Frimpong-Manso J, Obodai M, Dzomeku M, Apertorgbor MM (2011) Influence of rice husk on biological efficiency and nutrient content of Pleurotus ostreatus (Jacq. ex. Fr.) Kummer. Int Food Res J 18(1):249–254

García-García M, Rocha-Zavaleta L, Valdez-Cruz NA, Trujillo-Roldán MA (2014) Conservation of the mycelia of the medicinal mushroom Humphreya coffeata (Berk.) Stey. in sterile distilled water. MethodsX 1(1):19–22. https://doi.org/10.1016/j.mex.2014.04.002

Gregori A, Boh B, Wraber B, Habijanic J, Berovic M, Svagelj M, Pankl M (2015) Cultivation of medicinal mushrooms Ganoderma lucidum and Grifola frondosa mycelia and polysaccharides in bioreactors, April, 12–16. https://doi.org/10.15224/978-1-63248-055-2-03

Habtemariam S (2020) Polysaccharides in cancer therapy: targets and efficacy. Biomedicine 8:135

Hussain MH, Mohsin MZ, Zaman WQ, Yu J, Zhao X, Wei Y, Zhuang Y, Mohsin A, Guo M (2022) Multiscale engineering of microbial cell factories: a step forward towards sustainable natural products industry. Synth Syst Biotechnol 7(1):586–601. https://doi.org/10.1016/j.synbio.2021.12.012

Kachrimanidou V, Papadaki A, Papapostolou H, Alexandri M, Gonou-Zagou Z, Kopsahelis N (2023) Ganoderma lucidum mycelia mass and bioactive compounds production through grape pomace and cheese whey valorization. Molecules 28(17). https://doi.org/10.3390/molecules28176331

Kim HO, Yun JW (2005) A comparative study on the production of exopolysaccharides between two entomopathogenic fungi Cordyceps militaris and Cordyceps sinensis in submerged mycelial cultures. J Appl Microbiol 99(4):728–738. https://doi.org/10.1111/j.1365-2672.2005.02682.x

Lübeck M, Lübeck PS (2022) Fungal cell factories for efficient and sustainable production of proteins and peptides. Microorganisms 10(4). https://doi.org/10.3390/microorganisms10040753

Łysakowska P, Sobota A, Wirkijowska A (2023) Medicinal mushrooms: their bioactive components, nutritional value and application in functional food production—a review. Molecules 28(14). https://doi.org/10.3390/molecules28145393

Mag P, Ghatnur SM, Parvatam G, Balaraman M (2015) Culture conditions for production of biomass, adenosine, and cordycepin from Cordyceps sinensis CS1197: optimization by desirability function method. Pharmacogn Mag 11(Suppl 3):448–456

Mao X, Eksriwong T, Chauvatcharin S, Zhong J (2005) Optimization of carbon source and carbon/nitrogen ratio for cordycepin production by submerged cultivation of medicinal mushroom Cordyceps militaris. Process Biochem 40:1667–1672. https://doi.org/10.1016/j.procbio.2004.06.046

Oke MA, Afolabi FJ, Oyeleke OO, Kilani TA, Adeosun AR, Olanbiwoninu AA, Adebayo EA (2022) Ganoderma lucidum: unutilized natural medicine and promising future solution to emerging diseases in Africa. Front Pharmacol 13(August):1–26. https://doi.org/10.3389/fphar.2022.952027

Onyango BO, Palapala VA, Arama PF, Wagai SO, Gichimu BM (2011) Suitability of selected supplemented substrates for cultivation of kenyan native wood ear mushrooms (Auricularia auricula). Am J Food Technol 6(5):395–403. https://doi.org/10.3923/ajft.2011.395.403

Park JP, Kim SW, Hwang HJ, Yun JW (2001) Optimization of submerged culture conditions for the mycelial growth and exo-biopolymer production by Cordyceps militaris. Lett Appl Microbiol 33:76–81

Park JP, Kim YM, Kim SW, Hwang HJ, Cho YJ, Lee YS, Song CH, Yun JW (2002) Effect of agitation intensity on the exo-biopolymer production and mycelial morphology in Cordyceps militaris. Lett Appl Microbiol 34:433–438

Plehn S, Wagle S, Rupasinghe HPV (2023) Chaga mushroom triterpenoids as adjuncts to minimally invasive cancer therapies: a review. Curr Res Toxicol 5(November):100137. https://doi.org/10.1016/j.crtox.2023.100137

Rathore H, Prasad S, Kapri M, Tiwari A, Sharma S (2019) Medicinal importance of mushroom mycelium: mechanisms and applications. J Funct Foods 56(February):182–193. https://doi.org/10.1016/j.jff.2019.03.016

Regonesi G (2023) Bioreactors: a complete review. MicroalgaeX:38. https://doi.org/10.13140/RG.2.2.11630.79685

Roszczyk A, Turło J, Zagożdżon R, Kaleta B (2022) Immunomodulatory properties of polysaccharides from Lentinula edodes. Int J Mol Sci 23(16). https://doi.org/10.3390/ijms23168980

Sher H, Al-Yemeni M, Bahkali AHA, Sher H (2010) Effect of environmental factors on the yield of selected mushroom species growing in two different agro ecological zones of Pakistan. Saudi J Biol Sci 17(4):321–326. https://doi.org/10.1016/j.sjbs.2010.06.004

Song M, Kim N, Lee S, Hwang S (2007) Use of whey permeate for cultivating Ganoderma lucidum mycelia. J Dairy Sci 90(5):2141–2146. https://doi.org/10.3168/jds.2006-690

Spelman K, Sutherland E, Bagade A (2017) Neurological activity of Lion's Mane (Hericium erinaceus). J Restor Med 6(1):19–26. https://doi.org/10.14200/jrm.2017.6.0108

Spier MR, Vandenberghe LPDS, Medeiros ABP, Soccol CR (2011) Application of different types of bioreactors in bioprocesses. In: Bioreactors: design, properties and applications. Nova Science, pp 53–87

Venturella G, Ferraro V, Cirlincione F, Gargano ML (2021) Medicinal mushrooms: bioactive compounds, use, and clinical trials. Int J Mol Sci 22(2):1–31. https://doi.org/10.3390/ijms22020634

Webb C (2017) Design aspects of solid state fermentation as applied to microbial bioprocessing. J Appl Biotechnol Bioeng 4(1). https://doi.org/10.15406/jabb.2017.04.00094

Wu JY, Siu KC, Geng P (2021) Bioactive ingredients and medicinal values of grifola frondosa (Maitake). Foods 10(1). https://doi.org/10.3390/foods10010095

Yan J, Wang W, Wu J (2020) Since January 2020 Elsevier has created a COVID-19 resource centre with free information in English and Mandarin on the novel coronavirus COVID-19. The COVID-19 resource centre is hosted on Elsevier Connect, the company' s public news and information. J Infect Public Health 13:1645–1651

Yoon TJ, Yu KW, Shin KS, Suh HJ (2008) Innate immune stimulation of exo-polymers prepared from Cordyceps sinensis by submerged culture. Appl Microbiol Biotechnol 80(6):1087–1093. https://doi.org/10.1007/s00253-008-1607-y

Zou G, Li T, Mijakovic I, Wei Y (2024) Synthetic biology enables mushrooms to meet emerging sustainable challenges. Front Microbiol 15(February). https://doi.org/10.3389/fmicb.2024.1337398

Biotechnology for Submerged Cultivation of *Pleurotus ostreatus*

Layla Naim

Abstract *Pleurotus ostreatus* holds immense potential in the realm of biotechnology due to its nutritional value and medicinal properties. Submerged cultivation, a technique increasingly explored for its efficiency and capacity, presents a promising avenue for mass production of this valuable mushroom.

The process of submerged cultivation involves the growth of fungi in a liquid medium, offering several advantages over traditional solid-state cultivation, such as higher productivity and controlled environmental conditions. Various parameters crucial for submerged cultivation, including carbon and nitrogen sources, pH, temperature, agitation, and aeration, significantly influence the growth and metabolic activity of *P. ostreatus*.

Biotechnological advancements have facilitated the optimization of submerged cultivation conditions to enhance biomass and metabolite production in *P. ostreatus*. Techniques such as metabolic engineering, genetic manipulation, and optimization of culture conditions have been employed to improve the yield of bioactive compounds with pharmaceutical and nutraceutical applications.

Furthermore, the utilization of agro-industrial by-products as substrates for submerged cultivation not only addresses waste management issues but also reduces production costs, making the process economically viable. The application of bioreactors equipped with advanced monitoring and control systems has further revolutionized submerged cultivation by ensuring reproducibility and consistency in fungal biomass and metabolite production.

In conclusion, the integration of biotechnological approaches with submerged cultivation techniques offers a sustainable and efficient means of mass-producing *P. ostreatus* and its bioactive metabolites for various industrial applications, including pharmaceuticals, functional foods, and bioremediation. Continued research and development in this field hold promise for unlocking the full biotechnological potential of *P. ostreatus*.

L. Naim (✉)
Department of Agronomy, Faculty of Agronomy, University of Forestry, Sofia, Bulgaria

Department of Plant Production, Faculty of Agriculture, Lebanese University, Beirut, Lebanon

M. N. Owaid et al. (eds.), *Mushroom Production*, Interdisciplinary Biotechnological Advances, https://doi.org/10.1007/978-981-95-7793-4_3

Keywords *Pleurotus ostreatus* · Submerged cultivation · Bioreactor systems · Bioactive metabolites · Biotechnological optimization

1 *Pleurotus ostreatus*: Cultivation and Ecology

Pleurotus ostreatus, also known as the "oyster mushroom" (Barros et al. 2007) or "white rot fungi" (Bellettini et al. 2019), belongs to the Basidiomycota class within the Pleurotaceae family (El-Ramady et al. 2022). This fungus is classified as "white-rot fungi" due to its ability to produce ligninolytic enzymes like laccase and peroxidase, which degrade hemicellulose, cellulose, and lignin (Doroški et al. 2022). As a wood-destroying fungus-saprotroph (xylotroph) and primary decomposer commonly found in temperate regions, *P. ostreatus* can naturally colonize almost all hardwoods, especially dead and decaying tree stumps, fallen branches, and wet logs (Adebayo and Oloke 2017; Quiñónez-Martínez et al. 2022; Gafforov et al. 2023).

Initially originating from China, cultivation methods for *P. ostreatus* were developed in Germany during World War I and successfully applied on a large scale. This was a response to the need for new food sources due to the problem of hunger in Germany (Wojewoda 2003). Nowadays, *P. ostreatus* is distributed worldwide and is one of the three most extensively cultivated fungi globally (Royse et al. 2017; Li et al. 2020).

In their earlier work, Rajarathnam and Bano (1987) underscored a key factor behind the widespread cultivation success of *Pleurotus*: its rapid mycelial growth rate, facilitating swift substrate colonization and a high yield of fruit bodies. Additionally, *Pleurotus* is well recognized for its ability to thrive on various low-quality plant-origin agro/wood by-products, food and cellulose wastes, wheat straw bales, and sawdust as commercial substrates following straightforward treatments like pasteurization (Kibar and Pekşen 2008; Piska et al. 2017; Royse et al. 2017; Sałata et al. 2018; Barh et al. 2019; Bulam et al. 2019; Zięba et al. 2020; Galappaththi et al. 2021; Doroški et al. 2022; Lesa et al. 2022).

2 *Pleurotus ostreatus*: Nutritional Composition and Health Benefits

In today's market, consumers exhibit a high interest in bioactive foods that offer health-promoting and disease-reducing effects. Mushrooms, including *P. ostreatus*, are considered functional foods due to their nutritional value and health benefits (Assemie and Abaya 2022). Cultivating *P. ostreatus* offers numerous advantages over other edible mushroom varieties (Ejigu et al. 2022). Gastronomically, *P. ostreatus* has gained traction as a functional food ingredient because it is cholesterol-free and low in calories, carbohydrates, fat, and sodium (Sahoo et al. 2022).

Previous research by Bilal et al. (2010) reported that the digestibility of oyster mushroom protein ranges from 72% to 83%. Similarly, Akyuz et al. (2010) indicated protein content ranging from 17 g to 42 g per 100 g of dried fruit bodies, with digestibility influenced by strains, and physical and chemical differences in the growing medium. Moreover, *P. ostreatus* contains fat content ranging from 0.2 g to 8 g per 100 g of dried fruit bodies (Bilal et al. 2010; Akyuz et al. 2010).

Additionally, fresh *P. ostreatus* contains 85–95% moisture and approximately 100 different bioactive compounds, including various phenolic compounds such as phenolic acids, flavonoids, hydroxycinnamic acids, hydroxybenzoic acids, lignans, tannins, stilbenes, and oxidized polyphenols (Khan et al. 2008; Assemie and Abaya 2022). Nutritionally, *P. ostreatus* is rich in protein; fat-free; carbohydrates; vitamins B1, B2, B12, C, D, E, and K; minerals (potassium, iron, copper, zinc, and manganese); and dietary fibers (Sharifi-Rad et al. 2020).

From a medical standpoint, bioactive compounds like glucans, ergosterol, amino acids (arginine and glutamine), and proteoglycans present in *P. ostreatus* have been associated with antitumor activities, potentially inhibiting tumor-induced neovascularization (Mca et al. 2021). Moreover, *Pleurotus* species exhibit antibacterial properties against both Gram-negative and Gram-positive bacteria. The ethyl acetate extract of *P. ostreatus* has been investigated for its antimicrobial properties against *B. subtilis*, *S. aureus*, *E. coli*, *K. pneumoniae*, and *Proteus vulgaris* (Hrudayanath and Sameer 2014).

3 The Future of Mushroom Cultivation: From Traditional Methods to Submerged Mycelia

Mushroom usage has undergone significant evolution from its traditional roles in medicine and cuisine, motivating researchers to investigate their structure, biochemical composition, and cultivation methods (Dudekula et al. 2020).

Currently, commercial mushroom products primarily derive from the fruit bodies of solid-substrate cultivation, a time-consuming and labor-intensive process with numerous challenges (Tang et al. 2007). For instance, growing these fruiting bodies takes months, requires substantial substrate volumes and space, and often results in inconsistent product quality (Elisashvili 2012; Corrêa et al. 2016; Khan et al. 2018). Moreover, large-scale solid-state cultivation faces issues such as pH control, heat and mass transfer, water activity, substrate heterogeneity, environmental variability, and lengthy cultivation times, contributing to variable quality (Tang et al. 2007; Singh nee'Nigam 2009; Letti et al. 2018). Additionally, harmful radioisotopes can accumulate in the fruiting body from substrates during growth stages (Kalač 2012). Furthermore, year-round mushroom production fluctuates; low summer rates due to unfavorable conditions lead to higher market prices, while surplus winter production reduces prices for producers but increases agro-waste (García-Cruz et al. 2020).

Adopting submerged cultivation of mycelia as an alternative method is essential. Submerged cultivation involves growing mushrooms in a liquid medium where nutrients are dissolved, and oxygen supply is enhanced through agitation. This technique is more precise and faster compared to fruiting body cultivation (Bentil et al. 2018). Submerged cultivation ensures uniform substrate distribution, reduces contamination, and shortens growth cycles (García-Cruz et al. 2020). It also offers improved oxygen mass transfer and culture homogeneity. By using this method, physical factors (temperature, aeration, agitation), chemical factors (pH, medium composition), and biological factors (inoculum, morphology, rheology) can be controlled (Fazenda et al. 2008). As a reproducible technique for mycelium production, submerged cultivation achieves controlled yields of metabolites (Elisashvili 2012).

4 Submerged Cultivation of *Pleurotus ostreatus*: The Role of Strain Selection and Culture Medium Optimization

Submerged cultivation of edible mushrooms, particularly *Pleurotus ostreatus*, has gained significant interest due to its ability to produce high biomass and a diverse range of bioactive compounds, including proteins, enzymes, lipids, and carbohydrates, under controlled conditions in bioreactors (Bakratsas et al. 2021). This method offers a reliable alternative to traditional fruiting body cultivation by providing a controlled environment that enhances the safety and consistency of the produced bioactive compounds. The success of mushroom cultivation largely depends on the culture medium, which supplies the essential nutrients required for mycelium growth (Krupodorova et al. 2024). Additionally, selecting the appropriate strain is crucial, as strain-specific traits significantly impact both growth performance and the production of valuable metabolites (Sabri et al. 2019; Krupodorova et al. 2021). A thorough understanding of the cultural, morphological, physiological, ecological, and genetic variations among *P. ostreatus* strains is essential for effective strain selection. This approach ensures the successful commercial implementation of the most suitable strains for submerged cultivation (Krupodorova et al. 2024).

When selecting a *P. ostreatus* strain for submerged cultivation, several factors should be considered:

- Growth rate: Different strains of *P. ostreatus* exhibit varying growth rates under submerged conditions. Strains with faster growth rates can enhance production efficiency and reduce the time to harvest.
- Substrate utilization: Strains differ in their ability to utilize substrates in submerged fermentation. Selecting strains that effectively convert substrates into mycelium and fruit bodies is crucial for maximizing yield.
- Resistance to contaminants: Submerged cultivation conditions can promote microbial growth, making it essential to select strains with high resistance to contaminants to maintain culture health and ensure a successful harvest.

- Yield and quality: Strain-specific characteristics can influence both the quantity and quality of the mushrooms produced. Strains should be chosen based on desirable traits such as size, texture, and flavor of the fruiting bodies.
- Environmental adaptability: Some strains are better suited to specific environmental conditions within the submerged system, such as temperature, pH, and oxygen levels. Selecting strains that thrive under these conditions can lead to more stable and predictable production.

Table 1 provides an overview of commonly used *P. ostreatus* strains in submerged fermentation, highlighting their distinct characteristics and potential applications in biotechnology. This comprehensive approach to strain selection and cultivation conditions ensures the optimal use of *P. ostreatus* strains in submerged cultivation systems, enhancing their commercial viability and efficiency.

Strain selection and culture medium optimization are crucial for maximizing the efficiency of *Pleurotus ostreatus*-submerged cultivation. Variations in strain-specific growth, enzyme activity, and metabolite production necessitate strategic selection, as strain-medium interactions directly influence biomass yield and the synthesis of bioactive compounds and enzymes.

5 Optimization of Carbon and Nitrogen Sources for Enhanced *P. ostreatus* Production in Submerged Cultures

Basidiomycetes can be cultivated in liquid media without developing the fruiting body, facilitating exopolysaccharide (EPS) or biomass extraction (Rosado et al. 2003). To enhance the efficiency and cost-effectiveness of *P. ostreatus* production, optimizing the culture medium is crucial. Mycelial growth and secondary metabolite synthesis vary among strains due to differences in nutrient availability, including carbon, nitrogen, vitamins, and essential elements (Krupodorova et al. 2024; Dudekula et al. 2020).

- Carbon source: The selection of an appropriate carbon source is crucial for optimizing microbial metabolism and biosynthetic activity in basidiomycetes cultivation. A semisynthetic glucose-peptone medium is commonly used, but alternative carbon sources can significantly influence growth and metabolite production (Antontceva et al. 2018).

Studies have shown varying effects of carbon sources on *P. ostreatus* biomass production and secondary metabolite synthesis. Barakat and Sadik (2014) reported that while starch is optimal for mycelial biomass production, glucose enhances exopolysaccharide synthesis and antimicrobial and antioxidant activities. Similarly, Saskiawan (2009) compared multiple carbon sources and found that maltose yielded the highest mycelial biomass (7.06 g/L), followed by starch (5.12 g/L), lactose (4.66 g/L), and glucose (2.96 g/L). Although growth curves

Table 1 Characteristics of different *P. ostreatus* strains

Strain	Characteristics	Reference
P. ostreatus 2460	Rapid growth with dense, white mycelium on PDA; efficiently utilizes glucose and fructose; high antioxidant activity from phenolic compounds and flavonoids; strong laccase and cellulase activities; moderate antibacterial effects, especially against Gram-positive bacteria	Krupodorova et al. (2024)
P. ostreatus 2462	Fast-growing strain with high biomass in solid and liquid media; utilizes various nitrogen sources to boost growth and yield; exhibits antimicrobial activity against Gram-positive and Gram-negative bacteria; strong radical scavenging and anti-inflammatory polysaccharides	
P. ostreatus 551	Moderate growth with a preference for lignocellulosic substrates; dense, cottony mycelium with strong colonization; high protease and peroxidase activity; moderate antioxidant properties	
P. ostreatus 1685	Vigorous mycelial growth with high tolerance to temperature variations; enhanced growth in nitrogen-rich media; strong antioxidant potential	
P. ostreatus 2461	Fast-growing mycelium with a dense, filamentous structure on agar; efficiently utilizes complex carbon sources, boosting growth and enzyme production; strong antimicrobial activity, especially against fungal pathogens; high antioxidant activity with notable free radical scavenging capacity	
P. ostreatus M2191	Rapid, dense mycelial growth on PDA and other media; efficiently utilizes diverse carbon sources like glucose, maltose, and cellulose, enhancing adaptability; strong antioxidant properties; moderate antimicrobial activity, mainly against Gram-positive bacteria; high production of ligninolytic enzymes, including laccase and peroxidase	Vamanu (2012)
P. ostreatus PBS281009	Rapidly grows with dense, fluffy mycelium that colonizes agar and liquid media efficiently. Utilizes a wide range of nitrogen and carbon sources, boosting growth and metabolic activity. Shows strong antioxidant and antimicrobial properties against various pathogens and produces high levels of hydrolytic enzymes	
P. ostreatus ATHUM 4438	Adapts well to submerged fermentation, especially in stirred tank bioreactors. Efficiently uses diverse carbon sources, including agro-industrial by-products. Noted for high enzyme production and consistent bioactive metabolite yields across multiple fermentation cycles, with strong biomass accumulation and stable performance	Papaspyridi et al. (2012)
P. ostreatus 1801	Shows excellent growth and high biomass yield in submerged fermentation, producing significant bioactive metabolites. Efficiently utilizes various substrates, including agro-industrial residues. Performs well under controlled conditions with stable, consistent results across batches and good resistance to microbial contaminants	Hamza et al. (2024)
P. ostreatus ATHUM 5697	Produces mycelium rich in β-glucans and other bioactive metabolites. Efficiently uses food industry by-products as substrates and adapts well to submerged cultivation. Shows strong resistance to microbial contamination	Pilafidis et al. (2024)

(continued)

Table 1 (continued)

Strain	Characteristics	Reference
P. ostreatus LGAM 1123	Ideal for batch stirred tank bioreactors, with rapid, robust growth in submerged conditions. Highly resistant to contaminants and excels in producing mycoprotein rich in essential amino acids	Bakratsas et al. (2023b)
P. ostreatus	Grows quickly in submerged conditions, efficiently using pomelo peel powder as a carbon source. Resistant to microbial contamination and produces enzymes along with antioxidant-rich secondary metabolites	Majumder et al. (2020)
P. ostreatus	Adapts well to submerged fermentation, efficiently using various carbon sources and showing high polygalacturonase activity, ideal for industrial use. Exhibits stability and resistance to fermentation condition variations	Freixo et al. (2008)

Fig. 1 Concentrations of *P. ostreatus* biomass at different sources of carbon in the nutrient medium. Antontceva et al. (2018)

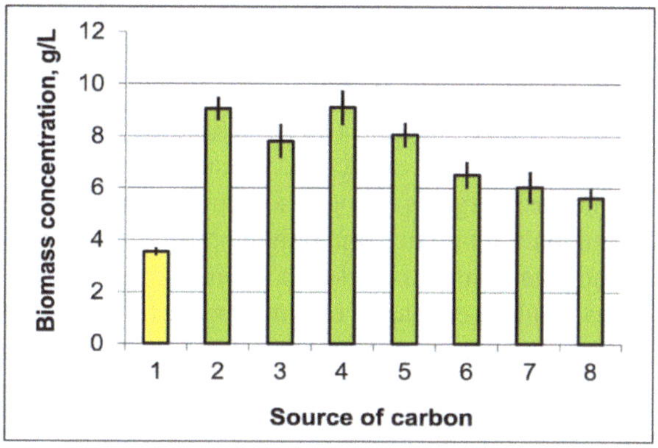

1 - glucose, 2 - rye whole grain flour, 3 - medium rye flour, 4 - light wheat flour, 5 - dark wheat flour, 6 - corn flour, 7 -potato starch, 8 - soy flour

showed minimal variation across carbon sources, starch accelerated the initial growth phase, making it ideal for early exponential development, while maltose proved most effective for liquid culture cultivation.

Expanding on these findings, Bakratsas et al. (2023b) demonstrated that *P. ostreatus* LGAM 1123 can efficiently utilize glucose- and xylose-based media, with their combination optimizing biomass yield, protein content, and amino acid profile. Meanwhile, Antontceva et al. (2018) explored cost-effective alternatives to glucose for improving economic efficiency in polysaccharide production, emphasizing the role of fungal biomass accumulation, which is strongly influenced by carbon source selection. The growth performance of *P. ostreatus* under different carbon sources is illustrated in Fig. 1.

More recently, Pilafidis et al. (2024) identified brewer's spent grain extract (BSGE) and diluted wine distillery effluent (WDE) as effective sole carbon sources for substantial mycelial biomass production. These agro-industrial by-products offer promising low-cost alternatives for large-scale cultivation.

- Nitrogen source: The effect of nitrogen supplementation on *P. ostreatus* mycelial growth varied with the nitrogen source. In peat extract-based media, yeast extract supported maximal growth, aligning with previous findings (Martin and Bailey 1985; Manu-Tawiah and Martin 1987). In synthetic media, urea and ammonium sulfate enhanced mycelial growth more effectively than ammonium phosphate (Manu-Tawiah and Martin 1988). Given these findings, yeast extract was selected as a suitable nitrogen source for further applications (Bakratsas et al. 2023b).

Regarding optimal nitrogen sources for submerged cultivation of *P. ostreatus*, the highest biomass yield was obtained when complex fertilizer, yeast extract, and peptone (in a 1:1 ratio) were used. Specifically, supplementation with 5 g/L of peptone and yeast extract resulted in the best mycelial biomass production (Vicenţiu-Bogdan et al. 2014).

Wu et al. (2008) tested various nitrogen sources (yeast peptone powder, keramine HD, hydrolyzed vegetable protein, Bacto peptone, $NaNO_3$, and $(NH_4)_2SO_4$) in a fructose-based medium and found that yeast peptone powder yielded the highest mycelial biomass and EPS production.

Nitrogen source selection also influenced enzyme activity. Organic nitrogen sources, particularly yeast extract and wheat bran, significantly increased laccase activity, whereas inorganic nitrogen sources (e.g., ammonium sulfate, ammonium nitrate, ammonium tartrate, ammonium phosphate, ammonium chloride, sodium nitrate, and potassium nitrate) led to lower enzyme activity. This suggests that *P. ostreatus* exhibits enhanced enzyme production in the presence of organic nitrogen sources, such as peptone, yeast extract, beer yeast, urea, malt extract, casein extract, soy extract/protein, and various amino acids (Hewage and Sanjeewani 2024).

Additionally, to maximize biomass and polysaccharide production, *P. ostreatus* showed the best mycelial mass accumulation when glucose was used at a concentration of 40 g/L and corn extract at 20 g/L as an organic nitrogen source (Gern et al. 2008).

In conclusion, the choice of carbon and nitrogen sources in submerged cultivation of *Pleurotus ostreatus* is primarily determined by the target metabolites intended for production.

6 Hydrodynamic and Environmental Factors Affecting *P. ostreatus* Growth in Submerged Bioreactors

Submerged liquid cultivation provides enhanced oxygen mass transfer and culture homogeneity, allowing precise control over physical (temperature, aeration, agitation), chemical (pH, medium composition), and biological (inoculum, morphology, rheology) factors (Fazenda et al. 2008).

The growth and productivity of *P. ostreatus* in submerged culture depend on strain selection, growth conditions, and nutritional requirements (Gregori et al. 2007). Pellet formation is influenced by impeller shear forces, inoculum concentration, pH, and medium composition (Treskatis et al. 1997). Due to the complex fungal morphologies in submerged systems, maintaining optimal agitation is crucial for efficient growth (Márquez-rocha et al. 1999). Increasing the energy dissipation/circulation function (EDCF) and aeration flow rate (AFR) significantly enhanced *P. ostreatus* biomass concentration and specific growth rate, with no signs of hydrodynamic stress (Tinoco-Valencia et al. 2014). Growth in mechanically agitated bioreactors was primarily constrained by oxygen availability in the culture medium. Previous studies have identified EDCF as a key hydrodynamic parameter for assessing the relationship between mechanical shear stress and fungal growth or process productivity (Li et al. 2002; Casas-López et al. 2005; Rocha-Valadez et al. 2005). *Temperatures* above the optimal value can be detrimental to protein expression and metabolite synthesis in microorganisms, while lower temperatures result in a reduced rate of cellular growth and product synthesis (Wang and Zhong 2007). For the genus *Pleurotus*, a study by Hoa and Wang (2015) identified 28 °C as the optimal growth temperature for both *Pleurotus ostreatus* and *Pleurotus cystidiosus*. However, the growth of *P. ostreatus* mycelium is most pronounced at 25 °C, indicating that substrate colonization is most efficient at this temperature (Akinyele et al. 2012).

Oxygen transfer, alongside temperature, is a critical factor for aerobic cell growth in bioreactors. Effective mixing ensures a sufficient supply of dissolved oxygen to the cells (Wang and Zhong 2007). Moreover, the oxygen concentration in the medium significantly influences fungal biomass biosynthesis, affecting both growth and metabolic activity (Wang et al. 2005). Antontceva et al. (2018) demonstrated that a higher oxygen dissolution rate leads to the highest yield of *P. ostreatus* biomass (Fig. 2).

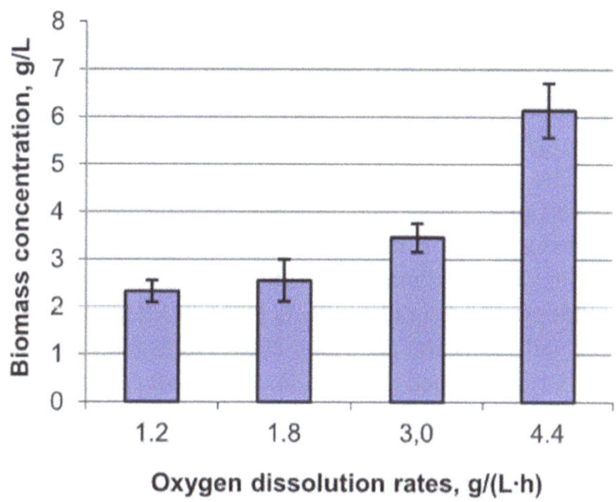

Fig. 2 Concentrations of *P. ostreatus* biomass at different dissolution rates of oxygen. Antontceva et al. (2018)

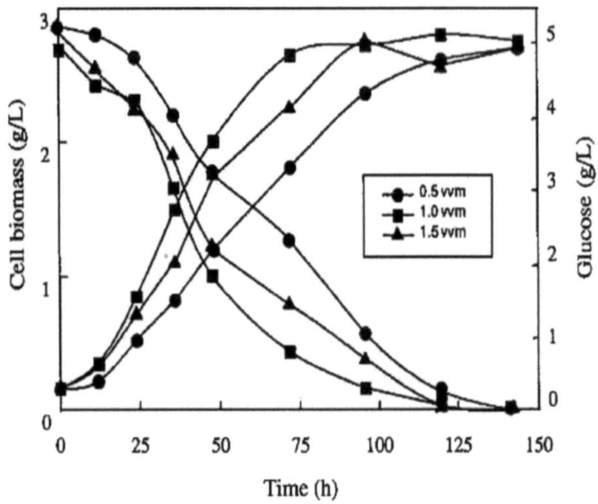

Fig. 3 Growth of *Pleurotus ostreatus* with different aeration rates. Márquez-Rocha et al. (1999)

Table 2 The effect of aeration on the growth characteristics of *P. ostreatus*

Aeration (vvm)	μ (h^{-1})	qs (gg^{-1}dry weight h^{-1})	Pellet diameter (mm) (±)
0.5	0.030	0.059	3.5 (0.35)
1.0	0.036	0.071	3.2 (0.41)
1.5	0.025	0.049	2.8 (0.31)
Culture condition: agitation 200 rpm, 26 C, 10 L working volume			

Márquez-Rocha et al. (1999)

Aeration plays a crucial role in fungal growth, as shown in Fig. 3, where different air flow rates influenced biomass development.

While air flow rates of 0.5 and 1.0 vvm resulted in no significant differences in pellet size, an increase to 1.5 vvm led to a notable reduction in both the specific growth rate (μ) and pellet size (Márquez-rocha et al. 1999) (Table 2). Despite this, Gern et al. (2008) identified 1.5 vvm as the optimal aeration rate for mycelial biomass accumulation. Additionally, Elisashvili et al. (2008a) reported that increasing aeration from 0.5 to 1.5 vvm significantly enhanced laccase enzyme activity, further highlighting the impact of oxygen availability on fungal metabolism.

The *pH* value plays a crucial role in *P. ostreatus* growth, exopolysaccharide (EPS) production, and biomass accumulation. Although *P. ostreatus* can grow across a broad pH range (4.0–8.0), the optimal pH for EPS and biomass production varies. The medium's pH, often overlooked, significantly influences fungal morphology, which in turn affects biomass yield and metabolite formation (Wang and McNeil 1995; Shu and Lung 2004; Gbolagade et al. 2006; Jonathan et al. 2009).

For submerged cultivation, *P. ostreatus* growth is generally optimal between pH 5.0 and 6.5, with most studies reporting peak growth at pH 5.5. Interestingly, its ability to produce EPS efficiently at pH 8 contrasts with other mushrooms, which typically prefer more acidic conditions (Kim et al. 2003).

Regarding biomass production, Bakratsas et al. (2024) found that *P. ostreatus* grew well at all tested pH levels (5–8), with the highest biomass (11.4 ± 2.0 g/L) and laccase activity (12.4 ± 1.5 U/mL) observed at pH 6. Other studies suggest that while biomass production peaks around pH 6.5, laccase production is generally favored at lower pH levels. Díaz et al. (2013) reported maximal laccase activity at pH 4.5 and the lowest at pH 8.5, mirroring findings from Prasad et al. (2005), who identified an optimal pH of 5.5 for laccase production within the 4.0–8.0 range.

7 Fermentation Process of *Pleurotus ostreatus* in Submerged Cultivation

- *Inoculation and Growth Phase*

The process begins with the inoculation of *Pleurotus ostreatus* spores or mycelial fragments into a sterile liquid nutrient medium. The medium typically contains a carbon source (glucose, sucrose, or lignocellulosic hydrolysates), nitrogen source (peptone, yeast extract, ammonium salts), minerals, and vitamins to support fungal growth (Wang et al. 2017). The mycelium initially exists as single hyphal fragments that aggregate over time to form dense, filamentous structures known as pellets (Papagianni 2004).

Upon inoculation, *P. ostreatus* mycelium initially exists as single hyphal fragments that aggregate over time to form dense, filamentous structures known as pellets. Pellet formation is influenced by several factors:

Agitation speed: It has been reported in the literature that agitation is one of the factors affecting pellet formation in terms of size, shape, and uniformity. Most studies indicate that the optimal range varies between 100–200 rpm. Moreover, agitation speed and duration consequently influence metabolite production and its concentration.

pH and oxygen levels: Maintaining pH at 4.5–6.5 and ensuring sufficient oxygen availability enhances pellet compactness and growth rate (Papagianni 2004).

Inoculum size: Higher inoculum concentrations (>10% v/v) tend to favor dispersed mycelial growth, while lower concentrations promote pellet formation (Bizukojc and Ledakowicz 2010).

- *Nutrient Uptake and Metabolism*

During submerged fermentation, *P. ostreatus* metabolizes available nutrients for growth and secondary metabolite production. The fungus secretes hydrolytic enzymes, such as cellulases and laccases, to degrade complex organic compounds into simpler molecules (Elisashvili et al. 2008b). Oxygen is a critical factor, as *P. ostreatus* is an obligate aerobe; thus, aeration and agitation are necessary to ensure sufficient oxygen transfer (Singh et al. 2016).

- *Optimization Strategies for Biomass Production*

Fed-batch fermentation: Gradual addition of carbon and nitrogen sources prevents substrate depletion and extends the growth phase (Xu et al. 2019).

Surfactant addition: Nonionic surfactants such as Tween 80 reduce surface tension and improve oxygen transfer, enhancing fungal biomass yield (Krull et al. 2013).

Metal ion supplementation: Adding trace metals like manganese and copper enhances enzyme activity and secondary metabolite production (Haddadin et al. 2009).

- *Importance of Monitoring and Control*
  To optimize the fermentation process, several key parameters must be monitored:

pH: The pH of the medium should be maintained within an optimal range (typically 4.5–6.5) to facilitate enzyme activity and mycelial growth (Liu et al. 2019).

Temperature: The ideal temperature for *P. ostreatus* growth in submerged culture is around 25–30 °C (Stajić et al. 2006).

Dissolved oxygen (DO): Adequate oxygenation is required, as oxygen limitation can lead to decreased biomass production and enzyme secretion. Stirring rates and aeration levels must be adjusted accordingly.

Nutrient concentration: Carbon and nitrogen source availability influences biomass yield and secondary metabolite synthesis (Sánchez 2010).

Agitation and shear stress: Excessive agitation can cause mechanical damage to mycelial structures, while insufficient mixing may lead to nutrient depletion zones (Papagianni 2004).

- *Harvesting and Downstream Processing*
  Once the mycelium reaches the desired growth phase, it is harvested through filtration or centrifugation. The biomass can be dried and processed for various applications, including functional food products, pharmaceuticals, and enzyme production (Sánchez 2010).

8 Optimizing *Pleurotus ostreatus* Cultivation: From Bioreactors to Industrial Utilization

Scaling up the submerged cultivation of *Pleurotus ostreatus* to industrial levels requires optimizing bioreactors, substrates, and process parameters. Stirred tank bioreactors with controlled aeration and agitation enhance biomass production (Bakratsas et al. 2024). Using agro-industrial by-products like aspen wood hydrolysate reduces costs and increases sustainability (Bakratsas et al. 2023b). Key factors such as temperature, pH, and light exposure influence yield and metabolite production (Bakratsas et al. 2024).

- *Bioreactor Selection and Optimization*

Bioreactor types: Stirred tank reactors (STRs) and airlift bioreactors are commonly used for fungal cultivation, with STRs offering better control over mixing and oxygenation (Zhang et al. 2021).

Aeration and agitation: Maintaining optimal dissolved oxygen levels (above 20% saturation) and controlled agitation (100–300 rpm) prevents excessive shear stress and promotes uniform pellet formation (Krull et al. 2013).

Scale-up strategies: Fed-batch and continuous fermentation strategies improve productivity by optimizing nutrient supply and minimizing substrate inhibition (Xu et al. 2019).

- *Substrate and Media Optimization*

Alternative carbon sources: Sugarcane bagasse, wheat bran, and corn stover hydrolysates provide cost-effective and sustainable carbon sources (Sun et al. 2021).

Nitrogen optimization: Organic nitrogen sources like yeast extract improve enzyme production, while ammonium salts provide cost-effective alternatives (González et al. 2019).

Metal ion supplementation: Trace metals such as manganese and copper enhance enzymatic activity and secondary metabolite production (Haddadin et al. 2009).

- *Applications of Cultivated Biomass*
  The cultivated biomass has diverse applications:

Food industry: *P. ostreatus* is a valuable protein source, with research exploring its use in single-cell protein production (Bakratsas et al. 2023a).

Pharmaceuticals: It contains bioactive compounds with immunomodulatory and antioxidant properties, making it relevant for drug development (Punthi et al. 2023).

Bioremediation: Its enzymatic capabilities enable pollutant degradation, making it effective in environmental cleanup (Moses 2024).

Sustainable waste management: Its ability to grow on textile and food waste highlights its role in circular economy initiatives (Sangosanya and Pistofidou 2024).

9 Submerged Cultivation of *Pleurotus ostreatus*: Limitations, Risks, and Future Perspectives

Despite its advantages, submerged cultivation of *Pleurotus ostreatus* presents multiple challenges, including oxygen transfer limitations, contamination risks, and substrate availability. Ongoing research is addressing these issues through innovative approaches such as alternative substrates, improved sterilization methods, and biotechnological applications like mycoremediation. Future advancements in strain development, process optimization, and sustainable cultivation models will further enhance the efficiency and commercial viability of this cultivation method.

Challenges and Limitations of Submerged Cultivation of Pleurotus ostreatus

Submerged cultivation of *Pleurotus ostreatus* (oyster mushroom) offers a controlled environment for producing mycelial biomass and bioactive metabolites. However, several challenges and limitations are associated with this method:

- *Technical and Process Challenges*

Oxygen supply: Maintaining adequate dissolved oxygen levels is crucial for fungal growth and metabolite production. However, the high oxygen demand of *P. ostreatus* can lead to suboptimal growth and product yields if not properly managed (Tang et al. 2007).

Shear stress and mixing: Efficient mixing ensures uniform nutrient distribution but can introduce shear forces that damage the delicate mycelial structures. Balancing mixing intensity is a significant challenge in bioreactor design (Tang et al. 2007).

Morphology and rheology: The morphology of *P. ostreatus* in submerged culture can vary, affecting the viscosity of the culture broth. Changes in viscosity impact mass and oxygen transfer rates, complicating process control and scalability (Tang et al. 2007).

Strain variability: Different strains of *P. ostreatus* exhibit varied growth rates and metabolite production profiles under identical conditions. This variability necessitates strain-specific optimization of cultivation parameters (Krupodorova et al. 2024).

- *Nutritional and Environmental Factors*

Nutrient requirements: Formulating an optimal nutrient medium is essential for maximizing biomass and metabolite yields. An inadequate or imbalanced nutrient supply can limit growth and product formation (Antontceva et al. 2018).

Temperature sensitivity: *P. ostreatus* grows best between 18 °C and 28 °C. Temperature deviations outside this range can adversely affect growth and enzyme activity, making precise control necessary (Díaz et al. 2011).

Substrate cost and availability: Traditional substrates can be expensive or seasonally unavailable. Research into alternative, cost-effective substrates, such as agricultural residues or invasive plant species like water hyacinth, has shown promise but requires further investigation to ensure consistency and efficiency (Ejigu et al. 2022).

- *Contamination Risks*

Fungal contaminants: The nutrient-rich environment of submerged cultures is highly susceptible to contamination by fungi such as *Trichoderma harzianum*, *Penicillium notatum*, *Sclerotium rolfsii*, and *Coprinus* spp. These contaminants can outcompete *P. ostreatus*, leading to reduced yields or complete culture loss. Higher contamination rates have been observed during warmer and more humid periods (Biswas 2014).

Mycotoxin contamination: Using mycotoxin-contaminated substrates poses risks. Although *Pleurotus* species have been studied for their potential to degrade certain mycotoxins, the safety and efficacy of such bioremediation processes require further assessment (Zapaśnik et al. 2024).

Bacterial contaminants: Bacteria can rapidly proliferate in liquid media, competing for nutrients and producing metabolites that inhibit fungal growth. Maintaining strict aseptic conditions during inoculation and cultivation is essential to minimize bacterial contamination.

Environmental factors: Temperature and humidity influence contamination rates. Higher temperatures and humidity levels increase the likelihood of microbial contamination, necessitating strict environmental control (Biswas 2014).

Ongoing Research Efforts and Future Directions

Current research focuses on improving the efficiency and sustainability of submerged cultivation methods for *Pleurotus ostreatus* through various innovative approaches:

- *Alternative Substrates and Resource Optimization*

Utilization of Agro-Industrial Wastes: Researchers are exploring agro-industrial residues such as palm oil fruit mesocarp and cocoa almond peels as cost-effective substrates. These materials support *P. ostreatus* growth while promoting agricultural waste recycling and reducing cultivation costs (Silva et al. 2020).

Single-cell protein production: Optimizing cultivation conditions for single-cell protein production using low-cost industrial by-products (fiber sludge from the pulp and paper industry) can provide sustainable protein sources for animal feed or human consumption (Bakratsas et al. 2023a).

- *Improved Sterilization and Cultivation Techniques*

Alternative pasteurization techniques: Traditional autoclaving is energy-intensive and costly. Submerged pasteurization, such as hot water immersion, offers an economical alternative while ensuring effective microbial control, making mushroom cultivation more accessible (Magaña and Shimizu 2022).

Process optimization: Further research is needed to optimize environmental parameters (pH, temperature, oxygen levels) and bioreactor designs to improve overall efficiency and scalability.

- *Sustainability and Environmental Applications*

Mycoremediation and bio-fermentation: *P. ostreatus* has demonstrated potential in degrading environmental pollutants and converting agro-industrial waste into valuable products such as biofertilizers and animal feed. Integrating these biotechnological applications can enhance the sustainability of mushroom production systems (El-Ramady et al. 2022).

Integration with circular economy models: Aligning submerged cultivation practices with circular economy principles by utilizing waste streams from other industries can enhance sustainability and reduce environmental impact.

References

Adebayo EA, Oloke JK (2017) Oyster mushroom (Pleurotus species); a natural functional food. J Microbiol Biotech Food Sci 7(3):254–264. https://doi.org/10.15414/Jmbfs.2017/18.7.3.254-264

Akinyele JB, Fakoya S, Adetuyi CF (2012) Anti-growth factors associated with Pleurotus ostreatus in a submerged liquid fermentation. Malays J Microbiol 8(3):135–140

Akyuz M, Onganer AN, Erecevit P, Kirbag S (2010) Nutritive value of edible wild and cultured mushrooms. Turk J Biol 23(2):125–130

Antontceva E, Shamtsyan M, Rabin A (2018) Selection of conditions for submerged cultivation of Pleurotus ostreatus. J Hyg Eng Des 96–100

Assemie A, Abaya G (2022) The effect of edible mushroom on health and their biochemistry. Int J Microbiol 2022., 8744788:7

Bakratsas G, Polydera A, Katapodis P, Stamatis H (2021) Recent trends in submerged cultivation of mushrooms and their application as a source of nutraceuticals and food additives. Future Foods 4:100086

Bakratsas G, Polydera A, Nilson O, Kossatz L, Xiros C, Katapodis P, Stamatis H (2023a) Single-cell protein production by Pleurotus ostreatus in submerged fermentation. Sustain Food Technol 1(3):377–389

Bakratsas G, Polydera A, Nilson O, Chatzikonstantinou AV, Xiros C, Katapodis P, Stamatis H (2023b) Mycoprotein production by submerged fermentation of the edible mushroom *Pleurotus ostreatus* in a batch stirred tank bioreactor using agro-industrial hydrolysate. Foods 12:2295. https://doi.org/10.3390/foods12122295

Bakratsas G, Tsoumanis C, Stamatis H, Katapodis P (2024) Exopolysaccharide production in submerged fermentation of *Pleurotus ostreatus* under red and green light. Fermentation 10(6):313. https://doi.org/10.3390/fermentation10060313

Barakat OS, Sadik MW (2014) Mycelial growth and bioactive substance production of Pleurotus ostreatus in submerged culture. Int J Curr Microbiol Appl Sci 3:1073–1085

Barh A, Sharma VP, Annepu SK, Kamal S, Sharma S, Bhatt P (2019) Genetic improvement in Pleurotus (oyster mushroom): a review. 3 Biotech 9:322. https://doi.org/10.1007/s13205-019-1854-x

Barros L, Baptista P, Correia DM, Morais JS, Ferreira ICFR (2007) Effects of conservation treatment and cooking on the chemical composition and antioxidant activity of Portuguese wild edible mushrooms. J Agric Food Chem 55:4781–4788

Bellettini MB, Fiorda FA, Maieves HA, Teixeira GL, Ávila S, Hornung PS, Júnior AM, Ribani RH, (2019) Factors affecting mushroom Pleurotus spp. Saudi J Biol Sci 26(4):633–646. https://doi.org/10.1016/j.sjbs.2016.12.005

Bentil JA, Thygesen A, Mensah M, Lange L, Meyer AS (2018) Cellulase production by white-rot basidiomycetous fungi: solid-state versus submerged cultivation. Appl Microbiol Biotechnol 102(14):5827–5839. https://doi.org/10.1007/s00253-018-9072-8

Bilal AW, Bodha RH, Wani AH (2010) Nutritional and medicinal importance of mushrooms. J Med Plants Res 4(24):2598–2604

Biswas MK (2014) Microbial contaminants in Oystermushroom (Pleurotus ostreatus) cultivation their management and role of meteorological factors. In Proceedings of the 8th International conference on mushroom biology and mushroom products (ICMBMP8), pp 567–575

Bizukojc M, Ledakowicz S (2010) The morphology of filamentous fungi in submerged cultures as a function of process variables. Chem Biochem Eng Q 24(2):157–166

Bulam S, Üstün NŞ, Pekşen A (2019) Evaluation of nutritional and medicinal values of edible wild and cultivated Pleurotus ostreatus. Turk J Agric - Food Sci Technol 7(12):2054–2061. https://doi.org/10.24925/Turjaf.V7i12.2054-2061.273

Casas-López JL, Sánchez-Pérez JA, Fernández-Sevilla JM, Rodríguez-Porcel EM, Chisti Y (2005) Pellet morphology, culture rheology and lovastatin production in cultures of Aspergillus terreus. J Biotechnol 116:61–77. https://doi.org/10.1016/j.jbiotec.2004.10.005

Corrêa RCG, Brugnari T, Bracht A, Peralta RM, Ferreira ICFR (2016) Biotechnological, nutritional and therapeutic uses of Pleurotus spp. (oyster mushroom) related with its chemical composition: a review on the past decade findings. Trends Food Sci Technol 50:103–117. https://doi.org/10.1016/j.tifs.2016.01.012

Díaz R, Alonso S, Sánchez C, Tomasini A, Bibbins-Martínez M, Díaz-Godínez G (2011) Characterization of the growth and laccase activity of strains of Pleurotus ostreatus in submerged fermentation. BioRes 6(1):282–290

Díaz R, Carmen MT, Martha S (2013) Influence of initial PH of the growing medium on the activity, production and genes expression profiles of laccase of Pleurotus Ostreatus in submerged fermentations. Electron J Biotechnol 16:6

Doroški A, Klaus A, Režek Jambrak A, Djekic I (2022) Food waste originated material as an alternative substrate used for the cultivation of oyster mushroom (Pleurotus ostreatus): a review. Sustainability 14:12509. https://doi.org/10.3390/su141912509

Dudekula UT, Doriya K, Devarai SK (2020) A critical review on submerged production of mushroom and their bioactive metabolites. 3 Biotech 10(8):337. https://doi.org/10.1007/s13205-020-02333-y

Ejigu N, Sitotaw B, Girmay S, Assaye H (2022) Evaluation of oyster mushroom (Pleurotus ostreatus) production using water hyacinth (Eichhornia crassipes) biomass supplemented with agricultural wastes. Int J Food Sci 2022:9289043. 9

Elisashvili V (2012) Submerged cultivation of medicinal mushrooms: bioprocesses and products (review). Int J Med Mushrooms 14(3):211–239. https://doi.org/10.1615/IntJMedMushr.v14.i3.10

Elisashvili V, Penninckx M, Kachlishvili E, Tsiklauri N, Metreveli E, Kharziani T, Kvesitadze G (2008a) Lentinus edodes and Pleurotus species lignocellulolytic enzymes activity in submerged and solid-state fermentation of lignocellulosic wastes of different composition. Bioresour Technol 99(3):457–462. https://doi.org/10.1016/j.biortech.2007.01.011

Elisashvili V, Penninckx M, Kachlishvili E, Asatiani M, Kvesitadze G (2008b) Use of white-rot basidiomycetes for the production of lignocellulolytic enzymes. FEMS Microbiol Lett 282(2):208–214

El-Ramady H, Abdalla N, Fawzy Z, Badgar K, Llanaj X, Tör˝os G, Hajdú P, Eid Y, Prokisch J (2022) Green biotechnology of oyster mushroom (Pleurotus ostreatus L.): a sustainable strategy for myco-remediation and BioFermentation. Sustainability 14:3667. https://doi.org/10.3390/su14063667

Fazenda ML, Seviour R, McNeil B, Harvey LM (2008) Submerged culture fermentation of "higher fungi": the macrofungi. Adv Appl Microbiol 63(07):33–103. https://doi.org/10.1016/S0065-2164(07)00002-0

Freixo M d R, Karmali A, Arteiro JM (2008) Production and chromatographic behaviour of polygalacturonase from Pleurotus ostreatus on immobilized metal chelates. Process Biochem 43(5):531–539. https://doi.org/10.1016/j.procbio.2008.01.010

Gafforov Y, Yamaç M, İnci Ş, Rapior S, Yarasheva M, Rašeta M (2023) *Pleurotus Eryngii* (DC.) Quél.; *Pleurotus Ostreatus* (Jacq.) P. Kumm. – *Pleurotaceae*. In: Olim K, Khojimatov, Gafforov Y, Bussmann RW (eds) Ethnobiology of Uzbekistan. Springer, Cham

Galappaththi MCA, Dauner L, Madawala S, Karunarathna SC (2021) Nutritional and medicinal benefits of oyster (Pleurotus) mushrooms: a review. Fungal Biotec 1(2):65–87. https://doi.org/10.5943/Funbiotec/1/2/5

García-Cruz F, Durán-Páramo E, Garín-Aguilar MA et al (2020) Parametric characterization of the initial pH effect on the polysaccharides production by *Lentinula edodes* in submerged culture. Food Bioprod Process 119:170–178. https://doi.org/10.1016/j.fbp.2019.10.016

Gbolagade JS, Fasidi IO, Ajayi EJ, Sobowale AA (2006) Effect of physico-chemical factors and semi-synthetic media on vegetative growth of Lentinus subnudus (Berk.), an edible mushroom from Nigeria. Food Chem 99:742–747

Gern RMM, Wisbeck E, Rampinelli RJ, Ninow JL, Furlan SA (2008) Alternative medium for production of *Pleurotus ostreatus* biomass and potential antitumor polysaccharides. Bio Res Technol 99:76–82

González JA et al (2019) Influence of organic and inorganic nitrogen sources on mycelial growth and enzyme production in *Pleurotus* spp. Appl Microbiol Biotechnol 103(12):4785–4793

Gregori A, Vagelf M, Pohleven J (2007) Cultivation techniquesand medicinal properties of Pleurotus spp. Food Technol Biotechnol 45(3):238–249

Haddadin MSY et al (2009) Role of trace elements in the growth and metabolism of *Pleurotus ostreatus*. Int J Biotechnol Res 2(4):88–94

Hamza A, Shankar MP, Chowdary US, Ghanekar S, Sahoo S, Krishnaiah CV, Kumar DS (2024) Submerged production of mycelium biomass and bioactive compounds from P. ostreatus in a controlled fermentation medium. Food Humanity 2:100302

Hewage R, Sanjeewani I (2024) Impact of nitrogen source on laccase activity in water by spawn pellets of Pleurotus ostreatus (oyster mushroom) and applications for water treatment. Second cycle, A2E. Alnarp: SLU, Dept. of Biosystems and Technology (from 130101)

Hoa HT, Wang CL (2015) The effects of temperature and nutritional conditions on mycelium growth of two oyster mushrooms (Pleurotus ostreatus and Pleurotus cystidiosus). Mycobiology 43(1):14–23. https://doi.org/10.5941/MYCO.2015.43.1.14

Hrudayanath T, Sameer KS (2014) Diversity, nutritional composition and medicinal potential of Indian mushrooms: a review. Afr J Biotechnol 13(4):523–545

Jonathan SG, Bawo DDS, Adejoye DO, Briyai OF (2009) Studies onbiomass production in Auricularia polytricha collected fromWilberforce Island, Bayelsa State, Nigeria. Am J Appl Sci 6(1):182–186

Kalač P (2012) A review of chemical composition and nutritional value of wild-growing and cultivated mushrooms. J Sci Food Agric 93:209–218. https://doi.org/10.1002/jsfa.5960

Khan MA, Amin SMR, Uddin MN, Tania M, Alam N (2008) Comparative study of the nutritional composition of oyster mushrooms cultivated in Bangladesh. Bangladesh J Mushroom 2(1):9–14

Khan AA, Gani A, Khanday FA, Masoodi FA (2018) Biological and pharmaceutical activities of mushroom β-glucan discussed as a potential functional food ingredient. Bioact Carbohydr Diet Fibre 16:1–13. https://doi.org/10.1016/j.bcdf.2017.12.002

Kibar B, Pekşen A (2008) Modelling the effects of temperature and light intensity on the development and yield of different Pleurotus species. Agric Trop Subtrop 41(2):68–73

Kim SW, Hwang HJ, Xu CP, Choi JW, Yun JW (2003) Effect of aeration and agitation on the production of mycelial biomass and exopolysaccharides in an enthomopathogenic fungus Paecilomyces sinclairii. Lett Appl Microbiol 36:321–326. https://doi.org/10.1046/j.1472-765x.2003.01318.x

Krull R et al (2013) Effects of surfactants on fungal growth morphology and oxygen transfer in submerged cultures. Bioprocess Biosyst Eng 36(2):219–231

Krupodorova T, Barshteyn V, Sekan A (2021) Review of the basic cultivation conditions influence on the growth of basidiomycetes. Curr Res Environ Appl Mycol 1:494–531

Krupodorova T, Barshteyn V, Tsygankova V et al (2024) Strain-specific features of *Pleurotus ostreatus* growth in vitro and some of its biological activities. BMC Biotechnol 24:9. https://doi.org/10.1186/s12896-024-00834-9

Lesa KN, Khandaker MU, Iqbal FMR, Sharma R, Islam F, Mitra S, Emran TB (2022) Nutritional value, medicinal importance, and health-promoting effects of dietary mushroom (Pleurotus ostreatus). J Food Qual 2022:2454180. https://doi.org/10.1155/2022/2454180

Letti LA, Vítola FM, de Melo Pereira GV, Karp SG, Medeiros AB, da Costa ES, Bissoqui L, Soccol CR (2018) Solid-state fermentation for the production of mushrooms. Curr Dev Biotechnol Bioeng. https://doi.org/10.1016/b978-0-444-63990-5.00014-1

Li Z, Shukla V, Werner K, Fordyce A, Pedersen A, Marten M (2002) Effects of increased impeller power in a production-scale Aspergillus oryzae fermentation. Biotechnol Prog 18:437–444. https://doi.org/10.1021/bp020023c

Li J, Han LH, Liu XB et al (2020) The saprotrophic *Pleurotus ostreatus* species complex: late Eocene origin in East Asia, multiple dispersal, and complex speciation. IMA Fungus 11(1):10. https://doi.org/10.1186/s43008-020-00031-1

Liu R, Zhang Y, Ding Z, Ma W, Zhang K (2019) Optimization of submerged culture conditions for mycelial growth and exopolysaccharide production by *Pleurotus ostreatus*. Biotechnol Appl Biochem 66(5):749–757

Magaña J, Shimizu N (2022) Submerged pasteurization of substrate for Pleurotus ostreatus cultivation in vitro, The XX CIGR World Congress, Kyoto International Conference Center, Kyoto, Japan

Majumder K, Paul B, Sundas R (2020) An analysis of exo-polygalacturonase bioprocess in submerged and solid-state fermentation by Pleurotus ostreatus using pomelo peel powder as carbon source. J Genet Eng & Biotechnol 18:1–8

Manu-Tawiah W, Martin AM (1987) Use of nitrogen supplemented peat extracts for the cultivation of Pleurotus ostreatus mushroom mycelium. Presented at the 13th Int. congress on the science and cultivation of edible fungi, Braunschweig, 20–26 Sept

Manu-Tawiah W, Martin AM (1988) Nitrogen sources and the growth response of Pleurotus ostreatus mushroom mycelium. Can Inst Food Sci Technol J 21(2):194–199. https://doi.org/10.1016/s0315-5463(88)70776-2

Márquez-rocha FJ, Guillén GK, Sánchez JE, Vázquez-duhalt R (1999) Growth characteristics of Pleurotus ostreatus in bioreactors. Biotechnol Tech 13:29–32. https://doi.org/10.1023/A:1008861432337

Martin AM, Bailey VI (1985) Production of mushroom mycelium in supplemented acid-extract from peat. Can Inst Food Sci Technol J 18(2):185–188

Mca G, Dauner L, Madawala S, Sc K (2021) Nutritional and medicinal benefits of oyster (Pleurotus) mushrooms: a review. Fungal Biol Biotechnol 1(2):65–87

Moses IO (2024) Oyster mushroom in bioremediation: a review of its potential and applications. J Multidiscip Sci MIKAILALSYS 3(1):61–70. https://doi.org/10.58578/mikailalsys.v3i1.4362

Papagianni M (2004) Fungal morphology and metabolite production in submerged mycelial processes. Biotechnol Adv 22(3):189–259

Papaspyridi L-M, Aligiannis N, Topakas E, Christakopoulos P, Skaltsounis A-L, Fokialakis N (2012) Submerged fermentation of the edible mushroom Pleurotus ostreatus in a batch stirred tank bioreactor as a promising alternative for the effective production of bioactive metabolites. Molecules 17(3):2714–2724. https://doi.org/10.3390/molecules17032714

Pilafidis S, Tsouko E, Sougleri G, Diamantopoulou P, Gkatzionis K, Ioannou Z, Sarris D (2024) Submerged cultivation of selected macro-fungi to produce mycelia rich in β-glucans and other bioactive compounds, valorizing side streams of the food industry. Carbon Resour Convers 7(2):100198

Piska K, Sułkowska-Ziaja K, Muszyńska B (2017) Edible mushroom Pleurotus ostreatus (oyster mushroom) –its dietary significance and biological activity. Acta Sci Pol Hortorum Cultus 16(1):151–161

Prasad KK, Mohan SV, Bhaskar YV, Ramanaiah SV, Babu VL, Pati BR, Sarma PN (2005) Laccase production using Pleurotus Ostreatus 1804 immobilized on PUF cubes in batch and packed bed reactors: influence of culture conditions. J Microbiol 43:301–307

Punthi F, Yudhistira B, Gavahian M, Chang CK, Husnayain N, Hou CY, Yu CC, Hsieh CW (2023) Optimization of plasma activated water extraction of Pleurotus ostreatus polysaccharides on its physiochemical and biological activity using response surface methodology. Foods 12(23):4347

Quiñónez-Martínez M, Peña-Avilés K, MartínezRuiz NR, Garza-Ocañas F, Nájera-Medellín JA, OlivasSánchez MP (2022) Production of Pleurotus ostreatus cultivated in substrates made from two invasive weeds. Agrociencia. https://doi.org/10.47163/agrociencia.v56i3.2796

Rajarathnam S, Bano Z (1987) Pleurotus mushrooms. Part I A. Morphology, life cycle, taxonomy, breeding, and cultivation. Crit Rev Food Sci Nutr 26(2):157–223. https://doi.org/10.1080/10408398709527465. PMID: 3322683

Rocha-Valadez JA, Hassan M, Corkidi G, Flores C, Galindo E, Serrano-Carreón L (2005) 6-pentyl-alpha-pyrone production by Trichoderma harzianum: the influence of energy dissipation rate and its implications on fungal physiology. Biotechnol Bioeng 91:54–61. https://doi.org/10.1002/bit.20489

Rosado FR, Germano S, Carbonero ER, Da Costa SMG, Iacomini M, Kemmelmeier C (2003) Biomass and exopolysaccharide production in submerged cultures of Pleurotus ostreatoroseus sing. and Pleurotus ostreatus "Florida"(Jack.: Fr.) Kummer. J Basic Microbiol 43(3):230–237

Royse DJ, Baars J, Tan Q (2017) Current overview of mushroom production in the world. In: Diego CZ, PardoGimenez A (eds) Edible and medicinal mushrooms. John Wiley, Sons, Ltd, pp 5–13. https://doi.org/10.1002/9781119149446.Ch2

Sabri MA, Shatha AS, Rukaibaa AC (2019) Utilization of agricultural and animal wastes in growth of novel Iraqi strains of edible mushrooms Pleurotus ostreatus and brown Agaricus bisporus. Plant Arch 19(Supplement 2):1188–1193

Sahoo S, Gayakwad T, Shahi S (2022) Medicinal value of edible mushrooms. Int J Health Sci 6:8760–8767

Sałata A, Lemieszek M, Parzymies M (2018) The nutritional and health properties of an oyster mushroom (Pleurotus ostreatus (Jacq. Fr) P. Kumm.). Acta Sci Pol Hortorum Cultus 17(2):185–197. https://doi.org/10.24326/Asphc.2018.2.16

Sánchez C (2010) Cultivation of *Pleurotus ostreatus* and other edible mushrooms. Appl Microbiol Biotechnol 85(5):1321–1337

Sangosanya A-O, Pistofidou A (2024) Textile and food waste biodegradation for the production of composite mycelium materials using Pleurotus ostreatus. Res Dir Biotechnol Des 2:e24. https://doi.org/10.1017/btd.2024.26

Saskiawan I (2009) Exopolysaccharide production and its bioactivities of the edible Pleurotus ostreatus in submerged culture. Biotropia 16(2):96–104

Sharifi-Rad J, Butnariu M, Ezzat SM et al (2020) Mushrooms-rich preparations on wound healing: from nutritional to medicinal attributes. Front Pharmacol 11:567518

Shu CH, Lung MY (2004) Effect of pH on the production and molecularweight distribution of polysaccharide by Antrodia camphorata in batchcultures. Process Biochem 39:931–937

Silva RMD, Carmo COD, Oliveira TASD, Figueirêdo VRD, Duarte EAA, Soares ACF (2020) Biological efficiency and nutritional value of Pleurotus ostreatus cultivated in agro-industrial wastes of palm oil fruits and cocoa almonds. Arq Inst Biol 87:e0852018

Singh nee'Nigam P (2009) Production of bioactive secondary metabolites. In: Singh nee'Nigam P, Pandey A (eds) Agro-industrial residues utilization. Springer, pp 133–134

Singh MP, Walia S, Bajaj BK (2016) Enhanced cellulase production by *Pleurotus ostreatus* under submerged fermentation and its application in bioethanol production. J Biosci Bioeng 122(3):351–359

Stajić M, Persky L, Friesem D, Hadar Y, Wasser SP, Nevo E, Vukojević J (2006) Effect of different carbon and nitrogen sources on laccase and peroxidases production by selected *Pleurotus* species. Enzym Microb Technol 38(1–2):65–73

Sun X et al (2021) Lignocellulosic hydrolysates as carbon sources for fungal biomass production: opportunities and challenges. Renew Energy Res J 36(4):542–556

Tang YZ, Zhu LW, Li HM, Li DS (2007) Submerged culture of mushrooms in bioreactorschallenges, current-state-of-the-art, and future. Food Technol Biotechnol 45:221–229

Tinoco-Valencia R, Gómez-Cruz C, Galindo E, Serrano-Carreón L (2014) Toward an understanding of the effects of agitation and aeration on growth and laccases production by Pleurotus ostreatus. J Biotechnol 177:67–73. https://doi.org/10.1016/j.jbiotec.2014.02.013

Treskatis SK, Orgeldinger V, Wolf H, Gilles ED (1997) Morphological characterization of filamentous microorganisms in submerged cultures by on-line digital image analysis and pattern recognition. Biotechnol Bioeng 53:191–201

Vamanu E (2012) Biological activities of the polysaccharides produced in submerged culture of two Edible Pleurotus ostreatus mushrooms. J Biomed Biotechnol 2012:1–8. https://doi.org/10.1155/2012/565974

Vicențiu-Bogdan H, Ana P, Georgiana PARFENE, Tiberius B (2014) Study of preliminary biotechnological conditions for Pleurotus ostreatus cultivation on submerged system. Innov Rom Food Biotechnol 15:58–62

Wang Y, McNeil B (1995) pH effect on EPS and oxalic acid production in culture of Sclerotium gluconicum. Enzyme Microb Technol 17:124–130

Wang SJ, Zhong JJ (2007) Bioreactor engineering. In: Bioprocessing for value-added products from renewable resources. Elsevier, pp 131–161. https://doi.org/10.1016/B978-044452114-9/50007-4

Wang L, Ridgway D, Gu T, Moo-Young M (2005) Bioprocessing strategies to improve heterologous protein production in filamentous fungal fermentations. Biotechnol Adv 23(2):115–129

Wang X, Bai P, Zhang J, Zhou Q, Fan L (2017) Submerged fermentation of *Pleurotus ostreatus*: evaluation of its antioxidant activity and extraction optimization of polyphenols. Process Biochem 62:144–150

Wojewoda W (2003) Checklist of polish larger basidiomycetes. W. Szafer Institute of Botany-Polish Academy of Sciences, Kraków

Wu CY, Liang ZC, Lu CP, Wu SH (2008) Effect of carbon and nitrogen sources on the production and carbohydrate composition of exopolysaccharide by submerged culture of Pleurotus citrinopileatus. J Food Drug Anal 16(2):6

Xu J et al (2019) Fed-batch fermentation strategies for maximized biomass production in *Pleurotus* spp. Biotechnol Prog 35(6):e2832

Zapaśnik A, Bryła M, Sokołowska B, Waśkiewicz A (2024) Pleurotus spp.—an effective way in degradation mycotoxins? A comprehensive review. Mycotoxin Res 41:1–13

Zhang Y et al (2021) Optimization of bioreactors for submerged cultivation of edible fungi. Mycol Res 24(5):349–365

Zięba P, Sękara A, Sułkowska-Ziaja K, Muszyńska B (2020) Culinary and medicinal mushrooms: insight into growing technologies. Acta Mycol 55(2):Article 5526. https://doi.org/10.5586/Am.5526

Controlled Cultivation of Mushrooms in Modular Robotic System

Shenbhagaraman Ramalingam, Jenefar Sudarson, Selvakumari Jeyaperumal, and Lakshmanan Govindan

Abstract Automated Robotic cultivation techniques in mushroom biotechnology has advanced, fast and significantly yielding technology for production of diverse mushrooms and bioactive compounds. Automated robotic methods helps to optimize fungal growth producing valuable molecules like proteins enzymes and polysaccharides. Temperature, agitation, pH, oxygen level, and inoculum quality must be precise controlled to maximize biomass and bioactive compounds yields. Automated modular robotic technique is upcoming technology, can be utilized across different scale for mushroom cultivation. Research aims to enhance yields and streamline production process via modular robotic technique offering promising avenue for nutrition, pharmaceuticals and industry. Modular robotic cultivation process significantly impacts fungal growth and bioactive compound synthesis using controlled modular robotic cultivation. Automated robotic composting technique helps to improve carbon source, organic and inorganic nitrogen sources and enhance the production of button mushroom especially in *Agaricus bisporus* strain. During automated modular robotic production conditions, the environmental oxygen level influence aerobic cell grow deeply and need proper oxygen transfers. pH adjustments is needed to keeps favorable growth condition, which affect mycelial growth and synthesis of metabolites. Proper oxygen transfer while minimizing stress conditions and there by enhancing mushroom yield through modular robotic

S. Ramalingam
Department of ENT, Saveetha Medical College and Hospitals, Saveetha Institute of Medical and Technical Sciences, Saveetha University, Chennai, India

J. Sudarson
Department of Biochemistry, Saveetha Medical College and Hospitals, Saveetha Institute of Medical and Technical Sciences, Saveetha University, Chennai, India

S. Jeyaperumal
National Centre for Disease Control, Kozhikode, Thiruvananthapuram Field Unit, Iranimuttam, Kerala, India

L. Govindan (✉)
Sri Lakshmi Narayana Institute of Medical Sciences (Affiliated to Bharath Institute of Higher Education and Research), Puducherry, India

M. N. Owaid et al. (eds.), *Mushroom Production*, Interdisciplinary Biotechnological Advances, https://doi.org/10.1007/978-981-95-7793-4_4

techniques. Modular robotic techniques maintains inoculum size and age, significantly influence mushroom yield and bioactive compound product. Thus, modular robotic cultivation offers a future direction for sustainable, scalable, and technology driven mushroom productivity system.

Keywords Modular robotic cultivation · Biomass · Automated storage system · Mushroom · Nutritional applications · Automated robotic composting

1 Introduction

Mushroom cultivation, once being a traditional agriculture practice, has evolved a lot with advance in technology (Kumar et al. 2023). One such innovation which hold big promise is the integration of modular robotic into cultivation process. This chapter explore the intersect of modular robotics and mushroom growing, showing how this combination help improving sustainability, efficiency and fungi productivity.

2 Enhanced Efficiency Through Automation

Traditional mushroom farming often involve labor-intense works like substrate preparation, inoculations, harvest, and maintaining (El-Ramady et al. 2022; Rukhiran et al. 2023). The addition of modular robotics helps streamline these task, reduces manual working, and cuts down operation cost. Automated system is able to measure and control environment like humidity, temperature, and then light more precisely, which optimize condition for mushroom growing (Wainaina and Taherzadeh 2023). Also, robot arms with sensors can detect and remove contamination better, ensuring more yields and good quality (Javaid et al. 2021). Even self-driving vehicle helps move materials and collect mushrooms around the facility, making operation more easy (Retsinas et al. 2023) (Table 1).

This table shows how automation systems in mushroom cultivation gives big advantage over old methods, like less labor needed, better precision, more control of environment, and easy to scale for different production goals. In mushroom growing, automation plays an important role in optimizing the growth condition, ensuring quality and also making many process more smooth (Guragain et al. 2024). Below is a detailed study about robotic automations used in mushroom production.

Table 1 Key differences among conventional and robotic schemes in mushroom farming (Suresh et al. 2021)

Aspect	Traditional methods	Automated systems
Labor intensity	High labor requirements for manual tasks	Reduced labor with automation
Substrate preparation	Manual mixing and preparation	Automated substrate mixing and delivery systems
Inoculation	Manual inoculation of spawn	Robotic arms for precise inoculation
Environmental control	Manual regulation of temperature and humidity	Automated environmental control systems
Contaminant detection	Visual inspection for contaminants	Sensors and cameras for automated detection
Harvesting	Manual harvesting and sorting	Robotic arms for automated harvesting
Transportation	Manual handling and transport within facility	Autonomous vehicles for material and mushroom transport
Data monitoring	Limited monitoring capabilities	Real-time monitoring and data analytics
Customization	Limited customization options	Flexible and customizable systems
Scalability	Limited scalability for large-scale production	Scalable to meet varying production requirements
Environmental impact	Resource-intensive and potentially wasteful	Resource-efficient and environmentally friendly

3 Robotic System for Mushroom Composting

Modular robotic can greatly improve the efficient, flexibility, and scale-up of compost making for mushroom farming (Petre and Teodorescu 2012). By employing modular robotic systems, each task in the composting process can be managed by specialized, interchangeable robotic units that can be reconfigured or replaced as needed.

4 Raw Material Preparation by Modular Robotic Units

Robotic conveyors and loaders equipped with modular conveyor belts and loaders can transport raw materials such as straw, manure, and gypsum to the processing area. Automated grinders and shredder units can grind and shred raw materials to the desired size for efficient composting (Szrek et al. 2022). Pre-wetting done by sprinkler robots equipped with adjustable nozzles can ensure precise water distribution for pre-wetting raw materials, and mixing done by robotic arms or rotating drums can thoroughly mix the materials to achieve a uniform composition. Robotic conveyors and loaders can transport thoroughly mixed raw materials to proper composting tunnel.

5 Role of Modular Robotic Units in Composting

5.1 Modular Robotic-Assisted Composting: Phase 1 (Preconditioning)

Modular robotic assisted composting process has been started by loading raw materials such as paddy straw, gypsum, calcium carbonate by the transport robots into composting tunnel. It will take care by Temperature and Humidity Sensing robots with Modular sensor units will be automatically inserted into the compost pile and it will starts continuously monitor environmental conditions (Han et al. 2021). During the composting process, aeration robotic units equipped with aeration systems can move through the compost pile, providing necessary oxygen and regulating temperature to ensure the proper conditioning of compost. Robotic turners can regularly turn the compost to ensure even decomposition and prevent anaerobic conditions (Alkoaik 2019). Once the composting process is completed, compost can be transferred to composting phase 2 for pasteurization.

5.2 Modular Robotic-Assisted Composting: Phase 2 (Pasteurization)

In this composting phase 2, the pasteurization process of compost was performed using modular robotic units with heating function, where they can inject steam or use heating elements to raise and maintain appropriate temperature for the compost for pasteurization (Vieira and Pecchia 2018). The robotics contain mobile units equipped with sensors to track temperature, humidity, and CO_2 levels (Nazir et al. 2014) during pasteurization. Cooling and moisture controlling unit contained (fans or ventilation) in modular robotic units cool down the compost and maintain optimal conditions for beneficial microorganisms. These units can maintain the desired moisture level in the compost (Lobo and Dorta 2019) (Fig. 1).

6 Modular Robotic Units in Quality Control and Testing of Compost

Autonomous robotic units can collect samples from different sections of the compost for testing and are equipped with lab-on-a-chip technology—these robots can analyze samples for pH, nutrient content, moisture, and microbial activity (Kitić et al. 2022).

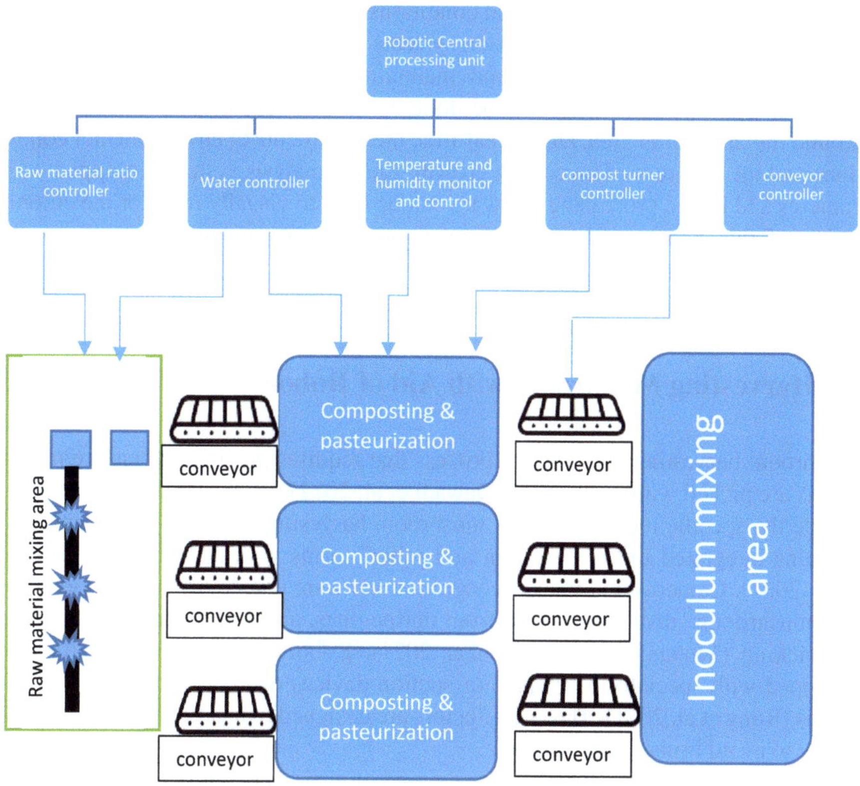

Fig. 1 Schematic diagram for robotic-controlled composting yard

7 Robotic Systems for Spawn Inoculation

Spawn inoculation involves introducing mushroom spawn (mycelium) into the prepared substrate. This process must be done in sterile conditions to prevent contamination (Sánchez 2004). Inoculation of spawn can be done using automated inoculation units equipped with sterilized needles or dispensing nozzles, robotic arms. These can precisely inject the spawn into the substrate and ensure uniform distribution of spawn into each substrate bag or tray (Krämer et al. 2016).

8 Robotic Systems for Spawn Running

Spawn running refers to the phase where the mycelium grows and colonizes the substrate (Aditya et al. 2024). This phase requires controlled environmental conditions to ensure optimal growth. The robotic climatic control units equipped with sensors can monitor and adjust temperature, humidity, and CO_2 levels in the incubation room to maintain optimal conditions for mycelium growth (Chong et al. 2023). Automated fans and ventilation systems can ensure adequate air circulation to prevent the buildup of CO_2 and maintain even temperature and humidity. Monitoring of spawn running can be done by mobile sensor units that can move around the incubation area, taking real-time measurements of environmental conditions and the progress of mycelium growth. Image analysis systems equipped with cameras and image processing software can be used to visually monitor the colonization of the substrate, identifying any issues such as contamination or uneven growth (Ghahraei et al. 2023).

9 Harvesting Mushroom with Aid of Robotics

Mushroom harvesting is a delicate process that requires precision to ensure mushrooms are picked without damage (Dawadi et al. 2022). Robotics can be tailored to handle the specific requirements of mushroom harvesting. Robotic harvesting unit contains integrated imaging system and robotic arms with end effectors (Huang et al. 2021). Cameras and image recognition software enable the robot to identify ripe mushrooms, distinguish them from unripe ones, and determine the best angle for picking (Mohanan and Salgaonkar 2021). Robotic arms with end effectors equipped with specialized grippers or suction devices that can gently pick mushrooms (Rong et al. 2021). The end effectors are designed to handle mushrooms delicately to avoid bruising and damage.

Harvesting robotics can be classified based on their function:

1. Autonomous mobile robotic units (AMRs): These robotic units can navigate the growing area, locate mushrooms using vision systems, and harvest them. They are equipped with sensors for obstacle detection and avoidance, ensuring safe operation in dynamic environments (Jou et al. 2022).
2. Guided picking robots: Similar to AMRs but operate along fixed paths or tracks within the growing area, providing a more controlled harvesting environment (Rong et al. 2021).
3. Multifunctional robot units: Robots that not only harvest but also sort mushrooms based on size, weight, and quality. This dual functionality streamlines the postharvest process, reducing the need for manual sorting (Kondoyanni et al. 2022) (Fig. 2).

10 Robotics in Postharvest Handling

After harvesting, mushrooms need to be handled carefully to maintain their quality until they reach the consumer (Janos Vetter 2003). Automatic-guided conveyor systems transport harvested mushrooms from the growing area to the packaging area.

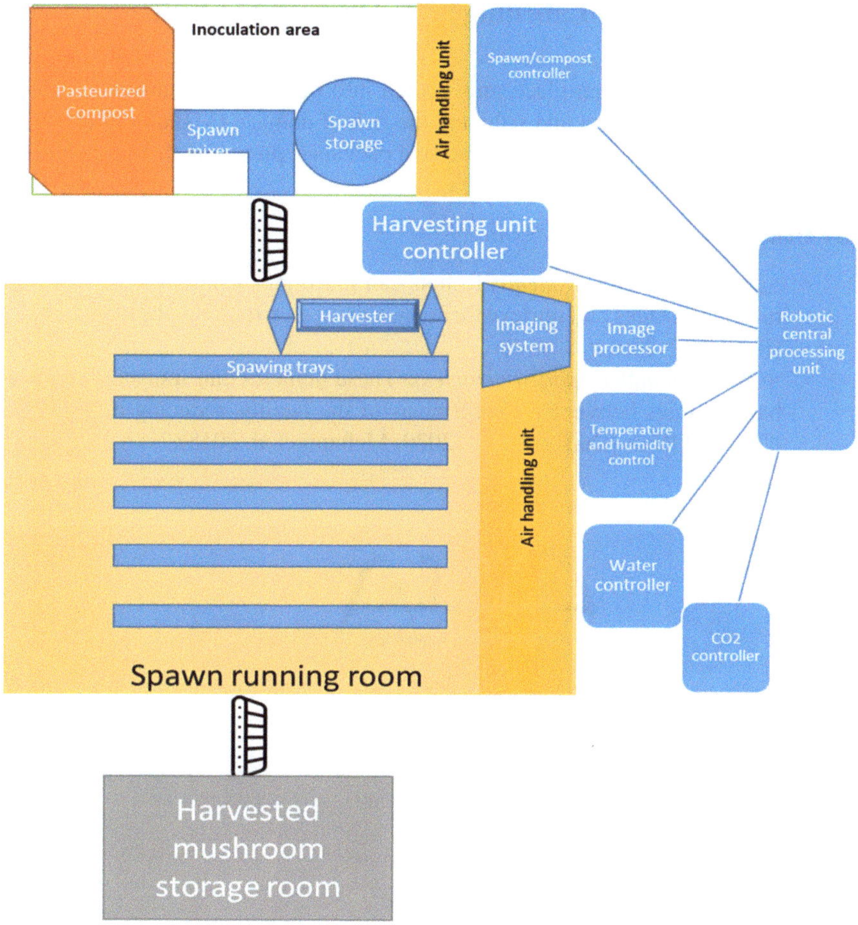

Fig. 2 Schematic diagram for robotic controlled mushroom growing unit

These systems can be integrated with sensors to ensure smooth and gentle movement (Jiang et al. 2022).

11 Sorting and Grading Robotic Units

Vision and sorting systems equipped with cameras and software can sort mushrooms by size, shape, and color, grading them for quality before packaging (Raman et al. 2021). Weight-based sorting weighs each mushroom and sorts them into categories based on predefined weight ranges (Valverde 2017).

12 Packaging and Labelling with Aid of Robot Units

Robotic packing units can place mushrooms into packaging containers, ensuring consistent packing density and reducing the risk of damage (Mandlik et al. 2021). Labelling and sealing systems label and seal packages, preparing them for shipping (Fig. 3).

(a) Handling and Transport Robots

- *Automated storage systems*: Robots can place inoculated bags or trays into incubation racks and retrieve them once the spawn running phase is complete (Reed et al. 2001).
- *AGVs (automated guided vehicles)*: These vehicles can transport substrate bags or trays between different areas of the facility, reducing manual handling and labor (De Ryck et al. 2020; Mehami et al. 2018).

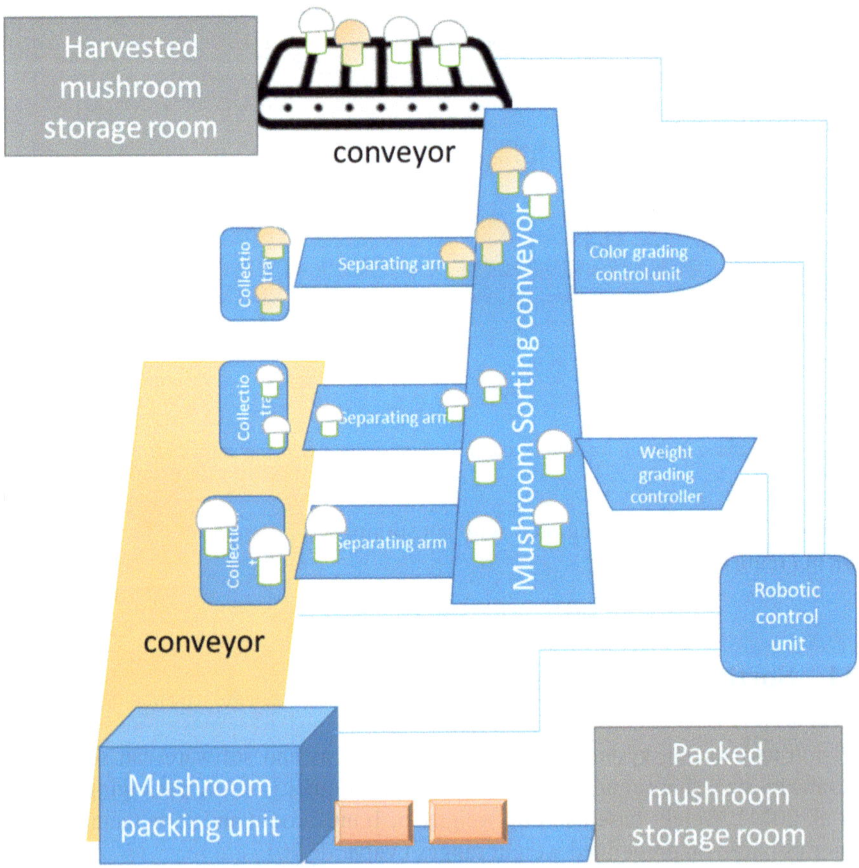

Fig. 3 Schematic diagram for robotic-controlled postharvest mushroom handling and packing unit

Integrating robotics into the mushroom harvesting process can significantly enhance efficiency, reduce labor costs, and improve the consistency and quality of the harvested mushrooms. Here's a comprehensive overview of how robotics can be utilized in mushroom harvesting:

13 Advantages of Modular Robotics in Cultivation of Mushrooms

- Modular systems can be easily reconfigured or upgraded to adapt to changing requirements or scale operations during compost preparation.
- Robots can operate continuously, reducing the time required for each phase of compost preparation.
- Automated systems ensure uniform processing, leading to higher quality compost.
- Robotic can operate continuously without fatigue, significantly increasing the harvesting speed compared to manual labor.
- Automated systems can harvest mushrooms with high precision, reducing damage and waste.
- Robotics ensure a uniform harvesting process, leading to more consistent product quality.
- Automation reduces the reliance on manual labor, lowering operational costs.
- Robots are capable of handling repetitive and physically demanding tasks, thereby minimizing physical strain on human workers.

14 Disadvantages

- High initial costs for purchasing and integrating robotic systems into existing operations.
- Regular maintain and tech support is important for keep robotic units working smooth.
- Staffs must be trained for use and handle robotic system, also for fix issues and do simple repairs.
- Existing workflows may need to be adapted to integrate robotic systems effectively.
- Modular systems can be expanded incrementally, allowing for scalable growth.

15 Autonomous Vehicles

Material transporting: In big mushroom cultivation setups, things like composts, substrate, and tray for growing need move around different section efficiently. Self-driving vehicles are guided on fixed paths or use sensor navigation, doing this work, cutting down manual works, and making workflow better.

Mushroom harvesting and moving: When mushrooms get mature, self-driving vehicles with soft harvesting tools pick them from the grow beds. These vehicles carry the picked mushrooms to the process or packing area, reducing damage from handling and keeping product good.

Maintenance and check: Autonomous vehicles also used for regular maintain jobs like cleaning, sanitizing, and checking the facility. They move through the grow space, keeping hygiene rules and finding any problems needing people's help.

By putting these parts into automation system, mushroom farming places can get more efficiency, the same quality, and better production while cutting labor costs and lowering chances of contamination or human mistake.

16 Precision Farming for Optimal Growth

Modular robot systems helps apply precision farming method suited for needs of different mushroom types. Using sensors and actuator, such systems able to watch and change things like nutrient amount, pH level, and air flows in real time (Pretto et al. 2021). This kind of precision ensures steady and best growing conditions, giving more yield and less wasting of resources (Matar et al. 2024). Also, robotic platform with machine learning can study data patterns to better farming method over time, changing with environment shifts and boost output (Kaczmarczyk et al. 2022). In mushroom farming, the good growth of fungi largely depends on keeping the right nutrient levels, pH balance, and air flow in the grow area (Ferreira et al. 2009). Using modular robotic to check and adjust these thing in real time can significantly improve the production and product quality. Below explains how each part can be handled:

17 Nutrient Levels

Monitoring: Sensors placed inside the growing media or nutrient liquid always measure how much nutrients like nitrogen, phosphorous, and potassium is there (Raman et al. 2021). These sensors may use methods like conductivity, ion-selective probe, or even spectroscopy for giving the right readings (Kaur et al. 2022).

Adjustment: Modular robotic systems can be equipped with mechanisms for dispensing nutrient solutions or additives constructed on the real-time statistics obtained from the radars. Precise dosing ensures that the mushrooms receive the necessary nutrients at all stages of growth, promoting healthy development and maximizing yields (Nicolescu et al. 2016).

1. *pH Balance*

   Monitoring: pH sensors monitor the acidity or alkalinity of the growing medium or nutrient solution (Rico 2020). Fluctuations in pH can impact nutrient availability and fungal growth. Real-time monitoring enables prompt detection of deviations from the optimal pH range.

   Adjustment: Modular robotics can incorporate pH adjustment systems, such as pH regulators or dosing pumps, capable of mixing alkaline or acid solutions to sustain the preferred pH level. Automated adjustments confirm that the pH ruins within the ideal choice for mushroom cultivation, optimizing nutrient uptake and metabolic processes.

2. *Airflow*

   Monitoring: Airflow sensors are placed around the grow facility measures how fast air moves and the way it circulates. Proper airflows are important for control temperature, humidity, and gas exchange, which directly affects mushroom development (Li et al. 2022).

   Adjustment: Modular robotic systems can run fans, vents, or air dampers to change airflow using real-time data from sensors. By changing how much and where air go, these system help spread fresh air same everywhere and take out extra moisture and CO_2, making a better grow condition for mushrooms.

Putting these monitoring and control parts into modular robots help apply precision farming in mushroom growing (Sulistyanto et al. 2018). By always tuning nutrient, pH, and airflow in real-time, farmers can get more yield, good mushroom quality, and save resources, which support sustainable and profitable farming (Balan et al. 2022).

18 Sustainability and Environmental Benefits

The integrating of modular robotics help promoting sustainability in mushroom farming by reducing resource use and impacting less on environment (Gregori and Pohleven 2007). Automated system use resources much efficient, giving water, nutrient, and power only where it needed. This focused method reduces the use of water, lessens fertilizer runoffs, and saves energy compared with normal farming (see Table 2). Also, by reducing need for chemicals and pest killers, robotic systems help better ecosystems and make food production more safe.

Table 2 Robotic mushroom cultivation reduces water usage, fertilizer runoff, and energy consumption compared to traditional methods (Pandey et al. 2022)

Aspect	Traditional methods	Robotic mushroom cultivation
Water usage	Often involves manual watering with imprecise control	Automated systems deliver precise amounts of water directly to mushrooms and substrate, minimizing waste
Fertilizer runoff	Excessive use of fertilizers can lead to runoff, contaminating water sources	Automated systems deliver nutrients directly to the substrate in controlled amounts, reducing the risk of runoff
Energy consumption	Manual labor and outdated equipment may require high-energy inputs	Robotic systems are designed for energy efficiency, utilizing advanced sensors and algorithms to optimize resource usage and minimize energy consumption

In robotic mushroom cultivation, the reliance on manual application of chemical inputs and pesticides is significantly reduced, leading to profound benefits for both ecosystems and food safety. Unlike older methods where chemicals often used without much control, robotic system allows precise and focus delivery of nutrients while using less pesticides (Balaska et al. 2023). Advance monitoring and detection system built into robot platforms help early find and solve threats, lowering the chance of pollution and environmental harm. So, less chemical use not just keep biodiversity and ecosystem safe but also makes mushroom more safe to eat with less leftover chemicals, helping environment and food safety both (Aditya et al. 2024). Giving nutrients precisely and controlling pests better also mean less bad effects like soil or water getting polluted (Hossain et al. 2022). This moving to sustainable way of growing not only helps nature and the health of people but also makes sure agriculture can stay strong in future. As tech is getting better, robot mushroom farming can significantly improve the use of resources, reduce environmental damage, and grow food in more sustainable ways all over the world.

19 Flexibility and Scalability

Modular robotics give flexibility and scalable solutions, letting mushroom growers adjust for different market needs and production demands. These robotic system can be modified and re-set up to fit with different farming setups, mushroom kinds, and production sizes. Whether it's small indoor farms or big commercial grow area, modular robotic platform are used for better space using, raising production, and adapting to market changing. Also, modular systems is easy to maintain, upgrade, and expand, making them more useful for long time and ready for future agriculture changes.

Customization and resetting in robotic mushroom farming means that automated system can change to different setup, grow different mushroom types, and scale up or down based on what needed. Not like traditional ways, which mostly have fixed structures and same processes, robotic systems give more design and working flexibility.

One part of customization is how robot tools can change with many grow setups. This includes different layout, size, and facility arrangements. Robots can be built to work in many environment, even in small grow room, large greenhouse, or reused buildings like containers or warehouses. This make possible for growers to better use the space, improve production, and deal with space limits.

Also, robotic system support many kinds of mushrooms with different grow needs. Various mushrooms need special climate like temperature, moist, and substrate mix. Robotic platforms using sensors and moving parts to check and fix these things in real time, giving correct condition for every type of mushroom. This helps grow better and get more mushrooms in different product types. Additionally, robotic mushroom cultivation offers scalability to meet varying production scales and demands. Whether it's a small-scale operation supplying local markets or a large-scale commercial farm serving regional or national distribution channels, robotic systems can be scaled up or down accordingly. Modular design and flexible configurations enable growers to expand or reconfigure their production facilities as needed, without significant disruption to operations. This scalability ensures that growers can adapt to changing market dynamics, seasonal fluctuations, and evolving consumer preferences while maintaining efficiency and profitability.

Totally, customize and reconfigure are important feature of robotic mushroom farming that helps growers to better their production, improve quality of mushrooms, and react good to market changes. By giving flexibility, multi-use, and ability to scale, robotic system lets farmers reach more efficient, more sustainable and strong production in mushroom farming, so shaping the future of farming innovation and food making.

20 Challenges and Future Directions

Even with many advantages, using modular robotics widely in mushroom farming still faces problems like high starting costs, tech difficulties, and rules to follow. But new research and developments is trying to solve these problem and push new ideas in this area. In future, new things like self-moving robots, swarm robotic systems, and better sensors may help more for efficiency, trust, and sustainability in mushroom farming.

Using robot systems in mushroom farms isn't without problems, including start-up money needs, tech hard to use, and rule issues. These make it harder for farmers who want to use automatic machines on their farms.

First, the cost to buy and setup robotic machines is very high. Buying robot parts, programs, and building the system, as well as putting them together, take a lot of money at start. Also, there are cost for keeping it working, fixing, or upgrading later, which makes the whole cost more over time. For small farmers or those with little money, this high first cost can be a big problem and need budget thinking or loans.

Second, using robotic systems is very technical and can be hard to learn for farmers. These systems need special knowledge in robot working, computer things, and engineering. Training worker to use and repair robot tools might take extra time and money, especially if they don't already know tech stuff. Making sure robots can work together with already present farm tools also make things more complicated.

Also, rules and laws are a big part in using robot tech in farming. Farms must follow safety, industry, and eco rules to make sure they're using robots legally and safely. Handling government permits or following complex laws can take lots of paper works and time for farmers. Following food safety rules and trackability is extra important in mushroom farms, because people eat them and need trust in quality.

While robots help make farming better, faster, and greener, these challenges high cost, tech difficulty, and law following can slow down their use. To get past these, growers, robot companies, and rule makers need to plan, invest, and work together. Even with all this, robotic mushroom farming still have big future to change farming and help with food supply and sustainability, so it's a great area for more new ideas and funding.

21 Conclusion

Integrating modular robotics shows a big change in mushroom growing, giving new chances to make things more efficient, more green, and more productive. By making hard jobs automatic, making grow condition better, and lowering bad effect on nature, robot systems help growers meet the rising need for healthy and sustainable foods. As tech keeps changing, modular robotics is going to be important for the future of mushroom farming, opening path to more strong and eco-friendly farming industry.

References

Aditya N, Jarial RS, Jarial K, Bhatia JN (2024) Comprehensive review on oyster mushroom species (Agaricomycetes): morphology, nutrition, cultivation and future aspects. Heliyon 10(5):e26539. https://doi.org/10.1016/j.heliyon.2024.e26539

Alkoaik FN (2019) Integrating aeration and rotation processes to accelerate composting of agricultural residues. PLoS One 14(7):e0220343. https://doi.org/10.1371/journal.pone.0220343

Balan V, Zhu W, Krishnamoorthy H, Benhaddou D, Mowrer J, Husain H, Eskandari A (2022) Challenges and opportunities in producing high-quality edible mushrooms from lignocellulosic biomass in a small scale. Appl Microbiol Biotechnol 106(4):1355–1374. https://doi.org/10.1007/s00253-021-11749-2

Balaska V, Adamidou Z, Vryzas Z, Gasteratos A (2023) Sustainable crop protection via robotics and artificial intelligence solutions. Machines 11(8). https://doi.org/10.3390/machines11080774

Chong JL, Chew KW, Peter AP, Ting HY, Show PL (2023) Internet of things (IoT)-based environmental monitoring and control system for home-based mushroom cultivation. Biosensors 13(1). https://doi.org/10.3390/bios13010098

Dawadi E, Magar PB, Bhandari S, Subedi S, Shrestha S, Shrestha J (2022) Nutritional and post-harvest quality preservation of mushrooms: a review. Heliyon 8(12):e12093. https://doi.org/10.1016/j.heliyon.2022.e12093

De Ryck M, Versteyhe M, Debrouwere F (2020) Automated guided vehicle systems, state-of-the-art control algorithms and techniques. J Manuf Syst 54:152–173. https://doi.org/10.1016/j.jmsy.2019.12.002

El-Ramady H, Abdalla N, Fawzy Z, Badgar K, Llanaj X, Törős G, Hajdú P, Eid Y, Prokisch J (2022) Green biotechnology of oyster mushroom (Pleurotus Ostreatus L.): a sustainable strategy for myco-remediation and bio-fermentation. Sustainability 14:3667

Ferreira ICFR, Barros L, Abreu RMV (2009) Antioxidants in wild mushrooms. Curr Med Chem 16(12):1543–1560

Ghahraei O, Kheshtzarrin M, Saghafinia A (2023) Development of a button mushroom harvesting robot using expert system and image processing in shelf cultivation method. Majlesi J Electr Eng 17(4):53–61. https://doi.org/10.30486/mjee.2023.1990853.1171

Gregori A, Pohleven J (2007) Cultivation techniques and medicinal properties of Pleurotus spp. Food Technol Biotechnol 45(3):236–247

Guragain DP, Shrestha B, Bajracharya I (2024) A low-cost centralized IoT ecosystem for enhancing oyster mushroom cultivation. J Agric Food Res 15:100952. https://doi.org/10.1016/j.jafr.2023.100952

Han J, Liu L, Zeng H (2021) Design and implementation of intelligent agricultural picking mobile robot based on color sensor. J Phys Conf Ser 1757(1):012157. https://doi.org/10.1088/1742-6596/1757/1/012157

Hossain ME, Shahrukh S, Hossain SA (2022) Chemical fertilizers and pesticides: impacts on soil degradation, groundwater, and human health in Bangladesh. In: Singh VP, Yadav S, Yadav KK, Yadava RN (eds) Environmental degradation: challenges and strategies for mitigation. Springer International Publishing, pp 63–92. https://doi.org/10.1007/978-3-030-95542-7_4

Huang M, He L, Choi D, Pecchia J, Li Y (2021) Picking dynamic analysis for robotic harvesting of Agaricus bisporus mushrooms. Comput Electron Agric 185:106145. https://doi.org/10.1016/j.compag.2021.106145

Javaid M, Haleem A, Singh RP, Suman R (2021) Substantial capabilities of robotics in enhancing industry 4.0 implementation. Cogn Robot 1:58–75. https://doi.org/10.1016/j.cogr.2021.06.001

Jiang F, Yang X, Wang Y, Yang L, Sun B (2022) Design of an Online Quality Inspection and Sorting System for fresh button mushrooms (Agaricus bisporus) using machine vision. Eng Lett 30(1):221–226

Jou R-Y, Li W-J, Shih H-D, Chiu H-C (2022) Research on autonomous Mobile intelligent IoT platform in mushroom cultivation. In: 2022 IEEE 5th international conference on knowledge innovation and invention (ICKII), pp 184–188. https://doi.org/10.1109/ICKII55100.2022.9983567

Kaczmarczyk V, Baštán O, Husák M, Benešl T, Bradáč Z (2022) Robotic platform equipped with machine learning. IFAC Pap OnLine 55(4):380–386. https://doi.org/10.1016/j.ifacol.2022.06.063

Kaur G, Kaur H, Mittal SK (2022) Electrochemical sensor based on ion-selective membrane of silica/polyaniline nano-composites for selective determination of uranyl ions. Talanta Open 6:100158. https://doi.org/10.1016/j.talo.2022.100158

Kitić G, Krklješ D, Panić M, Petes C, Birgermajer S, Crnojević V (2022) Agrobot Lala—an autonomous robotic system for real-time, in-field soil sampling, and analysis of nitrates. Sensors 22(11). https://doi.org/10.3390/s22114207

Kondoyanni M, Loukatos D, Maraveas C, Drosos C, Arvanitis KG (2022) Bio-inspired robots and structures toward fostering the modernization of agriculture. Biomimetics 7(2). https://doi.org/10.3390/biomimetics7020069

Krämer I, Federici M, Kaiser V, Thiesen J (2016) Media-fill simulation tests in manual and robotic aseptic preparation of injection solutions in syringes. J Oncol Pharm Pract 22(2):195–204. https://doi.org/10.1177/1078155214565123

Kumar S, Kumar V, Kumar D, Singh S, Kumar D, Verma S (2023) Mushroom cultivation: a sustainable approach to future agriculture to ensure quality food and nutritional security of current population in India. Int J Multidiscip Res Growth Eval 04(04):697–705

Li Y, Sun F, Shi W, Liu X, Li T (2022) Numerical simulation of ventilation performance in mushroom solar greenhouse design. Energies 15(16). https://doi.org/10.3390/en15165899

Lobo MG, Dorta E (2019) Utilization and management of horticultural waste. In: Postharvest technology of perishable horticultural commodities. Elsevier, pp 639–666. https://doi.org/10.1016/B978-0-12-813276-0.00019-5

Mandlik PSB, Abhishek P, Aishwarya A, Anuja M (2021) Automatic packaging machine. Int J Sci Res Eng Technol 8(3):342–347. https://doi.org/10.32628/ijsrset218349

Matar MAI, Radi M, Alzebda SA (2024) Advancements in smart modular farming systems for sustainable agriculture. In: Khamis R, Buallay A (eds) AI in business: opportunities and limitations, vol 2. Springer Nature Switzerland, pp 45–55. https://doi.org/10.1007/978-3-031-49544-1_4

Mehami J, Nawi M, Zhong RY (2018) Smart automated guided vehicles for manufacturing in the context of industry 4.0. Procedia Manuf 26:1077–1086. https://doi.org/10.1016/j.promfg.2018.07.144

Mohanan MG, Salgaonkar A (2021) Robotic mushroom harvesting by employing probabilistic road map and inverse kinematics. BOHR Int J Future Robot Artif Intell 1(1):1–10. https://doi.org/10.54646/bijfrai.001

Nazir M, Gb H, Gbani M, Shabeen K, Najeeb M (2014) Button mushroom cultivation, vol 10. Communication and Publication Centre, pp 1–33

Nicolescu FA, Marinescu DA, Avram GC (2016) Chapter 4 - Virtual robotic prototype for safe and efficient cultivation of mushrooms. In: Petre M (ed) Mushroom biotechnology. Academic Press, pp 49–67. https://doi.org/10.1016/B978-0-12-802794-3.00004-7

Pandey N, Kamboj N, Sharma AK, Kumar A (2022) An overview of recent advancements in the irrigation, fertilization, and technological revolutions of agriculture. In: Bahukhandi KD, Kamboj N, Kamboj V (eds) Environmental pollution and natural resource management. Springer International Publishing, pp 167–184

Petre M, Teodorescu A (2012) Biotechnology of agricultural wastes recycling through controlled cultivation of mushrooms. In: Petre M (ed) Advances in applied biotechnology. IntechOpen. https://doi.org/10.5772/29998

Pretto A, Aravecchia S, Burgard W, Chebrolu N, Dornhege C, Falck T, Fleckenstein F, Fontenla A, Imperoli M, Khanna R, Liebisch F, Lottes P, Milioto A, Nardi D, Nardi S, Pfeifer J, Popović M, Potena C, Pradalier C et al (2021) Building an aerial–ground robotics system for precision farming: an adaptable solution. IEEE Robot Autom Mag 28(3):29–49. https://doi.org/10.1109/MRA.2020.3012492

Raman J, Jang KY, Oh YL, Oh M, Im JH, Lakshmanan H, Sabaratnam V (2021) Cultivation and nutritional value of prominent Pleurotus Spp.: an overview. Mycobiology 49(1):1–14. https://doi.org/10.1080/12298093.2020.1835142

Reed JN, Miles SJ, Butler J, Baldwin M, Noble R (2001) AE—automation and emerging technologies: automatic mushroom harvester development. J Agric Eng Res 78(1):15–23. https://doi.org/10.1006/jaer.2000.0629

Retsinas G, Efthymiou N, Anagnostopoulou D, Maragos P (2023) Mushroom detection and three dimensional pose estimation from multi-view point clouds. Sensors 23(7):1–23. https://doi.org/10.3390/s23073576

Rico ALJ (2020) Automated pH monitoring and controlling system for hydroponics under greenhouse condition. ARPN J Eng Appl Sci 15(2):523–528. https://doi.org/10.36478/JEASCI.2020.523.528

Rong J, Wang P, Yang Q, Huang F (2021) A field-tested harvesting robot for oyster mushroom in greenhouse. Agronomy 11(6):1–19. https://doi.org/10.3390/agronomy11061210

Rukhiran M, Sutanthavibul C, Boonsong S, Netinant P (2023) IoT-based mushroom cultivation system with solar renewable energy integration: assessing the sustainable impact of the yield and quality. Sustainability 15(18). https://doi.org/10.3390/su151813968
Sánchez C (2004) Modern aspects of mushroom culture technology. Appl Microbiol Biotechnol 64(6):756–762. https://doi.org/10.1007/s00253-004-1569-7
Sulistyanto MPT, Harianto W, Nugroho DA, Retandi RE, Akbar AK, Tjahjanti PH (2018) The controlling and monitoring system in oyster mushroom cultivation using fuzzy logic through web technology integrated with internet of things. In: MATEC web of conferences, vol 197, p 15002. https://doi.org/10.1051/matecconf/201819715002
Suresh M, Srinivasan M, Gowri Shankar S, Karthikeyan D, Nakhul V, Naveen Kumar A, Sundar S, Maniraj P (2021) Monitoring and automatic control of various parameters for mushroom farming. IOP Conf Ser Mater Sci Eng 1055(1):012011. https://doi.org/10.1088/1757-899x/1055/1/012011
Szrek J, Jakubiak J, Zimroz R (2022) A mobile robot-based system for automatic inspection of belt conveyors in mining industry. Energies 15(1). https://doi.org/10.3390/en15010327
Valverde J (2017) Harvesting and processing of mushrooms. In: Edible and medicinal mushrooms. John Wiley & Sons, Ltd, pp 261–270. https://doi.org/10.1002/9781119149446.ch13
Vetter J (2003) Chemical composition of fresh and conserved Agaricus bisporus mushroom. Eur Food Res Technol 217:10–12. https://doi.org/10.1007/s00217-003-0707-2
Vieira FR, Pecchia JA (2018) An exploration into the bacterial community under different pasteurization conditions during substrate preparation (composting–phase II) for Agaricus bisporus cultivation. Microb Ecol 75(2):318–330. https://doi.org/10.1007/s00248-017-1026-7
Wainaina S, Taherzadeh MJ (2023) Automation and artificial intelligence in filamentous fungi-based bioprocesses: a review. Bioresour Technol 369:128421. https://doi.org/10.1016/j.biortech.2022.128421

Recent Prospects in Pathogen Control of *Agaricus bisporus* Farms

Funda Atila and Mehmet Cetin

Abstract *Agaricus bisporus*, widely known as the white button mushroom, is among the most commonly cultivated edible mushrooms worldwide. It holds both economic and nutritional importance, yet production often faces serious threats from fungal and bacterial diseases such as green mold, dry bubble, cobweb disease, wet bubble, and brown blotch. These infections lower yield and diminish market quality, resulting in notable financial losses. Traditional control still depends heavily on chemical fungicides. However, restrictions on their use, environmental concerns, and the rise of resistant strains have reduced their long-term reliability. In response, growers and researchers are turning toward more sustainable approaches. Promising options include biological control agents, nanotechnology-based delivery methods, and genetic improvement to increase resistance. This review highlights the major pathogens affecting *A. bisporus*, examines their economic impact, and summarizes recent developments in integrated disease management.

Keywords *Agaricus bisporus* · Integrated disease management · Biological control · Sustainable mushroom cultivation

1 Introduction

Agaricus bisporus (Lange) Imbach, widely known as the white button mushroom, is the most commonly cultivated mushroom species worldwide, representing approximately 35–45% of total production (Royse et al. 2017). In addition to being cultivated for food, the white button mushroom is a promising source of health-promoting and medicinal bioactive molecules (Atila et al. 2017). The industry faces major challenges from pests and pathogens, including viruses, insects such as flies and mites, nematodes, *Trichoderma* spp., *Lecanicillium fungicola*, and *Pseudomonas*

F. Atila (✉) · M. Cetin
Ege University, Bergama Vocational Training School, Greenhouse Program, Izmir, Turkey
e-mail: funda.atila@ege.edu.tr; mehmet.cetin@ege.edu.tr

M. N. Owaid et al. (eds.), *Mushroom Production*, Interdisciplinary Biotechnological Advances, https://doi.org/10.1007/978-981-95-7793-4_5

tolaasii (Sattar et al. 2021). In mushroom beds, fungal pathogens are among the most serious contaminants because they grow rapidly and compete with the crop for space and nutrients (O'Hanlon 2017). They may also release toxic secondary metabolites, volatile organic compounds, and extracellular enzymes, which reduce both the quality and yield of mushrooms (Ramakutoane et al. 2023). Management of fungal diseases of mushrooms is challenging and, at the same time, economically important.

Mushroom growers commonly use methods like strict hygiene and steam sterilization to control mycopathogens, but the primary method is chemical treatment (Carrasco et al. 2021). In Europe, only two fungicides (prochloraz and metrafenone) are approved for *A. bisporus*, while North America approves three (chlorothalonil, thiabendazole, and thiophanate-methyl) (Kredics et al. 2022). Prochloraz is effective but overuse of the it can prompt the onset of resistant strains in pathogens like *Lecanicillium* and *Cladobotryum* and environmental harm (Clarke et al. 2024). Many chemicals such as chlorothalonil, carbendazim, and benzimidazoles have been banned in Europe due to environmental concerns (Pereira et al. 2021). Moreover, there has been a consistent reduction in the number of approved artificially derived active substances in the European Union over the past few years, decreasing from 320 in 2017 to 234 in 2022. Projections indicate that up to half of the remaining substances may be further withdrawn or unsupported within the next decade (Marchand 2023).

Given these realities, the development of sustainable, low-toxicity approaches to fungal disease control has become a central focus in the field. Several promising directions have emerged. Examples include biological control methods using antagonistic organisms to suppress phytopathogenic fungi (Aslani et al. 2018; Stanojević et al. 2019; Altaf et al. 2022) and nanotechnology-based tools designed to improve the precision and efficiency of agricultural treatments (Pourbagher et al. 2021; Ramakutoane et al. 2023). Genetic improvement of mushroom strains to increase resistance to common pathogens is another approach gaining attention (Idrees et al. 2019; Huang et al. 2024).

This chapter explores these innovative strategies in the context of *A. bisporus* production, examining both recent technological advancements and integrated approaches that combine cultural, chemical, and biological tools for more effective disease management.

2 Major Mushroom Pathogens

Diseases caused by fungi and bacteria such as green mold (*Trichoderma aggressivum f. europaeum*), dry bubble (*L. fungicola*), cobweb disease (*Cladobotryum* spp.), wet bubble (*Mycogone perniciosa*), and brown blotch (*Pseudomonas tolaasii*) remain major challenges in mushroom cultivation. These pathogens cause more than cosmetic damage, as they can reduce both yield and the marketable quality of the harvest. In many cases, the resulting losses are considerable. In Western production systems, such infections are estimated to reduce annual yields by as much as 25% of total output (Quiroz et al. 2024). Because of the urgency of the problem, particularly for *A. bisporus* growers, this section examines the main fungal and bacterial threats affecting commercial mushroom farms. The focus is on pathogens that are both widespread and economically damaging, providing context for recent advances in disease control strategies.

2.1 Green Mold

Green mold, predominantly caused by *T. aggressivum*, poses a serious threat to button mushroom production. This pathogen occurs in two distinct biotypes: *T. aggressivum* f. *europaeum* (the European strain) and *T. aggressivum* f. *aggressivum* (the American strain). The disease was first reported in the British Isles and North America and has since spread to Western, Central, and Southeastern Europe, as well as parts of Central America and Australia (Quiroz et al. 2024). Infections can cause devastating yield losses, often ranging from 60% to 100%. A particularly severe outbreak occurred in Hungary in December 2015, where affected cultivation beds suffered nearly complete crop failure (Hatvani et al. 2017).

The disease typically presents as rapid colonization of the substrate or casing layer with dense white mycelium, which subsequently turns green as the fungus sporulates. *T. aggressivum* f. *europaeum* is especially efficient at colonizing mushroom compost, severely reducing yields by inhibiting the growth of *A. bisporus* (Kredics et al. 2022). In areas affected by green mold, mushroom production is often halted, and in advanced stages, red-pepper mites (*Pygmephorus* spp.) may feed on the pathogenic fungi (Prasad and Singh 2020). The disease can also spread via routine farm operations and pest flies acting as vectors (Sattar et al. 2021).

Compounds produced by *T. aggressivum*, the causative agent of green mold, are more harmful to hybrid off-white strains of *A. bisporus* than to commercial brown strains (Kredics et al. 2022). Effective management relies heavily on strict sanitation and hygiene measures, particularly during the post-crop steam-off and early spawn-run stages. Although carbendazim, a benzimidazole derivative, has demonstrated effectiveness, resistance among *Trichoderma* strains is increasingly common (Gea et al. 2021). Consequently, continuous research and improved sanitation methods are critical to combating this pervasive threat.

2.2 Dry Bubble

Dry bubble infection, induced by *Lecanicillium fungicola* (formerly *Verticillium fungicola*), accounts for about 20% of global button mushroom crop losses, translating to approximately €300 million annually in Europe (Gea et al. 2021). Symptoms typically first appear in the casing material after colonized compost is applied, rarely before harvesting when fructification is induced. Primary infections lead to the formation of undifferentiated masses of host tissue referred to as bubbles, accompanied by stipe enlargement, spotting, and sunken lesions on caps. Fruiting bodies showing symptoms become unmarketable, resulting in notable yield losses (Berendsen et al. 2010).

Infections often originate from the casing material, especially peat, but can also be introduced via infected mushrooms, phorid flies, or contaminated equipment (Ghimire et al. 2021). Airborne spores through ventilation systems further spread the disease. Effective chemical treatments are limited due to host sensitivity and resistance development in *L. fungicola* (Berendsen et al. 2010; Gea et al. 2021).

2.3 Cobweb Disease

Fungi belonging to the genus *Cladobotryum* are responsible for cobweb disease, a major threat to white button mushroom production (Liu et al. 2023). The disease appears as soft, cotton-like growths on the mushroom casing, ranging in color from white to yellowish or pink. There is a swift spread across the casing surface, accompanied by excessive mycelial growth on mushroom fruiting bodies, ultimately resulting in their deterioration. (Xu et al. 2020). Notably, *Cladobotryum dendroides* is identified by its coarse mycelium over the affected mushrooms, resembling a spider web, hence the name "cobweb disease" (Gea and Navarro 2017). This excessive growth results in blemishes on the mushroom caps, making them unfit for sale.

The occurrence of cobweb disease in commercial mushroom crops shows considerable variation, with reported prevalence ranging from 6.8% to 33% in countries like India, Turkey, and Spain. During the mid-1990s, the UK and Ireland experienced epidemic levels, resulting in crop losses up to 40% (Gea et al. 2021). Cobweb disease typically appears toward the end of the cultivation cycle (Carrasco et al. 2016a).

Although *Cladobotryum* spores do not persist in environments with low relative humidity (RH), the fungus forms microsclerotia—durable structures capable of germinating even under 0% RH conditions (Carrasco et al. 2017). Elevated humidity levels outside the cropping rooms promote the survival and spread of pathogenic conidia (Gea et al. 2021). The spores are lightweight, airborne, and readily dispersed by irrigation, spreading throughout growing rooms via ventilation systems (Clarke et al. 2024). Synthetic fungicides like prochloraz and metrafenone have been effective against mushroom pathogens (Carrasco et al. 2016b; Prasad and Singh 2020). As of June 2023, prochloraz has been banned in the European Union. At the same time, the heavy reliance on metrafenone has encouraged the emergence of resistant pathogen strains (Clarke et al. 2024).

2.4 *Wet Bubble Disease*

M. perniciosa is the causal agent of wet bubble disease, a serious issue in mushroom cultivation that leads to pronounced deformities in the fruiting bodies. Affected mushrooms develop large, irregular, nodular structures that closely resemble tumorous growths. A characteristic symptom of the disease is the "tear-drop" phenomenon, in which droplets of extracellular fluid form on the deformed tissues. These masses typically begin as white and spongy, but as the disease progresses, they darken to a brown color and may release amber-like exudates (Shi et al. 2020).

Infections by *M. perniciosa* can cause substantial yield losses in *A. bisporus* (Carrasco et al. 2019). Contamination usually stems from infected casing soil or spawn, and occasionally, poorly maintained ventilation systems contribute to the spread of spores throughout the cultivation area (Pieterse et al. 2005; Glamoclija et al. 2008).

Controlling the disease requires rigorous hygiene measures throughout the cultivation facility. Control measures commonly involve fumigating growing rooms, disinfecting surfaces with agents such as formalin or Lysol, sterilizing soil, and applying fungicidal formulations like prochloraz-manganese (Gea et al. 2010). For localized infections, some growers report success using simple but effective techniques like directly applying salt to infected patches and covering the area to prevent further spread (Pieterse 2005).

2.5 *Bacterial Brown Blotch Disease*

Brown blotch, caused by *P. tolaasii*, is among the most prevalent and economically damaging bacterial diseases of cultivated mushrooms (Evidente 2022). Global production losses attributed to this pathogen are estimated at 8–10% (Amin et al. 2021). The bacterium's pathogenicity is primarily linked to tolaasin, a lipodepsipeptide toxin. Tolaasin integrates into host cell membranes, compromising their structural integrity and ultimately leading to browning and collapse of the affected cells (Storey et al. 2020).

The disease produces visible blemishes on both the cap and stipe of button mushrooms, substantially reducing their market quality (Archana and Vaja 2022). *P. tolaasii* is frequently found in compost and casing materials, and its behavior can shift between pathogenic and nonpathogenic forms depending on environmental conditions and host factors (Sattar et al. 2021). High humidity and excessive surface moisture are known to accelerate symptom development, resulting in brown spots, pitting, and general discoloration of the pileus (Taparia 2021).

Management approaches combine chemical and nonthermal methods. In experimental and commercial trials, certain antibiotics have provided notable reductions in disease severity. Reported effective treatments include Terramycin (9 mg/ft^2), streptomycin (200 ppm), and oxytetracycline (300 ppm), while kasugamycin and kanamycin have also yielded positive outcomes (Prasad and Singh 2020). Chlorinated water, bacteriophage TO1, and tolaasin-detoxifying bacteria have been shown to reduce disease severity, but none are completely effective (Oh et al. 2000; Ghasemi et al. 2020).

3 Alternative Pathogen Control Methods on *Agaricus bisporus* Farms

3.1 Biocontrol Agents

Biological control refers to the method or technique through which an unwanted organism is managed using another organism that provides beneficial effects (Nega 2014). In recent years, the use of biological control agents (BCAs) has emerged as a promising alternative to synthetic fungicides for managing diseases in mushroom cultivation. Research has evaluated a wide spectrum of BCAs, ranging from essential oils and other plant-derived compounds (Benomari et al. 2018; Chang et al. 2022; Altaf et al. 2022) to beneficial microorganisms, including selected bacterial and fungal strains (Pirttilä et al. 2021; Roca-Couso et al. 2021), as well as aqueous extracts obtained from composted agricultural by-products (Orosz et al. 2021; González-Hernández et al. 2021). These agents employ diverse mechanisms such as direct parasitism, antibiosis through antimicrobial metabolite production, and competition for nutrients or colonization sites that suppress pathogen development or prevent their establishment (Lahlali et al. 2022).

Beyond their disease-suppressive potential, BCAs can help reduce dependency on chemical fungicides, thereby lowering environmental risks and supporting biodiversity within the production environment (Ons et al. 2020). Another important advantage is their comparatively low risk of inducing resistance in target pathogens, in contrast to many synthetic pesticides whose efficacy often declines with repeated use (Schmitt and Seddon 2005). Moreover, BCAs are generally recognized as posing minimal hazards to humans, animals, and nontarget organisms, offering a safer option from both food safety and occupational health perspectives (EFSA et al. 2020).

Integration of BCAs into broader integrated pest management (IPM) frameworks—combining biological tools with cultural practices and carefully targeted chemical applications—can further enhance control efficacy and sustainability (Raymaekers et al. 2020). While challenges remain, including variable performance under fluctuating environmental conditions and the complexities of regulatory approval, their role in commercial *A. bisporus* production is steadily expanding. With ongoing advances in biotechnology and refinement of field application strategies, BCAs have the potential to improve yields while supporting to the ecological sustainability of mushroom cultivation.

3.1.1 Microbial Biocontrol Agents

Microbial biocontrol agents (mBCAs) are attracting growing attention as sustainable alternatives to chemical fungicides, largely because of their minimal environmental impact. These beneficial microorganisms provide an ecologically sound means of suppressing fungal pathogens while maintaining food safety and protecting human health. Research on microbial antagonists has traditionally focused on *Bacillus* and *Trichoderma* spp., both well-known for their strong antifungal activity. Nevertheless, a wider variety of microbes has also shown potential for biocontrol, including *Pseudomonas fluorescens*, *Ampelomyces quisqualis*, *Agrobacterium radiobacter*, nonpathogenic strains of *Fusarium* spp., *Coniothyrium* spp., and atoxigenic isolates of *Aspergillus niger* (Panahi et al. 2020; Wang et al. 2024).

In mushroom cultivation, *Bacillus*-based products currently dominate the microbial biopesticide sector. Their prominence can be attributed to the diverse mechanisms these bacteria use to suppress pathogens (Khan et al. 2023). Supporting the proliferation of naturally occurring antagonists within the cultivation system can further enhance resource competition, thereby restricting the growth and spread of harmful fungi.

Bacillus strains are particularly adept at colonizing plant-derived substrates. Early establishment on these materials not only gives them a competitive advantage over pathogens but can also stimulate host development (Blake et al. 2021). They produce antimicrobial compounds—including antibiotics and volatile organic metabolites—that directly disrupt the life cycle of fungal pathogens. In addition, they compete with pathogens for nutrients and colonization sites, physically limiting their expansion (Cesa-Luna et al. 2020).

A key adaptive trait of *Bacillus* spp. is their ability to form endospores, enabling them to endure adverse environmental conditions and remain viable in the ecosystem for extended periods (Checinska et al. 2015). This trait also facilitates the production of stable, dry formulations with extended shelf life (Zhang et al. 2022). Furthermore, *Bacillus* spp. secrete a broad range of antimicrobial metabolites, enhancing their role as effective bioprotective agents (Caulier et al. 2019). Some *Bacillus* strains that naturally inhabit the casing layer have shown a targeted ability to suppress specific mycoparasites (Novikova and Titova 2023). Adding them to compost or casing soil has yielded promising results for the suppression of compost green mold (Milijašević-Marčić et al. 2017) and dry blister disease (Bhat et al. 2010) in *A. bisporus*.

Although *T. aggressivum* displayed resistance to many bacteria isolated from compost, certain strains proved effective. Stanojević et al. (2016) noted that *Bacillus* spp. strains suppressed the growth of *T. aggressivum* f. *europaeum*, *T. harzianum*, and *T. koningii*. In their study, Milijašević Marčić et al. (2017) tested 50 bacterial strains isolated from mushroom compost. Among these, *Bacillus subtilis* B-38 inhibited the growth of *T. harzianum* T54 by 48.08% and *T. aggressivum* f. *europaeum* T77 by 52.25% in both in vitro and in vivo assays. Stanojević et al. (2019) found that among all *Bacillus* spp. strains tested in vivo, *B. amyloliquefaciens* B-241 was particularly effective in suppressing compost green mold and dry bubble disease in *A. bisporus*. However, *Bacillus velezensis* QST 713 did not effectively control cobweb disease caused by *C. mycophilum* isolate 1546, while *B. velezensis* Kos exhibited moderate control, achieving 30–40% efficacy across two trials. Potočnik et al. (2018) conducted trials in mushroom cultivation rooms with *T. aggressivum* f. *europaeum* and found that prochloraz-Mn, *B. amyloliquefaciens* B-241, *B. velezensis* QST 713, *Bacillus pumilus* B-138, and *B. subtilis* B-233 were the most effective strains for suppressing the pathogen. Biofungicides formulated with *B. subtilis* and tea tree oil, or their combinations with prochlorase, demonstrated significant efficacy against *Cladobotryum* spp., albeit lower than prochlorase alone in single applications (Potočnik et al. 2010). *B. subtilis* used in combination with a fungicide showed reduced antagonistic activity than tea tree oil in its activity against the pathogen and reduced yield more than tea tree oil (Kosanović et al. 2013). However, Carrasco and Preston (2020) reported that *Bacillus* applications can reduce disease symptoms only under low disease pressure conditions, suggesting their effectiveness may be limited under high disease pressure.

3.1.2 Plant Extract and Essential Oils

Since ancient times, people have turned to plant extracts to combat bacteria, viruses, fungi, and even insects (Prakash et al. 2023). These extracts are often effective against a wide range of pests, and many are seen as appealing substitutes for synthetic pesticides because they pose fewer risks to human health and the environment. Studies have shown that plant-derived substances can work well against arthropod pests, disease vectors, and various other pathogens (Souto et al. 2021; Khursheed et al. 2022). Yet, despite this potential, their use in agriculture remains relatively limited.

There is limited research on the use of plant extracts and essential oils as pesticides in mushroom cultivation. Altaf et al. (2022) reported that treating compost and soil casing with botanicals or bacterial isolates can significantly increase mushroom yield by controlling green mold disease. According to their results, the ethanol extracts of *Juglans regia* and *Allium sativum* inhibited mycelium growth of *T. harzianum*, while minimal to no inhibition of *A. bisporus*. Similarly Shah et al. (2011) evaluated the antifungal activity of eight plant-derived substances, including *A. sativum*, *Urtica dioica*, *Lycopersicon esculentum*, *Datura stramonium*, *Mentha longifolia*, and *J. regia*, in the control of *T. harzianum* affecting white button mushrooms. They found that *J. regia* exhibited the highest inhibition at 52.9%, followed by *A. indica* at 34.1%, and *A. sativum* at 28.4%. *M. longifolia* demonstrated the weakest antifungal activity.

Essential oils, obtained by distillation from aromatic plants, contain a complex mixture of volatile and nonvolatile compounds. They have been applied to control fungi on several vegetable (Karkanis and Athanassiou 2021; Kačániová et al. 2022). In many previous studies, many of these natural compounds have been reported to show strong fungistatic effects against pathogens of edible fungi (Todorović et al. 2016; Luković et al. 2020; Dassanayake et al. 2021). Yet, employing these oils to manage diseases that impact the production of edible mushrooms may pose challenges due to the fact that both the antagonist and the crop of interest are fungi (Geösel et al. 2014; Mehrparvar et al. 2016; Gea et al. 2019).

Essential oils from plants, especially those of thyme (*Origanum vulgare*), oregano (*Thymus vulgaris*), and peppermint (*Mentha piperita*) and their main components, carvacrol, thymol, and menthol, have exhibited potent activity against various *Trichoderma* spp. and bacteria (Luković et al. 2020; Mutlu-Ingok et al. 2020; Greff et al. 2023). Kosanović et al. (2013) observed that when tea tree (*Melaleuca alternifolia*) essential oil was incorporated into button mushroom casing soil, the growth of *T. harzianum* was strongly suppressed.

Regnier and Combrinck (2010) found that lemon verbena and thyme oils could potentially be used as fungicides against *M. perniciosa* without inhibiting the growth of button mushrooms. In a commercial trial where these oils, along with their main components nerol and thymol at a concentration of 40 μL/L, were tested, neither treatment adversely affected the growth of *A. bisporus*, yielding similar overall yields as achieved with a commercial fungicide (Chronos 450 SC). However, the study also revealed that these essential oils were ineffective when used curatively to control wet blister pathogens and severe fungal infections. Glamoclija et al. (2006) reported that the essential oils of *Satureja thymbra* and *Salvia pomifera* showed in vitro antifungal activity against *M. perniciosa*, while Glamoclija et al. (2009) reported that the essential oil of *Crithmum maritimum* as well as the two main components of the oil, limonene and α-pinene, exhibited inhibitory activities against the pathogen. Essential oils are recognized for their biodegradability and low toxicity, making them suitable for disinfecting commercial casing soil. It is recommended to apply 2% oregano oil to disinfect the soil before applying it to germinated compost (Glamoclija et al. 2006).

Geranium (*Pelargonium graveolens*), cinnamon (*Cinnamomum verum*), and clove (*Eugenia caryophyllata*) oils have demonstrated significant efficacy against *L. fungicola*, *M. perniciosa*, and *Cladobotryum* spp. (Tanović et al. 2006, 2009). Santos et al. (2017) found that cinnamon, thyme, and clove oils were especially effective at suppressing the mycelial growth and conidial germination of *L. epidendrum*, with the strongest effects observed when the oils were applied after infestation. These findings suggest that essential oils or specific constituents derived from them could potentially be employed in future for mushroom cultivation, either as alternatives to synthetic fungicides or as supplementary preventive measures.

On the contrary, studies involving essential oils have not consistently yielded positive results. For example, cinnamon, geranium, and peppermint essential oils suppressed *C. dendroides* development, proving effective through both direct contact and vapor-phase action. However, the selectivity of these oils against the parasite diminished, causing damage to the crop, inhibiting pathogen mycelium growth, and inducing necrotic lesions on mushroom caps (Geösel et al. 2014). Tea tree oil, the active ingredient in the biofungicide Timorex 66 EC, significantly reduced cobweb disease in *A. bisporus*; nonetheless, its toxicity against *C. dendroides* mycelium in vitro was notably lower compared to prochloraz-Mn (Potočnik et al. 2010). Similarly, Potočnik et al. (2016) found that basil and peppermint oils exhibited potent activity against *T. aggressivum* f. *europaeum*, but some essential oils showed reduced selectivity and had the potential to harm the crop.

Gea et al. (2019) reported that when *A. bisporus* was treated with *T. vulgaris* and *Satureja montana* essential oils (1% concentration), results comparable to fungicide treatment with metrafenone were obtained in the control of cobweb disease; however, crop losses were recorded in the first flush. While the fungicidal potency of these oils is moderate, they can serve as a complement to commercial products for disease management, thereby reducing the quantity of fungicide employed.

3.1.3 Compost Tea

In addition to essential oils and plant extracts, researchers have explored the use of various aqueous extracts derived from composts made from agricultural wastes to control plant pathogens. These extracts, known as compost teas, have demonstrated antifungal effects against plant pathogens (On et al. 2015). Several compost teas produced from various organic waste sources, such as spent mushroom compost, grape marc, crop residues, and vermicomposted materials, have shown notable inhibitory effects against *L. fungicola* under laboratory conditions (Marín et al. 2013). These teas, rich in microbial activity and secondary metabolites, represent a promising direction in the search for biological disease control methods.

Gea et al. (2014) evaluated the potential of compost tea made from spent mushroom substrate (SMS) to suppress *L. fungicola*. The SMS, consisting mainly of peat-based casing material, was first steam-pasteurized at 70 °C for 12 h and then recomposted for 57 days. Once matured, it was steeped in water and allowed to ferment under aerobic conditions for 24 h, producing the compost tea. In vitro tests showed that this preparation markedly inhibited the mycelial growth of *L. fungicola*. Field trials in *A. bisporus* crops affected by dry bubble disease confirmed these findings, with treated plots showing a clear reduction in disease incidence. Importantly, no phytotoxic effects were detected, indicating that the treatment was safe for both crop health and yield.

The incorporation of SMS tea into the casing layer appears to influence the structure of the microbial community in ways that are unfavorable for *L. fungicola* proliferation. In this way, SMS tea serves not only as a direct antagonist to the pathogen but also as a modulator of the local microbiome (Navarro et al. 2011). The use of SMS tea in integrated pest management (IPM) systems can provide reliable protection against dry bubble disease while avoiding the pitfalls of chemical fungicide residues in *A. bisporus*-fruiting bodies.

As a strategy, SMS tea has several advantages. It is relatively easy to prepare, based on agricultural by-products, and environmentally benign. Nonetheless, its effectiveness can vary due to fluctuations in microbial and chemical composition between batches. Therefore, for broader adoption at the commercial scale, there is a clear need for standardized production protocols to ensure consistency and reliability in biocontrol performance.

3.1.4 Commercial BCA Products

The range of BCAs available for plant protection has expanded considerably in recent years, and some of them have been adapted for use in mushroom cultivation. Among the best-known examples is *Serenade®*, a biofungicide formulated with *Bacillus subtilis* strain QST 713. Since 2007, *Serenade®* has been applied to control *T. aggressivum* (Milijašević-Marčić et al. 2017) in European mushroom farms, particularly in France. Comparative trials assessing different BCAs have shown *Serenade®* to be among the most effective for protecting *A. bisporus*, especially when applied directly to the spawn before compost inoculation (Védie and Rousseau 2008).

Sjaarda et al. (2015) showed that *Serenade®* can suppress *T. aggressivum* and may also trigger defense responses in *A. bisporus*. In their study, treated cultures had higher laccase activity, which could be related to the breakdown of methyl-isocoumarin, a toxic compound produced by the pathogen. The product has been effective in French production systems, but results in other regions and cultivation methods have been less consistent (Kosanović et al. 2013).

Another formulation, *SerenadeMax®* (Bayer CropScience), contains *Bacillus velezensis* QST 713 as its active ingredient (Pandin et al. 2018). *Serenade® WP* (Bayer, Germany), also based on *B. velezensis* QST 713, and *Ekstrasol F SC* (BioGenesis d.o.o., Serbia), formulated with *B. subtilis* Ch-13 1, have likewise demonstrated activity against green mold in controlled experiments. When *T. aggressivum* f. *europaeum* T77 was introduced into mushroom beds, these BCAs were as effective as, and sometimes more effective than, prochloraz-manganese fungicides (Potočnik et al. 2019). Notably, trials combining *Serenade®* with prochloraz® in an 80:20 ratio revealed that *Serenade®* alone achieved better results than the mixed formulation. Although not consistently outperforming prochloraz-Mn across all trials, it proved more effective than some other biofungicides, such as those formulated with tea tree oil (Kosanović et al. 2013).

Tea tree oil, the active component of *Timorex 66 EC*, has also shown potential. In *A. bisporus* cultivation experiments, application at the manufacturer's recommended rates significantly reduced the incidence of cobweb disease (Potočnik et al. 2010). Nonetheless, further studies are needed to clarify how these biocontrol products interact with compost-associated microbial communities and to identify which formulations are best suited for integration into large-scale mushroom production systems.

3.2 Nanoparticles

In the biological sciences, nanotechnology has become an important multidisciplinary approach, applied in fields ranging from targeted drug delivery and vaccine transport to micronutrient supplementation and antimicrobial treatment. More recently, its use has expanded into agriculture, where the distinctive physicochemical characteristics of nanoparticles are being explored to develop innovative methods for controlling plant pathogens (Gupta et al. 2023). Nanoparticles, generally measuring between 1 and 100 nanometers, present a high surface area-to-volume ratio and notable reactivity, properties that enhance their effectiveness in antimicrobial interventions (Khan et al. 2019; Makvandi et al. 2020).

Among metal-based nanomaterials, those containing silver or copper have gained notable attention in plant disease management due to their strong antimicrobial properties (Javed et al. 2021; Maliki et al. 2022). Within this group, silver nanoparticles (AgNPs) are the most extensively investigated because of their broad-spectrum efficacy. While both physical and chemical synthesis techniques are available, interest has increasingly shifted toward green synthesis methods, valued for their lower environmental impact, cost efficiency, and potential for large-scale production (Kaabipour and Hemmati 2021).

The antimicrobial properties of AgNPs are closely linked to their extremely small size and high surface reactivity, which allow them to penetrate microbial cells and interfere with vital physiological functions. Various researches have shown that they can inhibit both spore germination and mycelial growth in several plant pathogenic fungi (Khan et al. 2021; Amna Ali et al. 2022; Kim et al. 2023). Reported susceptible species include *Fusarium oxysporum*, *Pythium spinosum*, and *Botrytis cinerea*. In one investigation, Ramakutoane et al. (2023) found that particles measuring 7–25 nm and applied at roughly 0.01 mg/mL yielded the most effective inhibition. Akpinar et al. (2021) also noted that reducing particle size and increasing concentration enhanced antifungal effect against *Fusarium oxysporum* f. sp. *radicis-lycopersici* (FORL).

Research on zinc oxide nanoparticles (ZnO NPs) has highlighted their potential as antifungal agents. For example, in lentil crops, Siddiqui et al. (2018) found that foliar applications of ZnO NPs reduced gall formation, suppressed nematode populations, and lessened the severity of wilt, blight, and leaf spot. In related research, Luksiene et al. (2020) observed that ZnO NPs, when activated by UV or visible light, provided better protection against *Botrytis cinerea* in strawberries than the commercial fungicide fenhexamid.

Copper nanoparticles (CuNPs) have been tested in a comparable way. Varympopi et al. (2020) reported that CuNPs outperformed conventional copper-based fungicides in vitro, pointing to their potential as a sustainable, cost-effective option. This aligns with the findings of Malandrakis et al. (2019), who found that AgNPs, ZnO NPs, and CuNPs reduced both the mycelial growth and the spore germination of *B. cinerea* under both laboratory and greenhouse conditions, with AgNPs and CuNPs giving the strongest disease suppression.

While there are numerous studies on the efficacy of nanoparticles synthesized from fungi (Klaus et al. 2020; Zhang et al. 2020; Owaid 2020), few studies have investigated the in vivo use of nanoparticles to control pathogens in the button mushroom. Ramakutoane et al. (2023) reported that different clay surfaces (zeolite, montmorillonite, and palygorskite) functionalized with AgNPs were very effective against mushroom phytopathogenic fungi. Moreover they observed that the small dispersion of AgNPs on clay surfaces greatly influences the overall antifungal activity, with small-sized NPs more efficient in controlling the pathogenic fungi. According to the results, AgNP-zeolite and AgNP-montmorillonite composites have effective antifungal properties and can be used as alternative fungicides against mushroom pathogens without affecting *A. bisporus* growth. In order for these nanoparticles, which have been found to be effective on pathogen development in plants, to be used in mushroom cultivation, a much larger number of in vivo and in vivo evaluations are needed.

3.3 Breeding for Disease Resistance

The development of strains with resistance to common pathogens through traditional breeding and biotechnology is one of the most important tools to reduce disease incidence and dependence on chemical controls (Das et al. 2023a). Efforts to develop pathogen-resistant fungal varieties often start by locating resistance traits within wild strains or germplasm that have adapted to specific environments. Once potential traits are identified, breeders test their heritability, make selective crosses, and study the offspring in detail. Large-scale cultivation trials under commercial conditions are then used to determine whether the resistance is both stable and useful in practice (Chakravarty 2011).

For *A. bisporus*, conventional breeding for disease resistance is hampered by both genetic and biological limitations. This mushroom's reproductive biology is complex, and cultivated strains today tend to have a narrow genetic range (Sonnenberg et al. 2017). Although some strains with varying degrees of resistance to certain pathogens have been identified, most available germplasm has yet to be studied in depth (Sattar et al. 2021). Wild relatives and locally maintained landraces could provide valuable resistance genes, but such resources are seldom incorporated into standard breeding programs (Savoie et al. 2013). As a result, many commercial cultivars are genetically uniform, making them more susceptible to emerging diseases.

Addressing these challenges calls for detailed study of the pathogens and the genetic basis of host resistance. In *A. bisporus*, resistance is often controlled by several genes, which can slow breeding progress. Recent advances in molecular tools and genomic analysis are helping to explain these mechanisms and may support the faster development of resistant hybrids.

A major recent advance in this area is the use of gene-editing systems, particularly CRISPR-Cas9. This technology allows highly targeted alterations to the genome, including the removal or inactivation of genes that make the host more susceptible to disease. For instance, Wan et al. (2021) showed that knocking out certain genes with CRISPR reduced the incidence of infections caused by *Pseudomonas tolaasii*, the bacterium responsible for brown blotch. These results point to the potential of genome editing as a powerful approach for breeding mushroom cultivars with improved resistance (Das et al. 2023b).

The combination of genomic and proteomic methods with mushroom pathology could further speed up the creation of resistant strains. By analyzing host-pathogen interactions in detail, researchers can pinpoint the genes and proteins linked to resistance and use this information to design targeted breeding or modification strategies. Such approaches could lead to fungal varieties that maintain productivity while withstanding disease pressures in commercial production systems.

4 Challenges in the Development of Alternative Control Methods for Mushroom Cultivation

The development of new BCAs for mushroom cultivation remains a demanding task for both researchers and industry practitioners. Demonstrating antimicrobial activity under laboratory conditions is a crucial first step; however, such in vitro successes often fail to translate directly into consistent performance in commercial settings. This makes it essential to conduct rigorous evaluations under conditions that closely mirror real-world production environments. At the earliest stages, careful screening to identify the most promising microbial strains or formulations is critical, as these candidates form the basis for subsequent field trials.

A major barrier to successfully translating laboratory findings into field performance is the limited knowledge of fungal crop ecology. Replicating the diverse and fluctuating conditions found on commercial farms within controlled laboratory setups remains difficult, and multiple factors can alter the results. Variables such as the timing of application, dosage levels, and the formulation of BCAs all play important roles in determining their success. To improve reliability, researchers are turning to more sophisticated methods. Molecular markers now allow monitoring of BCA persistence and activity during cultivation, while functional imaging techniques help reveal their impact on host plant physiology and the structure and behavior of the surrounding microbial community.

BCA performance is not determined by biological efficacy alone. Environmental fluctuations, pathogen diversity, and the practical aspects of product application all play a role. Developing stable, commercially viable formulations that retain consistent biological activity over time is an ongoing technical challenge. Equally important is optimizing delivery methods to accommodate different production systems and crop types. Furthermore, the regulatory pathways established to assess the safety and effectiveness of BCAs for both human health and the environment often involve extended timelines and significant costs. For these agents to be successfully incorporated into commercial mushroom cultivation, their introduction must be carefully coordinated so that they align with established management practices and enhance, rather than disrupt, ongoing production processes.

In parallel with these developments, nanopesticides have attracted attention as a novel approach to crop protection. These nano-formulations offer the potential for more precise application and reduced environmental impact in managing fungal pathogens and other agricultural pests. However, important uncertainties remain regarding their long-term safety for human health and their ecological consequences. Comprehensive, long-term studies will be required to evaluate not only their potential benefits but also any associated risks in fungal crop production systems.

The molecular mechanisms of action, risks and toxicities associated with nanopesticides are complex and uncertain. More toxicological research and risk assessment approaches are needed to understand nano-bio-eco interactions between nanomaterials and biotic and abiotic environments. Finally, companies developing disease-resistant varieties need to conduct costly research programs. The high cost of advanced genomic and proteomic technologies may limit access by smaller research institutions and commercial breeders. Therefore, there are very few varieties bred specifically for disease resistance and made available to breeders.

5 Conclusion

The cultivation of *A. bisporus* continues to encounter significant challenges from various pathogens. However, natural plant-derived fungicides, microbial biocontrol agents, compost teas, and nanoparticles provide promising avenues that are safe for human health and the environment to synthetic fungicides. These treatments can help avoid overreliance on single products, mitigating the risk of resistant strains. Further efforts should focus on generating extensive datasets through sequencing to gain insights into host-parasite relationships and develop breeding strategies for resistant varieties. Thorough sanitation of compost yards and cultivation facilities, incorporating bioactive substances and expanding genomic data through data mining are crucial for developing integrated disease management (IDM) programs. These programs should incorporate approved fungicides and best agricultural practices.

Moving beyond theoretical frameworks to practical deployment has become a priority, particularly in translating these advancements into on-farm practices for mushroom cultivation. The mushroom industry must overcome challenges such as the nature of the crop, restrictive legislative environments, and consumer demand for healthier and environmentally friendly cropping systems. Integrating these innovative solutions within an IPM framework can enhance pathogen control, fostering sustainable and profitable mushroom production. Continuous research and the adaptation of these strategies are vital to address emerging threats and secure the future of *A. bisporus* farming. Additionally, essential oil vapors could serve as a novel method to managing mold and bacteria in mushroom cultivation chambers, which are currently treated with alternative sanitizers. Further research is necessary to devise tailored, economical techniques for the feasible implementation of these commercial products, ensuring their availability to mushroom producers. The rise of fungicide-resistant strains and the limitations on pesticide use have heightened the need for resistant cultivars. Enhancing crop yield and quality, alongside developing disease resistance, should be the primary objectives for mushroom breeders and researchers.

References

Akpinar I, Unal M, Sar T (2021) Potential antifungal effects of silver nanoparticles (AgNPs) of different sizes against phytopathogenic Fusarium oxysporum f. sp. radicis-lycopersici (FORL) strains. SN. Appl Sci 3(4):506. https://doi.org/10.1007/s42452-021-04524-5

Altaf S, Jan SK, Basu U, Ahanger SA, Dave A, Kakraliya SS, Baazeem A, Mishra AK, Kumar A, Shah IA, Mushtaq M (2022) Sustainable management of green mold disease of white button mushroom using botanicals and biocontrol agents under temperate conditions. Horticulturae 8(9):768. https://doi.org/10.3390/horticulturae8090768

Amin Z, Wani FF, Gulzar H, Dar WA, Sheikh PA (2021) Diseases of white button mushroom (Agaricus bisporus)—a potential threat to mushroom industry. Int J Curr Microbiol App Sci 10(2):2076–2085. https://doi.org/10.20546/ijcmas.2021.1002.247

Amna Ali B, Azeem MA, Qayyum A, Mustafa G, Ahmad MA, Javed MT, Chaudhary HJ (2022) Bio-fabricated silver nanoparticles: a sustainable approach for augmentation of plant growth and pathogen control. In: Sustainable agriculture reviews, Nanoparticles: a new tool to enhance stress tolerance, vol 53. Springer, pp 345–371. https://doi.org/10.1007/978-3-030-86876-5_14

Archana TS, Vaja S (2022) Chapter 7—A review on diseases of button mushroom (*Agaricus* spp.): a potential hazard to mushroom cultivation. In: Recent advances in mushroom cultivation technology and its application, pp 135–149

Aslani MA, Harighi B, Abdollahzadeh J (2018) Screening of endofungal bacteria isolated from wild growing mushrooms as potential biological control agents against brown blotch and internal stipe necrosis diseases of *Agaricus bisporus*. Biol Control 119:20–26. https://doi.org/10.1016/j.biocontrol.2018.01.006

Atila F, Owaid MN, Shariati MA (2017) The nutritional and medical benefits of *Agaricus bisporus*: a review. J Microbiol Biotechnol Food Sci 7(3):281–286. https://doi.org/10.15414/jmbfs.2017/18.7.3.281-286

Berendsen RL, Baars JJ, Kalkhove SI, Lugones LG, Wösten HA, Bakker PA (2010) *Lecanicillium fungicola*: causal agent of dry bubble disease in white-button mushroom. Mol Plant Pathol 11(5):585–595. https://doi.org/10.1111/j.1364-3703.2010.00627.x

Blake C, Christensen MN, Kovács ÁT (2021) Molecular aspects of plant growth promotion and protection by *Bacillus subtilis*. Mol Plant-Microbe Interact 34(1):15–25. https://doi.org/10.1094/MPMI-08-20-0225-CR

Benomari FZ, Andreu V, Kotarba J, Dib MEA, Bertrand C, Muselli A, Djabou N (2018) Essential oils from Algerian species of Mentha as new bio-control agents against phytopathogen strains. Environ Sci Pollut Res 25(30):29889–29900

Bhat MA, Simon S, Munshi NA, Bhat ZA (2010) In vitro efficacy of casing and compost isolated bacterial inoculants against Verticillium fungicola (Preuss) Hassebrauk and Agaricus bisporus (Lange) Imbach. J Biol Control 24 (2): 137–141

Carrasco J, Navarro MJ, Santos M, Diánez F, Gea FJ (2016a) Incidence, identification and pathogenicity of *Cladobotryum mycophilum*, causal agent of cobweb disease on *Agaricus bisporus* mushroom crops in Spain. Ann Appl Biol 168:214–224. https://doi.org/10.1111/aab.12257

Carrasco J, Navarro MJ, Santos M, Gea FJ (2016b) Chemical control of mushroom cobweb disease caused by *Cladobotryum mycophilum*. In: Mushroom science XIX: science and cultivation of edible and medicinal fungi, Amsterdam, The Netherlands, p 448

Carrasco J, Navarro MJ, Gea FJ (2017) Cobweb, a serious pathology in mushroom crops: a review. Span J Agric Res 15(2):e10R01. https://doi.org/10.5424/sjar/2017152-10143

Carrasco J, Tello ML, de Toro M, Tkacz A, Poole P, Pérez-Clavijo M, Preston G (2019) Casing microbiome dynamics during button mushroom cultivation: implications for dry and wet bubble diseases. Microbiology 165(6):611–624. https://doi.org/10.1099/mic.0.000792

Carrasco J, Zied DC, Navarro MJ, Gea FJ, Pardo-Giménez A (2021) Commercial cultivation techniques of mushrooms. In: Advances in macrofungi. CRC Press, pp 11–40

Caulier S, Nannan C, Gillis A, Licciardi F, Bragard C, Mahillon J (2019) Overview of the antimicrobial compounds produced by members of the *Bacillus subtilis* group. Front Microbiol 10:435128. https://doi.org/10.3389/fmicb.2019.00302

Cesa-Luna C, Baez A, Quintero-Hernández, Cruz-Enriquez JDL, Castañeda-Antonio MD, Muñoz-Rojas J (2020) The importance of antimicrobial compounds produced by beneficial bacteria on the biocontrol of phytopathogens. Acta Biol Colomb 25(1):140–154

Chakravarty B (2011) Trends in mushroom cultivation and breeding. Aust J Agric Res 2(4):102–109

Chang Y, Harmon PF, Treadwell DD, Carrillo D, Sarkhosh A, Brecht JK (2022) Biocontrol potential of essential oils in organic horticulture systems: from farm to fork. Front Nutr 8:805138. https://doi.org/10.3389/fnut.2021.805138

Checinska A, Paszczynski A, Burbank M (2015) *Bacillus* and other spore-forming genera: variations in responses and mechanisms for survival. Annu Rev Food Sci Technol 6:351–369. https://doi.org/10.1146/annurev-food-030713-092332

Clarke J, McGuinness B, Fitzpatrick D, Kavanagh K, Grogan H (2024) Response of the mushroom pathogen *Cladobotryum mycophilum* to the fungicides prochloraz and metrafenone and two *Bacillus*-based biological control agents in mushroom crop trials. Crop Prot 177:106530. https://doi.org/10.1016/j.cropro.2023.106530

Carrasco J, Preston GM (2020) Growing edible mushrooms: a conversation between bacteria and fungi. Environ Microbiol 22(3):858–872

Das S, Ray MK, Panday D, Mishra PK (2023b) Role of biotechnology in creating sustainable agriculture. PLOS Sustain Transform 2(7):e0000069. https://doi.org/10.1371/journal.pstr.0000069

Das K, Ayim BY, Borodynko-Filas N, Das SC, Aminuzzaman FM (2023a) Genome editing (CRISPR/Cas9) in plant disease management: challenges and future prospects. J Plant Prot Res:159–172. https://doi.org/10.24425/jppr.2023.145761

Dassanayake MK, Chong CH, Khoo TJ, Figiel A, Szumny A, Choo CM (2021) Synergistic field crop pest management properties of plant-derived essential oils in combination with synthetic pesticides and bioactive molecules: a review. Foods 10(9):2016. https://doi.org/10.3390/foods10092016

EFSA Panel on Biological Hazards (BIOHAZ), Koutsoumanis K, Allende A, Alvarez-Ordóñez A, Bolton D, Bover-Cid S et al (2020) Scientific opinion on the update of the list of QPS-recommended biological agents intentionally added to food or feed as notified to EFSA (2017–2019). EFSA J 18(2):e05966. https://doi.org/10.2903/j.efsa.2020.5966

Evidente A (2022) Bioactive lipodepsipeptides produced by bacteria and fungi. Int J Mol Sci 23(20):12342. https://doi.org/10.3390/ijms232012342

Gea FJ, Tello JC, Navarro MJ (2010) Efficacy and effects on yield of different fungicides for control of wet bubble disease of mushroom caused by the mycoparasite *Mycogone perniciosa*. Crop Prot 29(9):1021–1025. https://doi.org/10.1016/j.cropro.2010.06.006

Gea FJ, Carrasco J, Diánez F, Santos M, Navarro MJ (2014) Control of dry bubble disease (*Lecanicillium fungicola*) in button mushroom (*Agaricus bisporus*) by spent mushroom substrate tea. Eur J Plant Pathol 138:711–720. https://doi.org/10.1007/s10658-013-0344-y

Gea FJ, Navarro MJ (2017) Mushroom diseases and control. In: Edible and medicinal mushrooms: technology and applications, pp 239–259

Gea FJ, Navarro MJ, Santos M, Diánez F, Herraiz-Peñalver D (2019) Screening and evaluation of essential oils from Mediterranean aromatic plants against the mushroom cobweb disease, Cladobotryum mycophilum. Agronomy 9(10):656. https://doi.org/10.3390/agronomy9100656

Gea FJ, Navarro MJ, Santos M, Diánez F, Carrasco J (2021) Control of fungal diseases in mushroom crops while dealing with fungicide resistance: a review. Microorganisms 9(3):585. https://doi.org/10.3390/microorganisms9030585

Geösel A, Szabó A, Akan O, Szarvas J (2014) Effect of essential oils on mycopathogens of *Agaricus bisporus*. In: 8th international conference on mushroom biology and mushroom products (ICMBMP8), p 534. Retrieved January 4, 2024, from https://www.researchgate.net/publication/268815181

Ghasemi S, Harighi B, Azizi A, Mojarrab M (2020) Reduction of brown blotch disease and tyrosinase activity in *Agaricus bisporus* infected by *Pseudomonas tolaasii* upon treatment with endofungal bacteria. Physiol Mol Plant Pathol 110:101474. https://doi.org/10.1016/j.pmpp.2020.101474

Ghimire A, Pandey KR, Joshi YR, Subedi S (2021) Major fungal contaminants of mushrooms and their management. Int J Appl Sci Biotechnol 9(2):80–93. https://doi.org/10.3126/ijasbt.v9i2.37513

Glamoclija J, Sokovic M, Vukojevic J, Milenkovic I, Van Griensven LJLD (2006) Chemical composition and antifungal activities of essential oils of *Satureja thymbra* L. and *Salvia pomifera* ssp. *calycina* (Sm.) Hayek. J Essent Oil Res 18(1):115–117. https://doi.org/10.1080/10412905.2006.9699404

Glamoclija J, Sokovic M, Ljaljevic-Grbic M, Vukojevic J, Milenkovic I, Van Griensven L (2008) Morphological characteristics and mycelial compatibility of different *Mycogone perniciosa* isolates. J Microsc 232:489–492. https://doi.org/10.1111/j.1365-2818.2008.02145.x

Glamoclija J, Sokovic M, Grubisic D, Vukojevic J, Milinekovic I, Ristic M (2009) Antifungal activity of *Crithmum maritimum* essential oil and its components against mushroom pathogen *Mycogone perniciosa*. Chem Nat Compd 45:96–97. https://doi.org/10.1007/s10600-009-9242-0

González-Hernández AI, Suárez-Fernández MB, Pérez-Sánchez R, Gómez-Sánchez MÁ, Morales-Corts MR (2021) Compost tea induces growth and resistance against *Rhizoctonia solani* and *Phytophthora capsici* in pepper. Agronomy 11(4):781. https://doi.org/10.3390/agronomy11040781

Gupta A, Rayeen F, Mishra R, Tripathi M, Pathak N (2023) Nanotechnology applications in sustainable agriculture: an emerging ecofriendly approach. Plant Nanobiol:100033. https://doi.org/10.1016/j.plana.2023.100033

Greff B, Sáhó A, Lakatos E, Varga L (2023) Biocontrol activity of aromatic and medicinal plants and their bioactive components against soil-borne pathogens. Plants 12(4):706

Huang Z, Nie Y, Huang Y, Liu L, Liu B (2024) Elucidating the role of monoacetylphlorogulcinol in the pathogenicity of *Pseudomonas 'gingeri'* against *Agaricus bisporus*. Pest Manag Sci 80(7):3057–3690. https://doi.org/10.1002/ps.8057

Hatvani L, Kredics L, Allaga H, Manczinger L, Vágvölgyi C, Kuti K, Geösel A (2017) First report of Trichoderma aggressivum f. aggressivum green mold on Agaricus bisporus in Europe. Plant Dis 101(6):1052

Idrees M, Yaqoob S, Leo SF, Khan AA, Sun L, Yang Y, Li D, Fu Y, Li Y (2019) A comparison of the physical, chemical, and structural properties of wild and commercial strains of button mushroom, *Agaricus bisporus* (Agaricomycetes). Int J Med Mushrooms 21(6):611–625. https://doi.org/10.1615/IntJMedMushrooms.2019030681

Javed B, Ikram M, Farooq F, Sultana T, Mashwani ZUR, Raja NI (2021) Biogenesis of silver nanoparticles to treat cancer, diabetes, and microbial infections: a mechanistic overview. Appl Microbiol Biotechnol 105:2261–2275. https://doi.org/10.1007/s00253-021-11171-8

Kaabipour S, Hemmati S (2021) A review on the green and sustainable synthesis of silver nanoparticles and one-dimensional silver nanostructures. Beilstein J Nanotechnol 12(1):102–136. https://doi.org/10.3762/bjnano.12.9

Kačániová M, Galovičová L, Borotová P, Vukovic NL, Vukic M, Kunová S, Hanus P, Bakay L, Zagrobelna E, Kluz M, Kowalczewski PŁ (2022) Assessment of *Ocimum basilicum* essential oil anti-insect activity and antimicrobial protection in fruit and vegetable quality. Plants 11(8):1030. https://doi.org/10.3390/plants11081030

Karkanis AC, Athanassiou CG (2021) Natural insecticides from native plants of the Mediterranean basin and their activity for the control of major insect pests in vegetable crops: shifting from the past to the future. J Pest Sci 94(2):187–202. https://doi.org/10.1007/s10340-020-01275-x

Khan I, Saeed K, Khan I (2019) Nanoparticles: properties, applications and toxicities. Arab J Chem 12(7):908–931. https://doi.org/10.1016/j.arabjc.2017.05.011

Khan M, Khan AU, Bogdanchikova N, Garibo D (2021) Antibacterial and antifungal studies of biosynthesized silver nanoparticles against plant parasitic nematode *Meloidogyne incognita*, plant pathogens *Ralstonia solanacearum* and *Fusarium oxysporum*. Molecules 26(9):2462. https://doi.org/10.3390/molecules26092462

Khan S, Zahoor M, Khan RS, Ikram M, Islam NU (2023) The impact of silver nanoparticles on the growth of plants: the agriculture applications. Heliyon. https://doi.org/10.1016/j.heliyon.2023.e16928

Kim DY, Patel SK, Rasool K, Lone N, Bhatia SK, Seth CS, Ghodake GS (2023) Bioinspired silver nanoparticle-based nanocomposites for effective control of plant pathogens: a review. Sci Total Environ:168318. https://doi.org/10.1016/j.scitotenv.2023.168318

Klaus A, Petrovic P, Vunduk J, Pavlovic V, Van Griensven LJLD (2020) The antimicrobial activities of silver nanoparticles synthesized from medicinal mushrooms. Int J Med Mushrooms 22(9):869–883. https://doi.org/10.1615/IntJMedMushrooms.2020035988

Kosanović D, Potočnik I, Duduk B, Vukojević J, Stajić M, Rekanović E, Milijašević-Marčić S (2013) *Trichoderma* species on *Agaricus bisporus* farms in Serbia and their biocontrol. Ann Appl Biol 163(2):218–230. https://doi.org/10.1111/aab.12048

Kredics L, Hatvani L, Allaga H, Büchner R, Cai F, Vágvölgyi C et al (2022) *Trichoderma* green mould disease of cultivated mushrooms. In: Advances in Trichoderma biology for agricultural applications. Springer, pp 559–606

Khursheed A, Rather MA, Jain V, Rasool S, Nazir R, Malik NA, Majid SA (2022) Plant based natural products as potential ecofriendly and safer biopesticides: a comprehensive overview of their advantages over conventional pesticides, limitations and regulatory aspects. Microb Pathog 173:105854. https://doi.org/10.1016/j.micpath.2022.105854

Lahlali R, Ezrari S, Radouane N, Kenfaoui J, Esmaeel Q, El Hamss H, Belabess Z, Barka EA (2022) Biological control of plant pathogens: a global perspective. Microorganisms 10(3):596. https://doi.org/10.3390/microorganisms10030596

Liu Z, Cong Y, Sossah FL, Lu Y, Kang J, Li Y (2023) Characterization and genome analysis of *Cladobotryum mycophilum*, the causal agent of cobweb disease of *Morchella sextelata* in China. J Fungi 9(4):411. https://doi.org/10.3390/jof9040411

Luković J, Đurović-Pejčev R, Đorđević T, Milijašević-Marčić S, Duduk N, Vico I, Potočnik I (2020) Antifungal and synergistic activity of five plant essential oils from Serbia against *Trichoderma aggressivum* f. *europaeum* Samuels & W Gams. Pestic Fitomed 35(3):173–181

Luksiene Z, Rasiukeviciute N, Zudyte B, Uselis N (2020) Innovative approach to sunlight activated biofungicides for strawberry crop protection: ZnO nanoparticles. J Photochem Photobiol B 203:111656

Makvandi P, Wang CY, Zare EN, Borzacchiello A, Niu LN, Tay FR (2020) Metal-based nanomaterials in biomedical applications: antimicrobial activity and cytotoxicity aspects. Adv Funct Mater 30(22):1910021. https://doi.org/10.1002/adfm.201910021

Maliki M, Ifijen IH, Ikhuoria EU, Jonathan EM, Onaiwu GE, Archibong UD, Ighodaro A (2022) Copper nanoparticles and their oxides: optical, anticancer and antibacterial properties. Int Nano Lett 12(4):379–398. https://doi.org/10.1007/s40089-022-00380-2

Milijašević-Marčić S, Stepanović M, Todorović B, Duduk B, Stepanović J, Rekanović E, Potočnik I (2017) Biological control of green mould on *Agaricus bisporus* by a native *Bacillus subtilis* strain from mushroom compost. Eur J Plant Pathol 148(3):509–519. https://doi.org/10.1007/s10658-016-1107-3

Mutlu-Ingok A, Devecioglu D, Dikmetas DN, Karbancioglu-Guler F, Capanoglu E (2020) Antibacterial, antifungal, antimycotoxigenic, and antioxidant activities of essential oils: an updated review. Molecules 25(20):4711. https://doi.org/10.3390/molecules25204711

Malandrakis AA, Kavroulakis N, Chrysikopoulos CV (2019) Use of copper, silver and zinc nanoparticles against foliar and soil borne plant pathogens. Sci Total Environ 670:292–299

Marchand E, Petit F, Alliot F, Blanchoud H, Costantini D, Guigon E, Goutte A (2023) Contrasted antibiotics and pesticides occurrence in fish exposed in situ to urban effluents: a 20-day caging experiment. Environ Toxicol Chem 43(4):701–711

Marín F, Santos M, Diánez F, Carretero F, Gea FJ, Yau JA, Navarro MJ (2013) Characters of compost teas from different sources and their suppressive effect on fungal phytopathogens. World J Microbiol Biotechnol 29(8):1371–1382

Mehrparvar M, Goltapeh EM, Safaie N, Ashkani S, Hedesh RM (2016) Antifungal activity of essential oils against mycelial growth of Lecanicillium fungicola var. fungicola and Agaricus bisporus. Ind Crops Prod 84:391–398

Novikova I, Titova J (2023) Antifungal activity of industrial *Bacillus* strains against *Mycogone perniciosa*, the causative agent of wet bubble disease in white button mushrooms. Microorganisms 11(8):2056. https://doi.org/10.3390/microorganisms11082056

Nega A (2014) Review on concepts in biological control of plant pathogens. J Biol Agric Healthc 4(27):33–54

Navarro MJ, Santos M, Diánez F, Tello JC, Gea FJ (2011) Toxicity of compost tea from spent mushroom substrate and several fungicides towards Agaricus bisporus. In: Mushroom Biology and Mushroom Products, Proc 7th Int Conf Mushroom Biology and Mushroom Products (Arcachon, France, 4–7 Oct 2011)

Oh SJ, Kim HK, Kim HK, Fermor TR (2000) Effect of sodium hypochlorite for controlling bacterial blotch on *Pleurotus ostreatus*. Mycobiology 28(3):123–126. https://doi.org/10.1080/12298093.2000.12015736

O'Hanlon R (2017) Fungi in the environment. In: Fungi: biology and applications, pp 333–353

Ons L, Bylemans D, Thevissen K, Cammue BP (2020) Combining biocontrol agents with chemical fungicides for integrated plant fungal disease control. Microorganisms 8(12):1930. https://doi.org/10.3390/microorganisms8121930

Orosz V, Tomócsik A, Demeter I, Aranyos TJ, Makádi M (2021) Control of plant pathogen *Fusarium* spp. with compost, compost tea application—a review. Columella J Agric Environ Sci *8*(2):55–71. https://doi.org/10.18380/SZIE.COLUM.2021.8.2.55

Owaid MN (2020) Biomedical applications of nanoparticles synthesized from mushrooms. In: Green nanoparticles: synthesis and biomedical applications. Springer, pp 289–303

On A, Wong F, Ko Q, Tweddell RJ, Antoun H, Avis TJ (2015) Antifungal effects of compost tea microorganisms on tomato pathogens. Biol Control 80:63–69

Panahi HKS, Dehhaghi M, Jouzani GS, Zare R, Aghbashlo M, Tabatabaei M (2020) Fungal biocontrol agents as a new source for bioethanol production. In: Fungi in fuel biotechnology. Springer, pp 69–104

Pandin C, Védie R, Rousseau T, Le Coq D, Aymerich S, Briandet R (2018) Dynamics of compost microbiota during the cultivation of *Agaricus bisporus* in the presence of *bacillus velezensis* QST713 as biocontrol agent against *Trichoderma aggressivum*. Biol Control 127:39–54. https://doi.org/10.1016/j.biocontrol.2018.08.022

Pereira PC, Parente CE, Carvalho GO, Torres JP, Meire RO, Dorneles PR, Malm O (2021) A review on pesticides in flower production: a push to reduce human exposure and environmental contamination. Environ Pollut 289:117817. https://doi.org/10.1016/j.envpol.2021.117817

Pieterse Z (2005) *Mycogone perniciosa*, a pathogen of *Agaricus bisporus* (Master's dissertation, University of Pretoria, South Africa)

Pirttilä AM, Mohammad Parast Tabas H, Baruah N, Koskimäki JJ (2021) Biofertilizers and biocontrol agents for agriculture: how to identify and develop new potent microbial strains and traits. Microorganisms 9(4):817. https://doi.org/10.3390/microorganisms9040817

Potočnik I, Vukojević J, Stajić M, Rekanović E, Stepanović M, Milijašević S, Todorović B (2010) Toxicity of biofungicide Timorex 66 EC to *Cladobotryum dendroides* and *Agaricus bisporus*. Crop Prot 29:290–294. https://doi.org/10.1016/j.cropro.2009.07.016

Potočnik IS, Todorović B, Đurović-Pejčev R, Stepanović M, Rekanović E, Milijašević-Marčić S (2016) Antimicrobial activity of biochemical substances against pathogens of cultivated mushrooms in Serbia. Pestic Fitomed 31(1–2)

Potočnik I, Rekanović E, Todorović B, Luković J, Paunović D, Stanojević O, Milijašević-Marčić S (2019) The effects of casing soil treatment with *Bacillus subtilis* Ch-13 biofungicide on green mould control and mushroom yield. Pestic Fitomed 34:53–60. https://doi.org/10.2298/PIF1901053P

Pourbagher R, Abbaspour-Fard MH, Sohbatzadeh F, Rohani A (2021) *In vivo* antibacterial effect of non-thermal atmospheric plasma on *Pseudomonas tolaasii*, a causative agent of *Agaricus bisporus* blotch disease. Food Control 130:108319. https://doi.org/10.1016/j.foodcont.2021.108319

Prakash B, Dubey NK, de São José JFB (eds) (2023) Plant essential oils: from traditional to modern-day application. Springer

Prasad D, Singh R (2020) Diseases, moulds, insect-pests and mites of mushroom. Int J Plant Prot 13:211–226. https://doi.org/10.15740/HAS/IJPP/13.2/211-226

Pieterse Z (2005) Mycogone perniciosa, a pathogen of Agaricus bisporus. University of Pretoria (South Africa), Master Thesis. https://repository.up.ac.za/server/api/core/bitstreams/6ddfbb39-e683-4b76-8e72-5eaa9c279160/content

Potočnik I, Todorović B, Rekanović E, Luković J, Paunović D, Milijašević-Marčić S (2018) Impact of Bacillus subtilis QST713 mushroom grain spawn treatment on yield and green mould control. Pesticidi i fitomedicina 33(3–4):205–211

Quiroz LF, Ciosek T, Grogan H, McKeown PC, Spillane C, Brychkova G (2024) Unravelling the transcriptional response of *Agaricus bisporus* under *Lecanicillium fungicola* infection. Int J Mol Sci 25(2):1283. https://doi.org/10.3390/ijms25021283

Ramakutoane TL, Roux-van der Merwe MP, Badenhorst J, Kesavan Pillai S, Ray SS (2023) Antifungal properties of nanosilver clay composites against fungal pathogens of *Agaricus bisporus*. ChemistrySelect 8(36):e202301696. https://doi.org/10.1002/slct.202301696

Raymaekers K, Ponet L, Holtappels D, Berckmans B, Cammue BP (2020) Screening for novel biocontrol agents applicable in plant disease management—a review. Biol Control 144:104240. https://doi.org/10.1016/j.biocontrol.2020.104240

Regnier T, Combrinck S (2010) *In vitro* and *in vivo* screening of essential oils for the control of wet bubble disease of *Agaricus bisporus*. S Afr J Bot 76(4):681–685

Roca-Couso R, Flores-Félix JD, Rivas R (2021) Mechanisms of action of microbial biocontrol agents against *Botrytis cinerea*. J Fungi 7(12):1045. https://doi.org/10.3390/jof7121045

Royse DJ, Baars J, Tan Q (2017) Current overview of mushroom production in the world. In: Edible and medicinal mushrooms: technology and applications, pp 5–13

Santos TLD, Belan LL, Zied DC, Dias ES, Alves E (2017) Essential oils in the control of dry bubble disease in white button mushroom. Cienc Rural 47:e20160780. https://doi.org/10.1590/0103-8478cr20160780

Sattar MN, Iftikhar S, Saeed F, Shaalan R, Hourani W, Bejjani R, Sebaaly Z, Afzaal S, Sassine YN (2021) Mushroom farm design, management of pests, and control of diseases. In: Mushrooms: *Agaricus bisporus*, p 297

Schmitt A, Seddon B (2005) Biocontrol of plant pathogens with microbial BCAs and plant extracts—advantages and disadvantages of single and combined use. In: Dehne HW, Gisi U, Kuck KH, Russell PE, Lyr H (eds) Proceedings of the 14th international reinhardsbrunn symposium: modern fungicides and antifungal compounds IV. BCPC, pp 205–225

Shah SS, Sahera Nasreen SN, Munshi NA (2011) Evaluation of some botanicals in controlling green mold (*Trichoderma harzianum*) disease in oyster mushroom cultivation. Int J Bot 7(3):209–215. https://doi.org/10.3923/ijb.2011.209.215

Shi N, Ruan H, Jie Y, Chen F, Du Y (2020) Sensitivity and efficacy of fungicides against wet bubble disease of *Agaricus bisporus* caused by *Mycogone perniciosa*. Eur J Plant Pathol 157:873–885. https://doi.org/10.1007/s10658-020-02047-0

Sonnenberg AS, Baars JJ, Gao W, Visser RG (2017) Developments in breeding of *Agaricus bisporus* var. *bisporus*: Progress made and technical and legal hurdles to take. Appl Microbiol Biotechnol 101:1819–1829. https://doi.org/10.1007/s00253-017-8102-2

Souto AL, Sylvestre M, Tölke ED, Tavares JF, Barbosa-Filho JM, Cebrián-Torrejón G (2021) Plant-derived pesticides as an alternative to pest management and sustainable agricultural production: prospects, applications and challenges. Molecules 26(16):4835. https://doi.org/10.3390/molecules26164835

Stanojević O, Berić T, Potočnik I, Rekanović E, Stanković S, Milijašević-Marčić S (2019) Biological control of green mould and dry bubble diseases of cultivated mushroom (*Agaricus bisporus* L.) by *Bacillus* spp. Crop Prot 126:104944. https://doi.org/10.1016/j.cropro.2019.104944

Storey N, Rabiey M, Neuman BW, Jackson RW, Mulley G (2020) Genomic characterisation of mushroom pathogenic pseudomonads and their interaction with bacteriophages. Viruses 12(11):1286. https://doi.org/10.3390/v12111286

Siddiqui ZA, Khan A, Khan MR, Abd-Allah EF (2018) Effects of zinc oxide nanoparticles (ZnO NPs) and some plant pathogens on the growth and nodulation of lentil (Lens culinaris Medik.). Acta Phytopathol Entomol Hung 53(2):195–211

Sjaarda CP, Abubaker KS, Castle AJ (2015) Induction of lcc2 expression and activity by Agaricus bisporus provides defence against Trichoderma aggressivum toxic extracts. Microb Biotechnol 8(6):918–929

Stanojević-Nikolić S, Dimić G, Mojović L, Pejin J, Djukić-Vuković A, Kocić-Tanackov S (2016) Antimicrobial activity of lactic acid against pathogen and spoilage microorganisms. J Food Process Preserv 40(5):990–998

Savoie JM, Foulongne-Oriol M, Barroso G, Callac P (2013) Genetics and genomics of cultivated mushrooms, application to breeding of agarics. In: Agricultural Applications. Springer, Berlin Heidelberg, pp 3–33

Tanović B, Potočnik I, Stanisavljević B, Đorđević M, Rekanović E, Marković M (2006) Response of *Verticillium fungicola* var. *fungicola*, *Mycogone perniciosa* and *Cladobotryum* sp. mushroom pathogens to some essential oils. Pestic Fitomed 21(3):231–237

Tanović B, Potočnik I, Delibašić G, Ristić M, Kostić M, Marković M (2009) *In vitro* effect of essential oils from aromatic and medicinal plants on mushroom pathogens: *Verticillium fungicola* var. *fungicola*, *Mycogone perniciosa*, and *Cladobotryum* sp. Arch Biol Sci Belgr 61(2):231–238. https://doi.org/10.2298/ABS0902231T

Taparia T (2021) A multifaceted analysis of bacterial blotch diseases of mushrooms (Doctoral dissertation, Wageningen University and Research)

Todorović B, Potočnik I, Rekanović E, Stepanović M, Kostić M, Ristić M, Milijašević-Marčić S (2016) Toxicity of twenty-two plant essential oils against pathogenic bacteria of vegetables and mushrooms. J Environ Sci Health B 51(12):832–839. https://doi.org/10.1080/03601234.2016.1208462

Varympopi A, Dimopoulou A, Theologidis I, Karamanidou T, Kaldeli Kerou A, Vlachou A, Skandalis N (2020) Bactericides based on copper nanoparticles restrain growth of important plant pathogens. Pathogens 9(12):1024

Védie R, Rousseau T (2008) Serenade biofungicide: une innovation mjeure dans les champignonnières françaises pour lutter contre Trichoderma aggressivum, agent de la moisissure verte du compost. La Lettre du CTC 21(1):2

Wan L, Wang Z, Tang M, Hong D, Sun Y, Ren J, Zhang N, Zeng H (2021) CRISPR-Cas9 gene editing for fruit and vegetable crops: strategies and prospects. Horticulturae 7(7):193. https://doi.org/10.3390/horticulturae7070193

Wang N, Sundin GW, Fuente LDL, Cubero J, Tatineni S, Brewer MT, Zeng Q, Bock CH, Cunniffe NJ, Wang C, Candresse T, Chappell T, Coleman JJ, Munkvold G (2024) Key challenges in plant pathology in the next decade. Phytopathology 114(5):837–842. https://doi.org/10.1094/PHYTO-04-24-0137-KC

Xu R, Liu X, Peng B, Liu P, Li Z, Dai Y, Xiao S (2020) Genomic features of *Cladobotryum dendroides*, which causes cobweb disease in edible mushrooms, and identification of genes related to pathogenicity and mycoparasitism. Pathogens 9(3):232. https://doi.org/10.3390/pathogens9030232

Zhang L, Wei Y, Wang H, Wu F, Zhao Y, Liu X, Wu H, Wang L, Su H (2020) Green synthesis of silver nanoparticles using mushroom *Flammulina velutipes* extract and their antibacterial activity against aquatic pathogens. Food Bioprocess Technol 13:1908–1917. https://doi.org/10.1007/s11947-020-02533-7

Zhang B, Xu L, Ding J, Wang M, Ge R, Zhao H, Zhang B, Fan J (2022) Natural antimicrobial lipopeptides secreted by *Bacillus* spp. and their application in food preservation: a critical review. Trends Food Sci Technol 127:26–37. https://doi.org/10.1016/j.tifs.2022.06.009

Genetic Progress for Sustainable Growing of *Agaricus bisporus* Strains

Manjit Singh

Abstract In the latter half of twentieth century, there have been vast improvements in technology for cultivation of button mushroom (*Agaricus bisporus*). The technology alone, however, can play a role only up to a point and for better and sustainable manifestation of technology, it was important to have proper genotypes (strains). In the twentieth century, the world mushroom production was very low, and button mushroom was the major contributor. Many other edible species of mushrooms have been brought under cultivation in the last few decades. At present, shiitake is the prime contributor to world mushroom production. Gains by technology are a function of genetic improvement. Genetic improvement in button mushroom initially was restricted to selection from available germplasm or use of random methods like multispore cultures and single spore selection. The life cycle of button mushroom was fully understood only in early 1970s, and the first button mushroom hybrid (U1) was developed and commercialized in 1981. Since then, a number of strains have been developed in the Netherlands, the USA, India, and various other countries. Genome of the both the homokaryons of first button mushroom hybrid Horst U1 has been sequenced. Work has been done on WRKY factors, RFLPs, RAPDs, ISSR, MAT locus, etc. These DNA markers for *A. bisporus* have helped in the identification of homokaryons and varieties, use of approaches like marker aided selection (MAS), and more in-depth study of the genetics. The carbon footprint of mushroom cultivation is low as compared to many other vegetables and animal products. Genetic improvement for producing more from less can further reduce environmental impact and contribute to sustainability. World mushroom production vis-à-vis genetic progress for sustainable growing of button mushroom is discussed in this chapter.

Keywords *Agaricus bisporus* · Genetic improvement · Strain development · Marker-assisted selection · Sustainable mushroom cultivation

---

M. Singh (✉)
ICAR-Directorate of Mushroom Research, Solan, India

M. N. Owaid et al. (eds.), *Mushroom Production*, Interdisciplinary Biotechnological Advances, https://doi.org/10.1007/978-981-95-7793-4_6

The sustainable growing of any crop requires improvement in agronomic practices along with targeted breeding and development of new cultivars aimed at improving yield, quality, resistance to diseases, adaptability to soil conditions/substrates, and environmental sustainability. We all know that phenotype is the outcome of genotype-environment interaction. It is important to understand the progress made in the development of varieties and also to understand the environmental impact of cultivation. Accordingly, this chapter has been subdivided into three sections. The first section deals with the changes in productivity and production of button mushroom, the second section deals with the strain development, and the last section deals with the environmental impact.

1 Progress in Productivity and Production of Button Mushroom

1.1 Increase in Productivity is the Key to Sustainability

Genetic improvements along with technological developments have contributed to produce enough food to meet the demands of the growing population. The world's population of 8.2 billion in the year 2024 is expected to continue to grow for another 50 or 60 years and is likely to be around 10.3 billion people in the mid-2080s (United Nations 2024). We have finite resources and limited land mass to feed this increasing global population. Genetic improvement of cultivars over time has helped us to produce more from limited land available. FAO data on cereal production demonstrates role of genetic improvement. For example, in the year 2023, the actual cropland area used for cereals was around 739 million ha and land spared due to crop yield improvements since 1961 was 1580 million ha (Fig. 1), implying that 1580 million ha additional land would have been needed to meet global cereal production if average crop yields had not increased since 1961. Thus, genetic improvement has helped us to produce desired quantity from 1/3rd of land that would have been required in the absence of genetic improvement. Genetic improvement has played similar role in other crops and mushrooms, though the extent may not be the same. A lot depends on the material we are improving and the efforts and investment being made to achieve this.

1.2 Next Revolution in Agriculture will Need Proper Management of Agro-Wastes

According to UNEP (2009), globally 140 billion tonnes of biomass is generated every year from agriculture. This biomass is generated by plants by converting solar energy into chemical energy, most of which is cellulose and lignin. For our survival on planet earth, we will have to learn to make use of this energy. Next revolution in

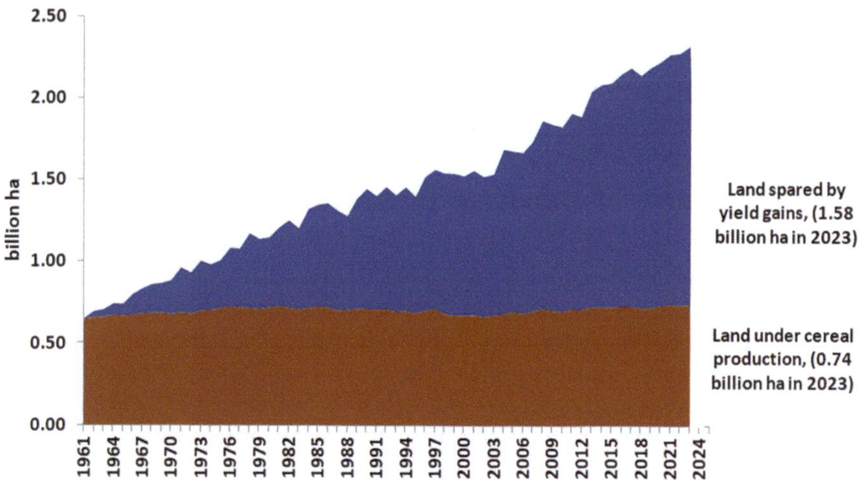

Fig. 1 Global land spared as a result of cereal yield improvements. (Data Source: FAO (2025)—with processing by 'Our World in Data' (https://ourworldindata.org/grapher/cereal-land-spared))

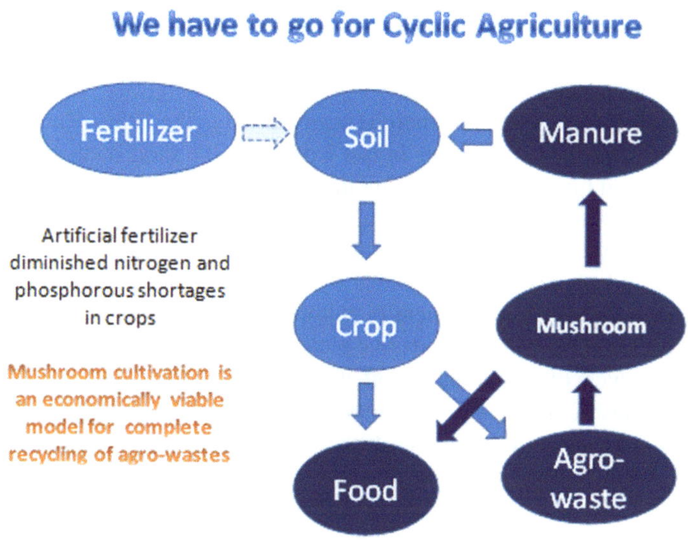

Fig. 2 Complete recycling of agro-wastes during mushroom cultivation

agriculture will depend on our ability to effectively recycle the agriculture biomass wastes and harvest this energy for our use as food and fuel. We can use the biomass waste to produce biogas, bioethanol, heat, or use it for making compost or as fodder (Chang and Wasser 2017). One of the novel ways is to use it for growing mushroom (Fig. 2). By doing so, we produce food from the waste and also completely recycle the agro-wastes as the material left after growing mushroom is recycled into fields as compost. This gain can be enhanced significantly using genetically superior strains. In fact, there has been increase in both production and productivity in different mushrooms in last few decades.

1.3 Productivity of Button has Increased Over Years

As we move from domestication to cultivation to commercialization, we can observe increase in productivity. Initial increase can be by improvements in cultivation techniques and selection of good strains available in nature. For subsequent increase in productivity as well as quality, genetic improvement involving combining of characters from different sources becomes important. Improvements in cultivation techniques and development of strains able to perform better using such techniques has led to increase in global mushroom production from 0.17 million tonnes in 1961 (Singh et al. 2017) to 47 MT in 2023. [FAOSTAT (2025) has reported 50.01 MT world mushroom production in 2023 with share of China as 47.15 MT. Actual production in China in 2023 as reported by Chinese Edible Fungi Association (CEFA 2025) is 43.34 MT. On the other hand, FAOSTAT has reported less production for many countries of Asia (e.g., South Korea, Taiwan) and many others like Bangladesh have not been included (Singh et al. 2017, 2020). The production value of 47.15 MT is the corrected value.]

Mushroom cultivation at commercial scale started in the 1950s. There have been lot of technological advancements in last 7 decades like short method of composting, bunkers and tunnels, phase III, automation in compost preparation, filling, and environment control. The technology alone, however, can play a role only up to a point and for better and sustainable manifestation of technology, it was important to have proper genotypes (strains). Gains by technology are a function of genetic improvement. Genetic improvement in button mushroom initially was restricted to selection or use of random methods like multispore cultures and single spore selection. The life cycle of button mushroom was fully understood only in 1971, and the first button mushroom hybrid (U1) was developed and commercialized in 1981. Since then, a number of strains have been developed in the Netherlands, China, India, and various other countries. In the present century, the number of workers contributed to the knowledge of the genome of *Agaricus*, and the genome of the both the homokaryons of first hybrid Horst U1 has been sequenced. Since then, work has been done on WRKY factors, RFLPs, RAPDs, ISSR, MAT locus, etc., and the DNA markers for *A. bisporus* have allowed use of approaches like marker-aided selection (MAS), identification of homokaryons, identification of varieties, and more in-depth study of the genetics.

There is no doubt that technological developments in cultivation methods of button mushroom have contributed to growth of button mushroom, but we need proper genotypes to take full advantage of these developments. The strains available in early days were simple selections. These were having better phenotypic stability as these could yield mushrooms under variable conditions. But the quality and quantity were limiting factors. In the last few decades, there have been genetic improvements and gains is evident from the increase in productivity, that is, biological efficiency (yield of fresh mushroom in kg/100 kg wet compost). Biological efficiency (BE) in the 1950s was 5–10 kg, and in the 1960s it was 10–12 kg. The current values of 30–35 kg biological efficiency mean that just like in case of cereals (Fig. 1), without this increase in productivity, we would have needed 3–4 times more mushroom farms if yield/100 kg compost remained same as in the 1950s. In fact, this value will be still higher if we consider the number of flushes harvested then and now. This increase in productivity happened early in developed countries while late in developing countries due to both late adoption of modern technology and availability of better strains.

For example, in the USA, yield levels have increased from 10 kg to 33 kg/m^2 from 1966–67 to 2018–19 (Fig. 3). Here, yields are expressed as kg/m^2 (USDA 2025). Normally, 90–100 kg compost is filled per square meter. Over the years, there has been an increase in the amount of filling, and this also contributed to the increase in yields. On the other hand, there has been a decrease in the number of flushes harvested. There is a decline in productivity in the last few years. COVID pandemic (Singh et al. 2021), changing global conditions, and labor costs seem to have contributed to this decline.

Similarly, in Canada, yield has increased from about 9 kg in 1964 to 26 kg per m^2 in 2006 and seems to have got stabilized around 25 kg (Fig. 4). There is a decline in the last few years (StatCan 2025). Global market conditions, labor costs, and COVID (Singh et al. 2021) seem to have contributed to this decline.

In the Netherlands, yield of about 5–10 kg/m^2 over 9 picking weeks in 1958 had increased to about 20 kg/m^2 over 6 picking weeks by 1980 (Fritsche 1981). Mushroom yield as kg/100 kg compost was 21.3 in 1990 that increased to 25.8 in 1996 (Van Griensven and Van Roestel 2004). The current yield is around 32–35 kg/100 kg compost. This increase is the outcome of use of phase III compost,

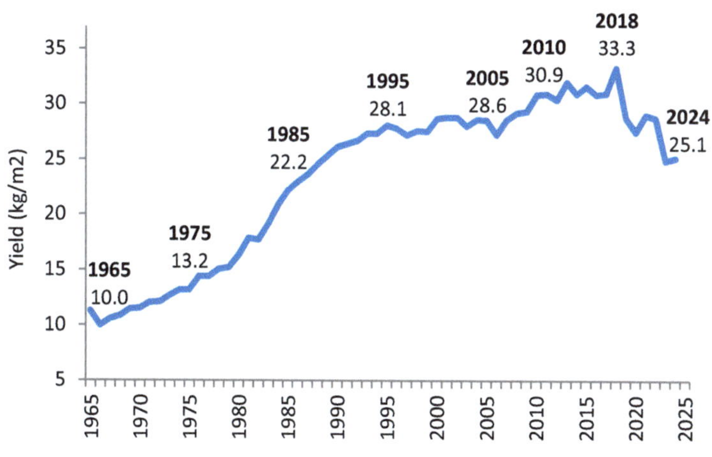

Fig. 3 Changes in productivity of button mushroom in the USA from 1965 to 2024. (Source: USDA 2025)

Fig. 4 Changes in productivity (yield kg/m²) of button mushroom in Canada from 1964 to 2024 (StatCan 2025)

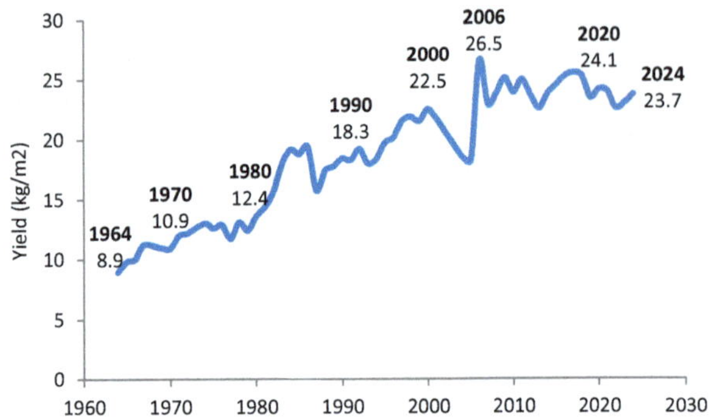

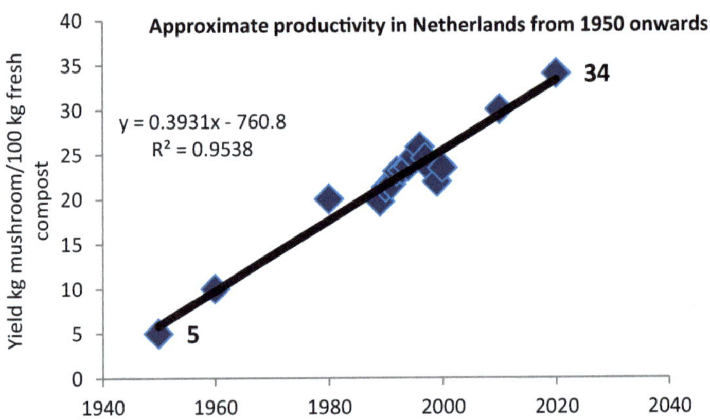

Fig. 5 Changes in productivity of button mushroom in the Netherlands from 1950s to 2023

automation in filling, environment control, etc., along with the use of better strains (Fig. 5).

In the 1950s, there was hardly any mechanization. Over time, there were improvements in composting technology and cultivation under controlled environmental conditions. After the 1980s, the use of hybrid strains has contributed to increase in the productivity. This increase is more in developed countries where the yields are in the range of 32–35 kg/100 kg compost as compared to about 25 kg in developing countries. For example, yields in India in the 1960s were only 2–3 kg (Shandaliya 2000), and present yields at various farms range from 24 to 26 kg/100 kg compost in two flushes. This gap in developed and developing countries reflects contribution of use of better strains, phase III compost, better environmental control, better casing materials, integrated pest and disease management, etc. Tropical conditions in countries like India where temperature goes up to 45 °C also contribute to relatively low productivity.

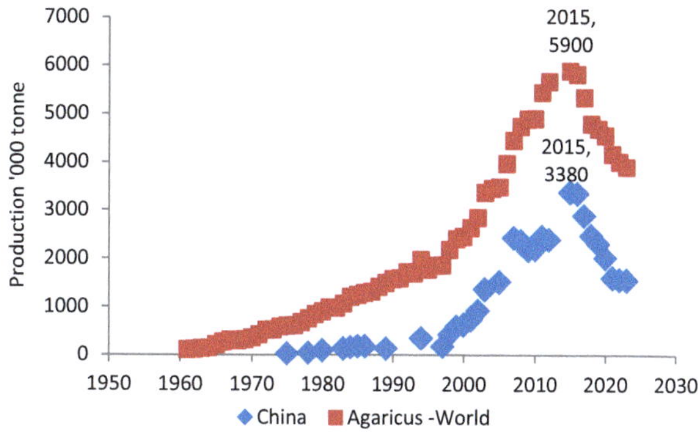

Fig. 6 Estimated button mushroom production in the world (1960–2023) and China (1978–2023)

1.4 Production of Button Mushroom has Increased, Share has Decreased

Commercial cultivation of button mushroom started in the 1950s. Production of button mushroom in the world has increased since 1960, and rate of increase appears to be more after 1990 till 2016 or so. There is declining trend after 2017 (Fig. 6).

At present, China accounts for over 90 percent of world mushroom production, but the share of button in the total mushroom production in China in 2023 was only 3.6%, while in 2010, it was over 10%. The decline in button mushroom production in China in last few years is the primary reason for decrease in global button mushroom production. After 2015, there was a decline in button mushroom production in China that has grossly affected the total world button mushroom production (Fig. 6). In the USA, Canada, Europe, and Australia, where button is the primary contributor to mushroom production, there have been stagnation or marginal decline in the production.

In the 1960s, button mushroom was the major mushroom cultivated in the USA, England, France, the Netherlands, etc., with small contribution of shiitake from Japan. Over the years, a number of other countries have started growing different types of mushrooms. There was major shift in share of button mushroom after 1980 when China started making progress in mushroom production. We can see that button mushroom was primary contributor in the 50s and its share in world mushroom production has decreased despite of the increase in production during these years (Fig. 7). Figure 7 is based on reports by various workers, and the values have been rounded off. The species-wise production in different countries will help in better understanding of the increase in button mushroom production vis- à-vis rapid increase in share of other species in total mushroom production.

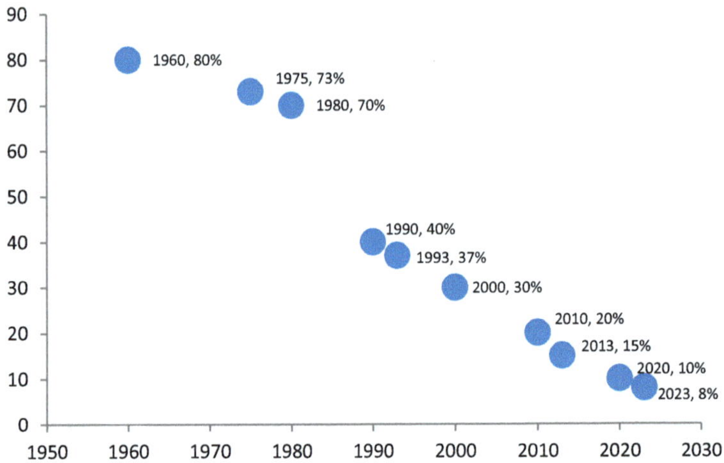

Fig. 7 Approximate share of button in total world production. (Share (rounded off by ±1%) is based on the world production data as reported by Chang and Miles 2004; Singh et al. 2018; FAOSTAT 2025)

1.5 Changes in Production of Agaricus and Other Mushroom Species in the Twenty-First Century

1.5.1 The USA

There was an increase in button mushroom production till 2015 or so. In the last few years, however, there is slight decline (Table 1). One noticeable change is the continuous increase in share of brown button mushroom in the USA. In 2024, 30% of button mushroom produced was brown button compared to 10% in 2000 (Fig. 8). Button mushroom is the main contributor to total mushroom production in the USA. Over the years, there is only marginal increase in the share of other mushrooms (USDA 2025) that account for about 2–3% of the US mushroom production (Table 1).

1.5.2 Canada

Mushroom production in Canada has shown increasing trend in the twenty-first century with fluctuations in some years like decline in 2002 and sharp increase in 2008 (StatCan 2025). The rate of growth in the last few years has decreased and the production seems to be stabilizing around 140 MT. Mainly button mushroom is cultivated and the share of exotic mushrooms in Canada is also 2–4% only (Beaulieu-Fortin 2019) (Table 2).

Table 1 Production (in '000 Tonne) of different mushrooms in USA

Year	Total button	Fresh	Processed	White button	Brown button	Shiitake	Oyster	Others	% share of exotics	Total
2000–01	383.83	314.17	69.66	343.92	39.92	4.05	1.65	0.60	1.61	390.1
2001–02	376.98	312.96	64.02	334.08	42.90	3.64	1.83	0.65	1.60	383.1
2002–03	379.38	316.27	63.12	329.17	50.22	3.20	1.62	0.58	1.41	384.8
2003–04	381.54	319.10	62.44	337.55	44.00	3.41	1.90	0.75	1.56	387.6
2004–05	380.15	315.84	64.30	334.55	45.60	3.91	2.33	0.60	1.77	387.0
2005–06	378.15	319.59	58.56	324.94	53.21	3.49	2.07	0.54	1.57	384.2
2006–07	369.16	316.02	53.13	324.16	44.99	3.17	2.29	0.96	1.71	375.6
2007–08	361.67	308.30	53.37	313.93	47.74	4.39	1.93	0.60	1.88	368.6
2008–09	364.64	308.59	56.05	314.26	50.38	4.29	2.29	0.62	1.93	371.8
2009–10	352.47	303.89	48.58	297.90	54.57	2.91	2.65	1.44	1.96	359.5
2010–11	383.72	325.91	57.81	321.47	62.25	3.32	3.51	1.24	2.06	391.8
2011–12	400.00	349.91	50.09	333.74	66.26	3.63	3.19	1.40	2.01	408.2
2012–13	394.46	344.24	50.21	325.64	68.82	3.75	3.16	1.53	2.09	402.9
2013–14	400.10	348.81	51.29	330.98	69.12	4.06	3.25	0.74	1.98	408.2
2014–15	411.49	370.62	40.88	338.58	72.91	4.20	3.50	1.66	2.24	420.9
2015–16	416.86	373.81	43.05	341.95	74.91	4.59	4.54	1.95	2.58	427.9
2016–17	411.22	368.15	43.08	330.03	81.19	4.90	5.03	2.21	2.88	423.4
2017–18	404.26	368.79	35.46	319.38	84.88	4.59	4.92	2.29	2.85	416.1
2018–19	368.53	339.34	29.19	285.17	83.36	3.05	3.91	1.77	2.32	377.3
2019–20	361.22	336.02	25.21	275.63	85.59	3.18	3.86	2.03	2.45	370.3
2020–21	334.48	304.24	30.24	247.09	87.40	3.27	3.04	3.02	2.71	343.8
2021–22	308.39	279.35	29.04	216.51	91.88	3.51	2.27	4.42	3.20	318.6
2022–23	318.92	274.42	44.50	221.04	97.88	2.56	3.49	3.70	2.98	328.7
2023–24	291.58	255.92	35.67	209.43	82.15	**1.74**	1.69	3.73	2.42	**298.7**
2024–25	**296.48**	**270.14**	**26.33**	**207.39**	**89.09**	**2.42**	**1.56**	**3.41**	**2.43**	**303.9**

1.5.3 India

Button mushroom production in India started in the 1960s, and cultivation was restricted to hilly regions only. Over the decades, button mushroom production has spread all over the country, and there has been increase in the production and productivity of button mushroom (Fig. 9). Initially, the focus was on button mushroom only. In subsequent years, production of other mushrooms also started and a number of other mushrooms are now being cultivated (Fig. 10). At present, the share of button in total Indian production is about 70% (Sharma and Bijla 2024). However, only button mushroom is produced throughout thc year under controlled conditions, and the production of mushrooms like oyster, paddy straw mushroom, and milky mushroom is a seasonal activity.

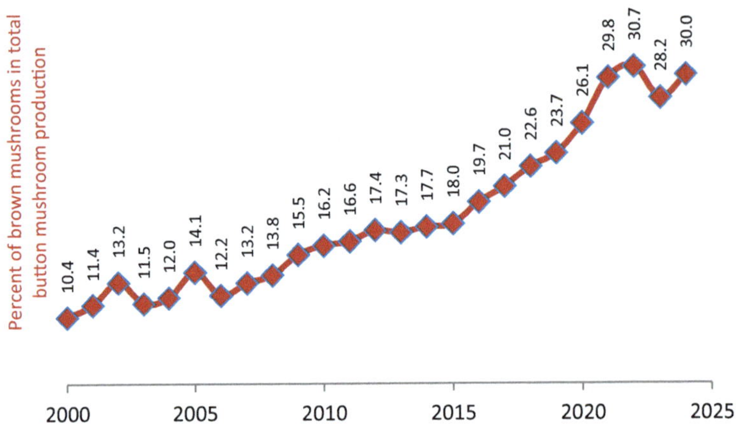

Fig. 8 Percent share of Brown (Portabella & crimini) in total button mushroom production in the USA from the year 2000 to 2024

Table 2 Mushroom production in Canada

	Quantity (Tonne) sold fresh	Quantity (Tonne) sold processed	Total Production	% processed
2000	72,176	8,065	80,240	10.1
2001	77,279	9,078	86,357	10.5
2002	64,726	10,349	75,075	13.8
2003	77,609	10,329	87,938	11.7
2004	75,261	9,421	84,682	11.1
2005	73,724	6,348	80,073	7.9
2006	77,966	11,313	89,279	12.7
2007	73,258	10,032	83,289	12.0
2008	1,03,516	12,838	1,16,354	11.0
2009	1,03,217	11,243	1,14,459	9.8
2010	1,02,798	12,097	1,14,895	10.5
2011	1,10,560	11,614	1,22,173	9.5
2012	1,13,921	10,905	1,24,826	8.7
2013	1,09,923	10,232	1,20,155	8.5
2014	1,12,425	9,632	1,22,057	7.9
2015	1,11,718	2,965	1,14,683	2.6
2016	1,11,729	6,983	1,18,711	5.9
2017	1,13,208	7,046	1,20,253	5.9
2018	1,17,916	7,649	1,25,565	6.1
2019	1,24,533	7,582	1,32,114	5.7
2020	1,21,044	11,545	1,32,589	8.7
2021	1,25,862	11,934	1,37,796	8.7
2022	1,27,105	11,986	1,39,090	8.6
2023	**1,29,371**	**12,638**	**1,42,009**	**8.9**
2024	**1,35,434**	**13,135**	**1,48,569**	**8.8**

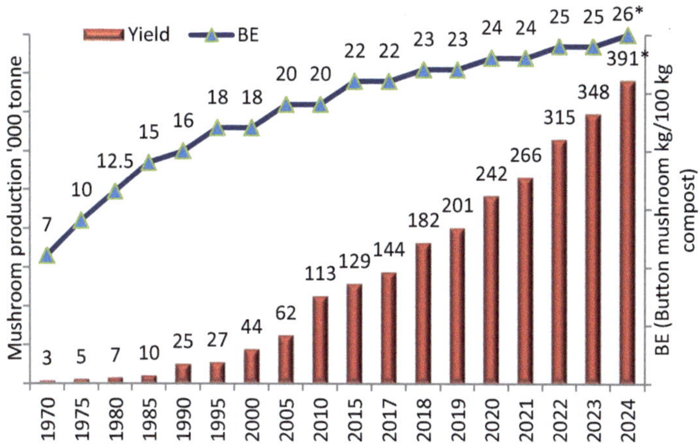

Fig. 9 Total mushroom production (in thousand tonnes) and approximate productivity of button mushroom as fresh mushroom (kg) per 100 kg wet compost, (*estimated value). (Data Source: Sharma and Bijla 2024)

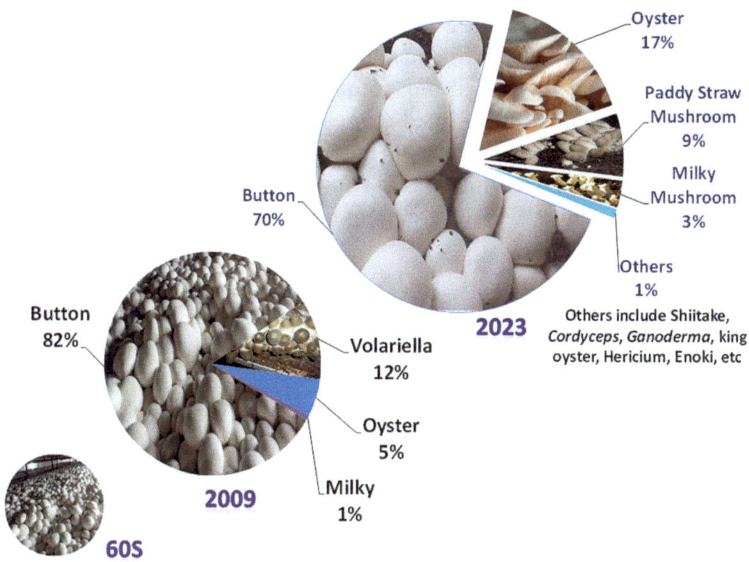

Fig. 10 Changes in contribution of different species to total mushroom production in India

1.5.4 China

In the year 1978, mushroom production in China was just 0.0586 million tonne (MT) that accounted for 5.7% of the world mushroom production. By 1994, the production was 2.641 MT that accounted for over 50% share of the world mushroom production (Singh et al. 2018). The growth story continued, and the production has reached 44.2 MT by 2024 (CEFA 2025); however, growth was slow in the last century. There was rapid increase in mushroom production in first decade of the twenty-first century. The annual rate of growth in last 10 years has slowed down to around 3%. Major share of increase is due to increase in production of Shiitake. In 2015, button mushroom accounted for about 10% of mushroom production in China that has declined to 3.3% in 2024 (Table 3, Figs. 11 and 12).

1.5.5 Japan

Mushroom production in Japan has stagnated around 0.45 MT. As we can see in Table 4, there is no contribution of button mushroom in the mushroom production of the country and major share is of *Flammulina* and *Hypsizygus*. Shiitake, once accounted for maximum production in Japan, is now at number three (Singh et al. 2018).

1.5.6 Button Production in South Korea

During the 1970s, the focus in South Korea was on button and cultivation was mainly for export purpose. With increasing competition from China and later from India for export of canned mushrooms, the share of button mushroom in total mushroom production declined (Fig. 13). In the twenty-first century, the average total production of mushrooms in South Korea was 190.7 million kg and that of button was only 14.4 million kg (Yoo et al. 2016; Soylu and Kang 2016; Personal communication Won Sik Kong for data from 2016–19; Singh et al. 2020) (Table 5).

1.5.7 Taiwan

Agaricus bisporus has been grown in Taiwan since 1958. For some years, Taiwan was the biggest supplier of canned mushrooms in the world (Ku and Lee 1989; Singh et al. 2018). Figure 14 is based on FAOSTAT, and it possibly represents production of button mushroom in Taiwan. The actual production of all mushrooms is much higher (Singh et al. 2018) though exact yearly data for the same is not available.

1.5.8 Global Picture

The world mushroom production in the first decade of this century was increasing at about 10%; it started declining after 2011, and in last few years, it is increasing at only 2–3% growth rate. Based on FAOSTAT data (corrected for mushroom production in China using data from CEFA), it can be seen in Fig. 15 that average increase

Table 3 Production of shiitake, wood ear, oyster, and button mushroom in China ('000 tonne) since 2010

	Shiitake	Wood ear mushroom		Oyster mushroom		Button	Enoki		
Species	*Lentinula edodes*	*Auricularia auricula*	*Auricularia polytricha*	*Pleurotus ostreatus*	*Pleurotus eryngii*	*Agaricus bisporus*	*Flammulina velutipes*	Others	Total
2010	4312	2896	1258	5599	426	2207	1848	3466	22,012
2011	5018	3461	1435	5633	523	2462	2493	4693	25,718
2012	6355	4754	1261	5330	729	2184	2401	5266	28,280
2013	7103	5564	1309	5948	673	2377	2729	5993	31,696
2014	7691	5791	1526	5458	1254	2301	2513	6167	32,701
2015	7667	6337	1826	5902	1365	3380	2613	5674	34,764
2016	8983	6795	1834	5381	967	3352	2669	5984	35,965
2017	9865	7519	1686	5464	1597	2895	2479	5615	37,120
2018	10,432	6740	1898	6435	1956	2483	2576	5369	37,889
2019	11,159	7018	1683	6865	2035	2314	2590	5675	39,339
2020	11,882	7064	1892	6830	2135	2022	2279	6510	40,614
2021	12,957	7034	2207	6113	2052	1610	2146	7220	41,339
2022	12,955	7490	2231	6157	1516	1573	2025	8280	42,227
2023	13,037	7145	2852	6132	1784	1562	1874	8955	43,341
2024	**13,093**	**7134**	**2883**	**6253**	**1765**	**1451**	**2141**	**9479**	**44,199**

Source: CEFA, 2025. https://bigdata.cefa.org.cn/output.html

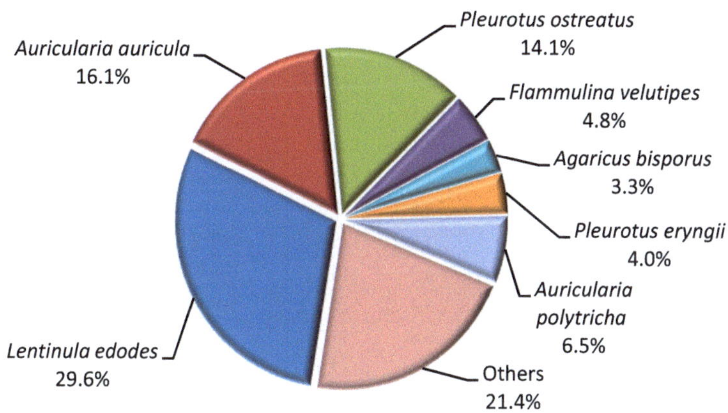

Fig. 11 Relative contribution of different mushroom species in total mushroom production in China in 2024

from 2002 to 2012 was 11.1% while during 2013 to 2023, it declined to 3.5%. At this rate, world mushroom production may reach 100 MT or so by 2050. Mushroom production in major mushroom producing countries is stagnant or is even showing declining trends in some. For example, in the largest mushroom producing country China, the growth rate in last few years is only 2–3%. Japan is the second largest producer, and here production has almost stabilized around 0.46 MT. Production has

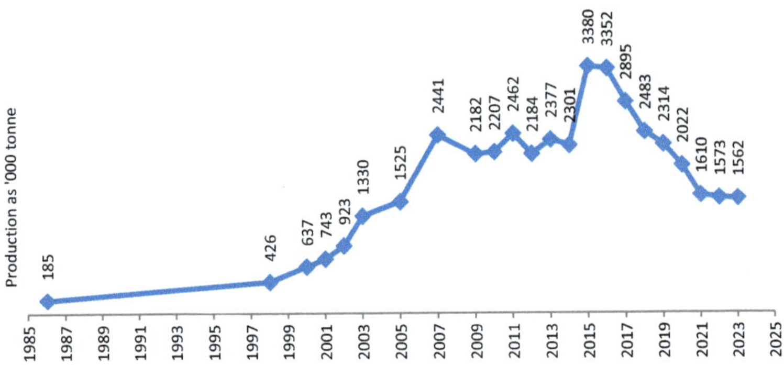

Fig. 12 Button mushroom production in China

Table 4 Production ('000 tonne) of different mushroom species in Japan

	Shiitake	Nameko	Enokitake	Oyster	Buna Shimeji	Maitake	Eringi	Others	Total
2001	100.9	23.8	108.4	6.8	86.6	44.0	10.1	0.1	380.7
2002	95.6	24.8	110.4	5.8	83.8	46.8	19.5	0.1	386.8
2003	94.1	25.1	110.2	5.2	84.4	45.8	29.9	0.1	394.8
2004	94.8	25.8	113.0	4.7	88.1	46.0	33.3	0.2	405.9
2005	93.8	24.8	114.5	4.1	99.8	45.1	34.3	0.1	416.6
2006	93.4	25.6	114.6	3.4	103.2	46.0	36.4	0.2	422.8
2007	92.1	25.8	129.8	3.0	109.0	43.6	38.3	0.2	441.8
2008	97.4	25.9	131.1	2.6	108.1	43.4	38.2	0.3	447.0
2009	100.2	26.1	138.5	2.4	110.7	41.0	37.2	0.2	456.4
2010	101.7	27.3	141.0	2.5	110.5	43.4	37.5	0.4	464.3
2011	97.1	25.4	143.2	2.1	118.0	44.5	38.1	0.7	469.0
2012	92.4	25.8	134.1	1.9	122.3	43.3	38.2	0.8	458.7
2013	92.4	23.4	133.6	2.3	117.4	45.5	40.2	0.8	455.6
2014	89.1	21.8	135.9	2.3	115.8	49.5	39.6	0.9	455.0
2015	86.3	22.9	131.7	3.3	116.2	48.9	39.7	1.3	450.1
2016	88.2	22.9	133.3	3.4	116.3	48.5	40.5	1.3	454.5
2017	86.8	22.9	135.6	3.8	117.7	47.7	39.1	1.7	455.4
2018	88.2	22.8	140.0	4.0	117.9	49.7	39.4	2.4	464.4
2019	88.0	23.3	129.0	3.9	118.6	51.1	37.6	2.3	453.8
2020	86.4	22.8	127.9	3.8	122.8	55.0	38.5	3.2	460.4
2021	86.6	24.1	129.6	4.5	119.5	54.5	38.3	3.1	460.2
2022	82.0	23.7	126.3	4.5	122.8	56.8	37.8	3.0	457.0
2023	76.1	23.8	117.5	4.6	117.9	55.3	**35.9**	2.9	**434.0**
2024	**73.0**	**23.9**	**120.7**	**4.0**	**117.5**	**56.6**	**34.6**	**3.1**	**433.4**

Source: MAFF, 2025. https://www.e-stat.go.jp/dbview?sid=0003446939

shown declining trend in the USA (the third largest producer) in the last few years. Mushroom production, on the other hand, is increasing in developing countries. Production in India has almost tripled in the last decade. In addition to increase in total mushroom production in developing countries, there is increase in mushroom

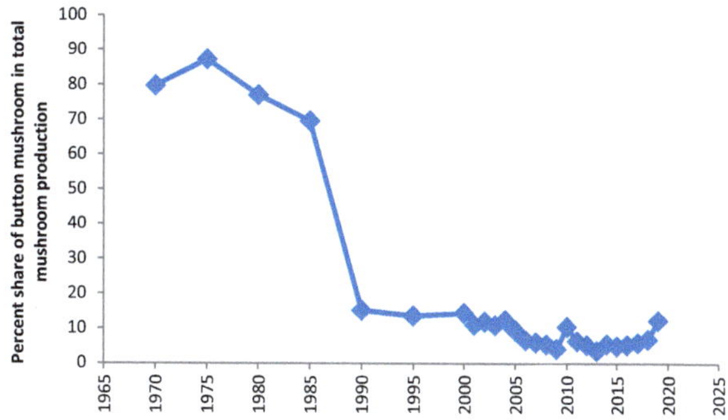

Fig. 13 Percent share of *A. bisporus* in total mushroom production in South Korea (1970–2019)

Table 5 Production of different mushroom species in South Korea

Year	*Pleurotus ostreatus*	*Pleurotus eryngii*	*Lentinula edodes*	*Flammulina velutipes*	*Agaricus bisporus*	Others	Total
2003	61,965	18,358	36,203	41,232	19,790	4,280	1,81,828
2004	52,211	32,736	38,040	32,796	24,053	15,195	1,95,031
2005	56,866	43,230	38,936	40,161	18,985	3,578	2,01,756
2006	45,782	43,256	37,900	34,400	11,892	6,472	1,79,702
2007	45,967	46,357	39,556	36,864	11,150	6,506	1,86,400
2008	40,071	45,906	39,466	55,231	10,822	6,713	1,98,209
2009	39,159	36,808	43,747	61,057	8,174	9,618	1,98,563
2010	45,191	44,361	39,997	53,187	22,635	10,905	2,16,276
2011	46,598	54,820	36,642	43,098	13,052	9,527	2,03,737
2012	51,991	50,605	30,971	50,841	10,996	13,537	2,08,941
2013	66,039	44,098	26,871	33,416	6,678	11,519	1,88,621
2014	76,389	47,814	25,058	33,259	11,493	14,194	2,08,207
2015	62,467	46,530	30,537	37,554	9,732	11,640	198,460
2016	58,784	48,588	28,652	38,092	10,173	7,199	1,91,488
2017	53,532	50,635	29,297	28,535	10,638	7,302	1,79,939
2018	39,675	49,136	26,890	28,532	11,348	8,072	1,63,653
2019	48,327	46,816	25,032	31,818	21,913	4,440	1,78,346
Mean	52,413	44,121	33,753	40,004	13,737	8,865	1,92,892

Based on Yoo et al. (2016); Per. Comm Won Sik Kong for data from 2016–19

diversification. Button mushroom has dominated in India since 1960, and with time, contributions from oyster and paddy straw mushroom have also increased (Fig. 10). In the last few years, many farmers are coming forward to cultivate shiitake, loin's mane, king oyster, *Ganoderma*, *Cordyceps militaris*, morels, etc.

Relative contribution of different species to the total world mushroom is changing continuously (Fig. 16). In the year 2023, the share of Shiitake at global level is estimated to be 28.4%, *Auricularia* spp. (*A. polytricha*+ *A. auricula*) 21.6%, *Pleurotus* spp. (*P. ostreatus*+ *P. eryngii*) 17.9%, button mushroom 7.8%, *Flammulina* 4.4%, and others 19.9%. It can be seen that despite increase in production of button mushroom, its share in total production is on decline, and share

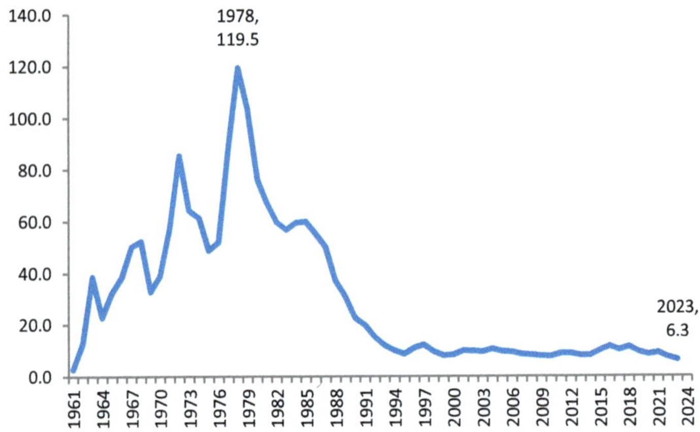

Fig. 14 Mushroom production in Taiwan (1961–2023) (FAOSTAT 2025)

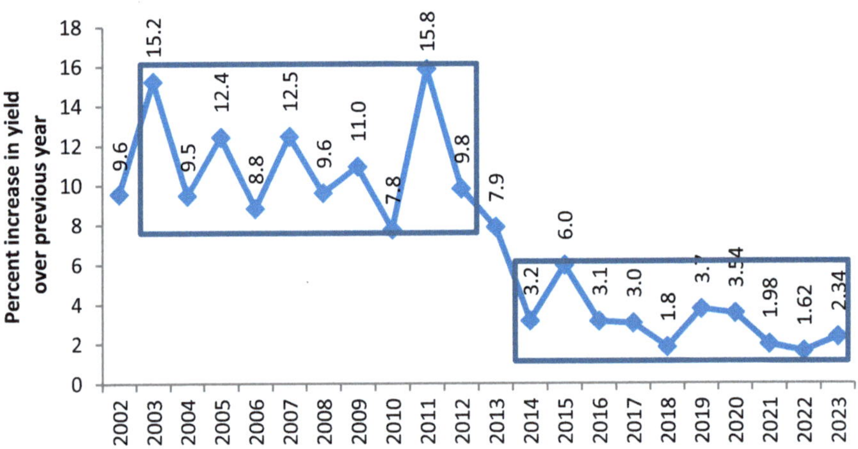

Fig. 15 World production using combined data of FAO and CEFA. (China production data in FAOSTAT 2025 replaced with CEFA 2025)

of other mushrooms is rising continuously as a number of other species are now commercially cultivated.

2 Progress in Genetic Improvement

The first cycle of selection of suitable varieties was in the seventeenth century by mushroom growers when the French started cultivation of button mushroom in the caves. Till 1926, all cultivated button mushrooms were brown types. It was in 1926 when a pure white mutant was selected and propagated by Lambert (1960). The

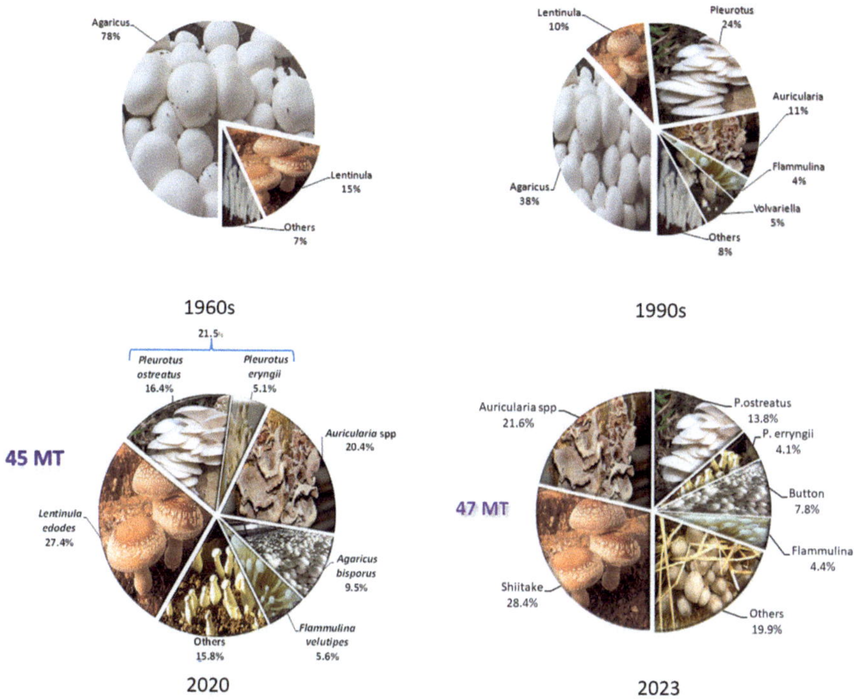

Fig. 16 Relative contribution of different species to the world mushroom production in 1960, 1990, 2020, and 2023

strains available at present can be broadly grouped as "white strains, off-white strains," "hybrid strains" (i.e., hybrid between white and off-white strains), and "brown strains" (Fritsche and Sonnenberg 1988). During initial cultivation in lime tunnels, only the strains that were able to grow better became the seed source for next crop. The first important step or the starting step for genetic improvement was development of pure culture techniques in 1894 and 1902 and subsequent method of spawn production on grains (Sinden 1932). The second step was development of short method of composting in 1950s and subsequent refinements in compost preparation, infrastructure (Phase II & III tunnels, bunkers), and environmental control in cultivation rooms. The role of genetic improvement in first half of last century was limited to improvement by selection from the available germplasm. Subsequently, there were efforts to improve strains by multispore cultures and single spore selection, and efforts were made to hybridize by mixing strains, using auxotroph markers, etc. (Singh and Kamal 2017). Such random approaches continued till understanding of life cycle of button mushroom in the beginning of the 1970s (Elliott 1971; Raper et al. 1972). This understanding lead to the start of systematic work on development of hybrids, and the first hybrid U1 was developed and commercialized in 1981 (Fritsche 1981, 1991a). Single spore isolates from this hybrid have been used to develop a number of other strains that have high genetic similarity with parent due to suppression of recombination (Sonnenberg et al. 2016).

Table 6 Homo- and heterokaryotic spores form bi-, tri-, and tetrasporic basidia (based on Eliot 1971)

Type of basidia	No of SSI picked	Type of spore	Number[a]	
Bisporic	20	Diad type I (WB + WB) Diad type II (WW + BB)	11 (55%) 5 (25%)	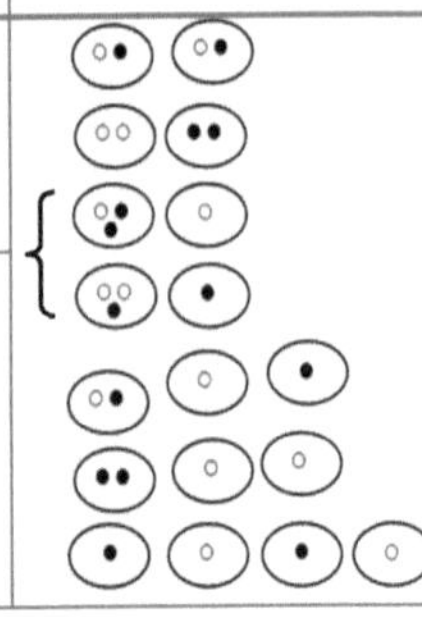
		Diad type III (WWB + B or BBW + W)	4 (20%)	
Trisporic	11	Triad type I (WB + W + B) Triad type II (BB + W + W or WW + B + B)	3 (27%) 8 (73%) 1 (100%)	
Tetrasporic	1[b]+ 19	Tetrad (B + W + B + W)		

W (white) for mating type 1 and B (black) for mating type 2. Even though sample size is small, it is sufficient to conclude that migration of nuclei is not a random process as most of the spores are carrying nuclei of different mating type

[a]Number of different types from single spore isolates (SSI)

[b]All four spores from one tetrasporic basidia developed into cultures

Development of U3 line is the first example of this approach of making selection from fertile single spores of U1 (Singh and Kamal 2017).

Till the understanding of life cycle and development of first hybrid, there was accumulation of knowledge on various aspects. Lambert (1950) based on cytological studies and comparison of monosporous cultures opined that improved commercial strains of the white mushroom variety can be obtained by single spore isolations or by combination of these isolates. Bretzloff et al. (1962) reported differences in colony growth rate of five multispore cultures from a white strain after few sub-culturing.

It was realized that a prerequisite for the genetic improvement of any organism needs understanding of its natural breeding system. Elliott (1971) isolated single spores from two-, three-, and four-spored basidia using micromanipulator, cultured them, and evaluated for fruiting. The study clearly established that *Agaricus bisporus*, like many other bisporic fungi, has secondary homothallism (Table 6) also sometimes referred as intramixis. This finding has since been confirmed and the life cycle thoroughly investigated (Raper et al. 1972).

Even though breeding system was clearly understood, it was realized that *Agaricus bisporus* is a difficult organism to manipulate through breeding due to its unusual biology. Unlike other *Basidiomycete* fungi, there are no clamp connections in heterokaryotic mycelia and hence no way to differentiate homo- from heterokaryotic mycelia. We know that frequency of nonfertile spores is very low. Moreover, cells are coenocytic. Kilgman (1943) observed 7–25 nuclei in the apical cell of the growing mycelium. The only method is to evaluate all cultures from the single spores for fruiting and treat non-fruiting ones as homokaryotic. The most common approach is to take spore print, go for serial dilution and plate the spores, and pick individual colonies. The cultures so generated are used to make spawn and evaluate these for fruiting. Efforts have been made to develop methods for easier identification of homkaryotic mycelia. Some of these are as follows:

- Growth in Petri plates and downward linear growth on compost (Singh and Kamal 2011a, b, 2012). Slow growth of single spore may indicate homokaryotic nature, but needs confirmation by growing and checking for lack of fruiting.
- Micromanipulation is a sure way to pick up monokaryotic spores from tetrasporic basidia, but the technique is cumbersome (Elliott 1971).
- Collection of germplasm and identification of strains producing higher number of homokaryotic spores. *A. bisporus* var *bisporus* mainly produces two-spored basidia, whereas *A. bisporus* var *brunetti* mainly produces four-spored basidia. Homokaryons from these two subspecies can be crossed but the average quality of most *brunetti* strains is poor (Sonnenberg et al. 2017).
- Study of factors promoting higher frequency of tri and tetrasporic basidia. Counts of two-, three-, and four-spored basidia have been shown to vary according to the age of the mushroom and the age of the crop (Pelham 1967). Fruit bodies from later flushes or growing at high temperature and abnormal climatic conditions may produce more tetrasporic basidia. Here too, final test is fruiting.
- Regeneration of homokaryons from promising heterokaryotic strains using protoplast techniques. These homokaryons will have same cytoplasmic background and are thus not exactly similar to the original parent spores.
- Molecular approaches to identify homokaryotic basidiospore cultures that include use of isoenzymes, RAPD, RFLP, ISSR, IRAPs, REMAPs markers (Nazrul and YinBing 2011; Lee et al. 2018; Gupta et al. 2019). Different approaches to identify homokaryons have been reviewed by various workers (Sonnenberg et al. 2017; Singh and Kamal 2017; Kamal et al. 2019), after sequencing of genome, use of primers for MAT locus seems a better method to identify homokaryons (Choi et al. 2023). This further helps in the identification of compatible homokaryons.

First hybrids were commercialized by Darlington in November 1981 (Fritsche 1991b). It can be presumed that till then the major impact for increase in productivity was due to refinements in cultivation methods. Thereafter breeding work started in many countries and many private spawn producers like Lambert, Sylvan, Amycel, and Somycel started developing their own hybrids and strains. In most of the cases, there was similarity due to limited germplasm and suppression of recombination (Sonnenberg et al. 2011).

In the 1960s, when button mushroom cultivation was introduced in India, strains like S-11, S-310, S-649, RRL 89, and TM 79 were popular. Breeding work in India started in 1985 at National Centre for Mushroom Research and Training (now Directorate of Mushroom Research), Solan, by isolating single spores, classifying these to fertile and infertile and intermating of non-fertile isolates (Bhandal and Mehta 1989; Thakur and Bhandal 1992). S-11 remained popular for many decades and was suitable for compost made using long method (Bhandal and Mehta 1989). Varieties like NCS 100, NCS 101, and NCH 102 were developed and released in 1997. NCH 102 was the first hybrid developed in India. Subsequently, by single

Fig. 17 NBS-5—a non-browning strain

spore isolations and hybridization, a number of button mushroom varieties of white and brown button have been notified by ICAR-Directorate of Mushroom Research, Solan, some of which are non-browning strains like NBS-1 and NBS-5 (Singh et al. 2016) (Fig. 17).

The Tamil Nadu Agricultural University (TNAU) released two button mushroom strains: Ooty 1 and Ooty (BM) 2. These strains were released in 1999 and 2002, respectively, by scientists at the Horticulture Research Station of TNAU in Ooty (India). Genetic improvement work was also done at Agriculture University CSK HPKVV, Palampur, by Kapoor et al. (2016), where they isolated single spores from commercial strains.

Breeding work was also done in South Korea at National Institute of Horticultural and Herbal Science, RDA, Eumseong. The first Korean button mushroom cultivar 'Sae-Ah' was developed in 2010, cultivar 'Saejeong' was developed in 2011, and cultivar, 'Saedo,' was developed in 2012. This new cultivar was selected by genetic analysis from crossing the heterokaryon line A175 and the homokaryon line ASI1346-15 (Oh et al. 2016).

Breeding work on mushrooms has been initiated in Iran (Gordan and Farsi 2005), China (Yating et al. 2022), and many other countries. In Europe and the USA, the work has been done by private companies, and details of their work are not easily available except the details of varieties developed by such companies. Sonnenberg et al. (2017) demonstrated that many of the hybrids/strains that appeared in the market after 1980 (Table 7) were derived from U1.

Sonneberg et al. (2017) genotyped over 70 varieties (Table 7) that included hybrids, traditional white, off white, and brown cultivars using SNP markers derived from several re-sequenced lines and demonstrated that the traditional and present day cultivars cluster into the expected groups with a very low genetic variation within each group and are essentially derived varieties. A brief of some of the current varieties and others as based on sources from internet is listed in Table 8.

Table 7 Varieties genotyped using SNP markers by Sonnenberg et al. (2017)

White hybrids (4)	Le Champion 308	Som 608	Sylvan A15	Sylvan U1
Traditional white cultivars (31)	Claron A1 Claron A6 Claron A8.8 Darmycel 21/2 Dawnston D649 Italspawn F5 Italspawn 286	Le Lion B14 Le Lion B74 Le Lion B8 Le Lion B80 Les Miz 35 Les Miz 66 Les Miz 96 Les Miz 98	Les Miz Y217 Royal 70 Scamyco 135 Scamyco 144 Sinden A1 Sinden A61 Somycel 153 Somycel 191	Somycel 22 Somycel 32 Somycel 32 Somycel 459 Somycel 53 Somycel 85 Somycel 91 Somycel BX105
Traditional off-white cultivars (23)	Claron A3.6 Claron A4 Claron A4.6 Darlington 735 ItalSpawn F2	Le Lion B62 Le Lion B86 Le Lion C63 Le Lion H1 Les Miz 36 Les Miz 54	Royal 20A Royal 30A Sinden A4 Somycel 11 Somycel 14 Somycel 176	Somycel 735 Somycel 76 Somycel 87 Somycel 9 Somycel X135 Somycel9.2
Traditional brown cultivars (12)	Claron C4 Hauser C3-7 Hauser C38	Italspawn FB30 Le Champion 101 Le Lion C9	Les Miz 23 Les Miz 28 Royal 101	Somycel 456 Somycel 856 Sylvan 856

However, Sylvan and Amycel are two of the leading global producers of mushroom spawn, offering a variety of strains for button mushrooms (*Agaricus bisporus*), both white and brown.

Sylvan, Inc. (often referred as Sylvan) is a global leader. It has facilities in the USA, Huai'an, China, Langeais, France, and Horst, the Netherlands. Sylvan has done various acquisitions over the years to strengthen its market position, particularly in Europe and beyond. Somycel, Europe's leading spawn producer was acquired in 1991. [Somycel (France) and Darmycel (UK) were sister companies since Dec 1973 (Mushroom News 1973, 21(12), p2), and these were acquired by Sylvan foods in Oct 1991.] Hauser Champignonkulturen AG was acquired in 1992. This expanded Sylvan's presence in the European spawn market. Mispaj Micelios was acquired in June 2020, and Gurelan Mycelium was acquired in September 2022, and these acquisitions consolidated Sylvan's position in the Spanish market. Sylvan acquired French agaricus mushroom spawn producer EuroMycel from Bonduelle (BON.PA) as announced on October 14, 2024 (www.sylvan.bio/news/). This move further strengthened Sylvan's position in Europe. In 2018, Sylvan became a global player as it executed an agreement with the Chinese group Yuguan (Jiangsu Yuguan Modern Agriculture Science & Technology Ltd) (https://mushroombusiness.com/news/sylvan-acquired-by-yuguan/). Yuguan Group is supported by the investment firm KKR (a US-headquartered private equity fund). This provided entry into the vast Chinese market as China is the largest producer of mushrooms accounting for over 90 percent of world production. Sylvan is expected to continue operating as a separate division under its current management in the USA. In April 2022, Sylvan (under Yuguan) opened the "Dennis C. Zensen Campus" spawn facility in Huai'an, Jiangsu—further cementing its China presence.

Table 8 Some of the varieties developed/sold by different agencies

Name of strain	Key characteristics
Sylvan Inc (https://www.sylvan.bio/food/)	
A15	A white hybrid, heavy first flush, commonly used strain in high tech farms, grown mainly for fresh market, high and consistent yields, adaptable to automation.
737	An intermediate hybrid strain
Tuscan	A Sylvan strain
Bavarian brown	Has rich golden-brown (chestnut) color and longer shelf life. It is designed for both hand-picking and machine harvesting. Can produce both small mushroom (cremini) and large mushrooms (Portobello)
Amycel (https://www.amycel.com/mushroom-spawn)	
Delta	Off white hybrid, moderate pinner. Can grow in warm season
Phoenixx	Off-white hybrid that produces large heavy-bodied mushrooms
Triple X	Most vigorous off-white hybrid, faster in all growing phases, excellent quality and density, well suited for a wide range of growing conditions
Exxcalibur	A new off-white hybrid that combines features of Phoenixx and Triple X, vigorous colonizing strain, produces large heavy-bodied mushrooms
Heirloom	Strain BR06, genetically unique and Patented (EP 1 993 350) and widely grown brown hybrid for both portabella and crimini; known for color, shape, density, and productivity
Brawn	Produces a darker cap than Heirloom, an alternative to Heirloom and a variety of size mixes. Patent protected; it cannot be cross-CAC'ed with Heirloom or any other brown strain on the market
EuroMycel (now Sylvan Europe) (https://www.euromycel.com/strains/)	
Euromycel E-58 Premium	Off-white mushroom strain, medium size, suitable for fresh market
E58 HD	This off-white, hybrid strain, bulbous stipe, suitable, for both manual and mechanical harvest
E21	Smooth, white, hybrid, mechanical harvesting variety. Uniform pinhead formation and a straight cylindrical stem. Suitable for the canning industry
E11	A white hybrid strain that has a higher pin density, Cap size 18–45 mm. suitable for canning, as well as the freezing process. A mechanical harvesting variety
Hollander Spawn (https://www.hollanderspawn.com/strains/)	
F41	Big Hybrid F41 produces a large, white, dense mushroom with a well-rounded cap of 4–6 cm across. Suited for growing large closed mushrooms for the fresh-market. Has some tendency for scaling on the first flush
F599	Intermediate Hybrid F599 an easy pinning, prolific strain. Mushrooms are of medium size; smooth, extremely dense white with a diameter of the cap of 3 to 6 cm. Produces high yields of medium size and big snow-white mushrooms having long shelf life
F59	Intermediate hybrid F59 is a medium size, smooth, extremely dense white mushroom with a diameter of the cap of 3 to 6 cm

(continued)

Table 8 (continued)

Name of strain	Key characteristics
FB33	Brown FB33 mushroom are solid, medium to large-sized (3 to 7 cm), brown colored cap, and white stem. Good shelf life and distinctive flavor. Suitable for Portabella production
Mycelia (https://mycelia.be/mushroom-species/agaricus-bisporus/)	
M7205	White, large, smooth to slightly scaly. Specific conditions for good fructification
M7206	White, smooth, small to average in size, a short, thick stem. Wide adaptability to growing conditions and the compost quality
M7215	Medium-size white button mushroom strain; suitable for the fresh market
M7218	White, smooth, medium-size mushroom strain. Suitable for fresh market
M7219	Medium-sized smooth and round mushroom
M7243	A brown variety, suitable for both open and closed forms. Good shelf life and taste
Lambart Spawn (https://www.lambertspawn.com/products/agaricus-strains)	
901	Off-White strain close to U1, high stability, suitable for fresh market
927	Off-White strain, also close to U1, good performer in Phase-3
950	Off-White strain, Rounder cap, larger sized mushroom suitable for gourmet market
931	Off-White strain, High yields, suitable for slicing
805	Brown strain, good for production of both crimini and portabella
845	Brown strain, more suitable for production of crimini
Indian strains/hybrids	
DMR button-03	A white variety notified by ICAR-DMR Solan, India, in 2013, suitable for fresh market. Cap slightly scaly. Veil opening very late and gills are light pink (Sharma et al. 2017)
DMR button-04	A pure white variety notified by ICAR-DMR Solan, India in 2013, suitable for fresh market. Cap dome shaped, slightly scaly. Veil opening very late (Sharma et al. 2017)
DMR button-06	A brown variety notified by ICAR-DMR Solan, India, in 2013, high yields, dome-shaped cap. Veil opening very late (Sharma et al. 2017)
NBS-1	A non-browning white strain notified in 2014 by ICAR-DMR, Solan, India (Singh et al. 2016; Sharma et al. 2020)
NBS-5	A non-browning white strain notified in 2014 by ICAR-DMR, Solan, India (Singh et al. 2016; Sharma et al. 2020)
DMR button-14	A high–yielding, rapidly growing white to off-white strain(Sharma et al. 2020)
DMR button-59	A medium–sized, high yielding white to off-white strain, requires low CO_2 (Sharma et al. 2020)
Others/old	
Horst U1 (Horronda)	First real hybrid developed by Horst (Holland), commercialized by Darlington Mushroom Labs, UK (Darmycel) in 1981 (Fritsche 1991b). This hybrid has been widely used in breeding programs. Suitable for cultivation under controlled conditions

(continued)

Table 8 (continued)

Name of strain	Key characteristics
Horst U3 (Horwitu)	White high-yielding strain developed by Horst (Holland), commercialized by Darlington Mushroom Labs, UK (Darmycel) in 1981(Fritsche 1991b) Single spore isolate from U1 (Singh and Kamal 2017)
Saedo	A Korean hybrid developed in 2012 (Oh et al. 2016).
Somycel-11	White, smooth, scales with strong air circulation, high dry matter content. One of first introduction in India where it is commonly referred as S-11
Somycel 12	White hybrid, able to grow at higher temperatures and higher carbon dioxide
Somycel-92	White, thick fruit-bodies, fleshy round pileus, few scales depending on conditions, short strong stipes, heavy and very good quality mushrooms, high yields and protein content
Somycel-653	Snow-white, middle-sized fruit-body with tendency to form small fruit-bodies, straight cylindrical thin stipes
Somycel 516	Strains from the U-3 group, best for small fruit bodies suitable for processing. Used in Poland

Amycel LLC along with Spawn Mate is a notable player in the mushroom industry, particularly in the development of hybrid brown button mushrooms like their patented "Heirloom" strain (BR06) and Brawn strain. Their white and off-white strains like Delta, Phoenixx, Triple X, and Exxcalibur are mainly for the fresh mushroom market.

Horst U1 was the first hybrid developed in button mushroom by intermating of homokaryotic strains H39 (single-spore isolate from the traditional white line Somycel 53) and H97 (single-spore isolate from the traditional off-white line Somycel 9.2) (Sonnenberg et al. 1999). There are other old varieties like Somycel 11, Somycel 91, Somycel 209, and Somycel 516; Hauser A3.2, Hauser A6, Hauser A6.5, and Hauser A8.8. There are many other relatively new varieties reported from Poland like Amycel-29, Amycel-2000, SP-251, SP-252, SP-312, ItalspawnF59, Polmycel-29, and Polymycel-31 (Sobieralski et al. 2014). It is difficult to make a list of all the strains used in different countries, and it is still more difficult to say that how many of these are originals. Considering that it is easy to vegetatively propagate mushroom cultures or make culture from fruit body or its spores, it is not clear how much spawn producers are respecting the breeder's rights. There are a number of small suppliers of different strains in many countries that have given new names to the commonly used varieties by simple subculturing or selecting a good mushroom from the crop and culture it. As any variety can be easily multiplied and there are hardly any morphological characteristics to clearly identify different varieties, there are ongoing debates on breeder's rights and considering single spore isolates as essentially derived varieties (Sonnenberg et al. 2011, 2017). Molecular markers are being used by many companies to protect and patent their varieties (Table 8).

Patenting mushroom varieties, particularly hybrids, unique breeding techniques, and novel parent strains, is crucial for companies to get return on investment for developing new varieties. Some of the patents related to these aspects are as in Table 9. However, there is no uniformity in laws for patenting and variety protection

Table 9 Some patents on button mushroom hybrids, lines, techniques, etc., in different countries

Patent number	Status	Year of grant	Assignee	Inventor(s)	Jurisdiction	Key characteristics/comments
CA2896208A1	Application (published)	2014	Sylvan America, Inc.	Ronald W. Kerrigan, et al.	CA (Canada)	New hybrid *Agaricus bisporus* mushroom strains having improved resistance to brown blotch disease (*Pseudomonas tolaasii*) aims to reduce losses from this common and economically significant mushroom disease.This is a patent application in Canada. It corresponds to an international application, likely WO2015/127210A1
CN108967041B	Granted	2021	Institute of Edible Fungi, SAAS	Li Gang, et al.	China	A new *Agaricus bisporus* variety 'Jinhe No. 1' and its breeding method, focusing on high yield and good quality.This patent also covers both a new strain and its breeding method
CN110352790B	Granted	2021	Fujian Province Academy of Agricultural Sciences	Wei Dong, et al.	China	A cultivation method for an *Agaricus bisporus* variety named 'Tianhong 3', focusing on high yield and quality. It describes both a strain and its cultivation method
CN115838821A	Application (published)	2023	Shanghai Academy of Agricultural Sciences	Li Gang, et al.	China	*Agaricus bisporus* germplasm strain "Shangnong Baiyu" and its breeding method, characterized by white color and good quality.This is a patent application, not a granted patent

(continued)

Table 9 (continued)

Patent number	Status	Year of grant	Assignee	Inventor(s)	Jurisdiction	Key characteristics/comments
EP1993350B2	Granted	2016	Amycel LLC		EP (Europe)	BR06 (commercialized as Heirloom) Hybrid brown *Agaricus bisporus* strain; significantly higher yields. This patent focuses on *Agaricus bisporus* mushroom strains with improved resistance to bacterial blotch caused by *Pseudomonas tolaasii*. Amycell won the court case in EPO
EP3247787B1	Granted	2022	Somycel Wageningen Universiteit	Anton Sonnenberg, et al.	EP (Europe)	New hybrid *Agaricus bisporus* mushroom strains (e.g., CNCM I-4945, I-4946) with improved characteristics like higher average weight and suitable phenotype for mechanical harvesting and smooth white-cap color.This patent relates to WO2016116632A1 and aims to address the lack of genetic diversity in commercial strains and improve profitability
ES2999373T3	Granted	2023	Wageningen Universiteit	Anton Sonnenberg, et al.	ES (Spain—national validation of EP)	New hybrid *Agaricus bisporus* mushroom strains with improved characteristics (likely high average weight, good for mechanical harvesting, smooth white-cap color) as a national validation of EP 3247787B1

KR20250028230A	Application (published)	2025	Not readily available	Not readily available	South Korea	As this is a 2025 application, specific details may not be fully public yet, but it would relate to a new mushroom variety or cultivation method
US 10051831 B2	Granted	2018	Sylvan America, Inc.	Kerrigan; Richard W. et al.	The USA	Describes methods to produce hybrid *Agaricus bisporus* (button mushroom) strains with greatly reduced or absent sporulation (i.e., sporeless or low-sporing mushrooms. Useful for cleaner cultivation environments and reducing allergen or contamination risk in commercial mushroom production
US-4996390-A	Granted	1991	The Pennsylvania State University	Lee C. Schisler, et al.	The USA Relevant for Canada	Novel interspecific mushroom strains (H1-SSI strains, e.g., H1-041), specifically cellular hybrids of *Agaricus bisporus* and *Agaricus bitorquis* through protoplast fusion. Strain HB3 mentioned, showing intermediate growth temperature and distinct isoenzyme patterns. It aims to combine desirable traits from two different *Agaricus* species, such as improved thermotolerance from *A. bitorquis* and high yield from *A. bisporus*

(continued)

Table 9 (continued)

Patent number	Status	Year of grant	Assignee	Inventor(s)	Jurisdiction	Key characteristics/comments
US-6343435-B1	Granted?	1998	Institut National de la Recherche Agronomique Paris	F. Moquet et al	The USA	The invention concerns intraspecific hybrids of *Agaricus bisporus* capable of bearing fruit at relatively high temperature. Made by cross-breeding between *A. bisporus*, var bisporus and var *burnettii*
US-10440930-B1	Granted	2019	Sylvan America, Inc.	Kerrigan; Richard W. et al	The USA	Relates to a hybrid *Agaricus bisporus* culture—specifically strain J15987 (deposited NRRL 67646). Covers methods for creating and utilizing hybrid cultures and homokaryotic lines derived from this strain
US20040144020A1	Application (published)	2004	Sylvan America, Inc.	Paul A. Horgen, et al.	The USA	This is a patent application that describes a new strain "Sylvan 149" and methods for its identification using molecular markers (RAPD)

US20100154079A1	Application published	2010	Sylvan America, Inc.	Kerrigan, et al.	The USA	Hybrid mushroom strain J9277, ITS descendants, and related methods. The strain itself is identified as "J10165" within the patent. It represents Sylvan's efforts to introduce new genetic material into the *Agaricus bisporus* gene pool, aiming for improved yield and quality beyond the traditionally dominant "U1" lineage. The key characteristics of this new strain include the following: abundant production of mushrooms. Mushrooms with smooth, bright white caps, which is a highly desirable trait for commercial white button mushrooms
US20100212042A1	Application (Published)	2010	Amycel LLC	Christopher William Robles, Stephen Christopher Lodder	The USA	Hybrid *Agaricus bisporus* mushroom strains for commercial production, specifically brown mushrooms (Portabella type) with improved characteristics such as increased productivity, darker cap color, and thicker cap

(continued)

Table 9 (continued)

Patent number	Status	Year of grant	Assignee	Inventor(s)	Jurisdiction	Key characteristics/comments
US-20100218294-P1	Application	2010 2010	Sylvan America Inc.	(Richard W. Kerrigan, Mark P. Wach)	The USA	*Agaricus bisporus* strain named as J10165. Smooth, bright white caps; cultural incompatibility with "U1" lineage (genetic disease barrier against viruses); unique ITS DNA sequence fingerprint for identification. Hybrid of "J9277" and "B7023"
US-PP17396-P3	Granted (Plant Patent)	2007	Monterey Mushrooms, Inc.	Sonnenberg; Anton.	The USA	Mushroom variety named "Broncoh." A new and distinct variety of *Agaricus bisporus* mushroom plant named "HORST U-8," characterized by its white cap and improved shelf life.This is a Plant Patent (PP), specifically for a new plant variety (mushroom is considered a plant for this purpose). "HORST U-8" suggests a lineage from the well-known Horst strains
US4996390-A	H1-SSI strains (e.g., H1-041)	1991	(Canadian research context)	Dahlberg; Kurt R.	The USA (relevant for Canada)	Interspecific hybrid with rapid compost colonization, smooth white cap, equivalent/higher yields, high-temperature tolerance (32–33 °C)

WO1994000006A1 EP0707443	Application (Published)	1994	Sylvan America, Inc.	Peter A. Horgen, et al.	WO (International)	This facilitates more efficient breeding of new hybrid strains with improved traits like productivity, disease resistance, and aesthetics.This patent addresses a key challenge in *A. bisporus* breeding (obtaining stable homokaryons for crosses)
WO2015127210A1 EP3107374	Application (published)	2015	Sylvan America, Inc.	Ronald W. Kerrigan, et al.	WO (International)	New hybrid strains of *Agaricus bisporus* having improved resistance to brown blotch disease (*Pseudomonas tolaasii*). This international application has national counterparts
WO2018102290A1 EP3548605	Application (published)	2018	Monterey Mushrooms, Inc.	John H. M. P. C. Peeters, et al.	WO (International)	Homokaryotic *Agaricus bisporus* mushroom fungus line J14756-ss3, and F1 hybrids (e.g., J15051) derived from it. Aims for improved yield, quality, and uniformity. This is the international publication corresponding to US 10051831 (which was granted)

(continued)

Table 9 (continued)

Patent number	Status	Year of grant	Assignee	Inventor(s)	Jurisdiction	Key characteristics/comments
WO2016116632A1 EP3247787	Published (a PCT application proceeded to patents like EP 3 247 787 B1 and US10440930)	2016	Somycel & WUR	Anton Sonnenberg, et al.	WO (World Intellectual Property Organization—PCT International Application)	This international application describes new heterokaryotic hybrid *Agaricus bisporus* mushroom strains that exhibit improved characteristics beneficial for growers, processors, and consumers. The invention aims to address the limited genetic diversity in commercially cultivated white *Agaricus bisporus* strains, which largely stem from a single lineage (like "Horst U1"). This is an international application which serves as the foundation for granted patents in various regions (e.g., EP 3 247 787 B1 in Europe, US10440930B2 in the USA, ES2999373T3 in Spain)
CRISPR-edited *Agaricus bisporus* (browning resistance)	Not explicitly patented	2016 (First approval)			The USA, Canada	Gene-edited to resist browning by knocking out polyphenol oxidase gene; not regulated as GMO as no foreign DNA introduced. Not explicitly patented in snippets, but its regulatory status is noted
JP Patent No. 7632898	Registered	2025	Toei Shinyaku Co., Ltd		Japan	Method and Kit for Detection and Quantification of β-glucan in mushrooms

of mushrooms (fungi) in different regions, even though we know that fungi are neither plants nor animals. Rules differ in different parts of the world, for example:

- In Europe, fungi are patentable and are excluded from the provisions of Article 53(b) EPC that excludes plant and animal varieties from patent protection (refer legal case surrounding patent by Amycel EP 1993350 B2). (https://www.juve-patent.com/cases/amycel-scores-win-as-upc-decides-on-fungi-patentability/ (updated 5 August 2024).
- In the USA, utility patents for novel *Agaricus bisporus* strains are granted, and Sylvan and Amycel have number of mushroom related patents in the USA (Table 9). Going a step further, the USA has approved gene editing of white button mushroom to resist browning by knocking out polyphenol oxidase gene using CRISPR technology, and this technology was not regulated under GMO rules, as this process did not introduce any foreign DNA.
- Australia explicitly includes fungi and algae under its Plant Breeder's Rights scheme, and new varieties of fungi can be protected by PBR for up to 20 years. (https://www.ipaustralia.gov.au/plant-breeders-rights/what-are-plant-breeders-rights)
- In Canada, Plant Breeders' Rights provide exclusive rights for new plant varieties for up to 20 years but the Canadian Food Inspection Agency clearly states that "Algae, fungi and bacteria cannot be protected under plant breeders' rights." Protection for *Agaricus bisporus* strains would thus fall under general patent law.
- In India, there is a system of release of plant varieties by Central Variety Release Committee (CVRC). Mushroom varieties are not being considered for release by CVRC due to lack of spawn standards and detailed protocols for variety identification. Another government agency, Protection of Plant Varieties and Farmers Right (PPVFR) also does not include mushrooms for protection of breeder's rights. As of now, the varieties developed are notified by research institutes and growers within India can get spawn or culture of these varieties from such institutions.
- China produces over 90% of the mushrooms of the world and has a different intellectual property strategy. China's intellectual property strategy focuses on protecting cultivation methods alongside new strains. This reflects a comprehensive approach to agricultural innovation and addressing challenges in genetic material protection. This differs from Western patents that often focus on the genetic makeup of the strain. China has recently updated its Regulations on the Protection of New Plant Varieties (effective June 1, 2025) (https://mmlc-group.com/pvr-2025-regs/), but it is not clear if these include fungi also.
- Japan has a Plant Variety Protection and Seed Act. *Agaricus bisporus* is listed as a botanical taxon. There is a patent by Toei Shinyaku Co., Ltd. for a β-glucan quantification method for *Agaricus* mushrooms. Major focus in Japan is, however, on enoki, nameko, oyster, eryngii, shimeji, and maitake varieties.

There is stagnation/decrease in button mushroom production due to the following:

- Availability of alternate mushroom species, some of which have better nutritional properties, flavor and nutraceutical value. Such mushroom species also have higher preference in China and many other Asian countries. As a result, the share of button mushroom in total production is on decline in China, South Korea, Taiwan, etc. As shown in Fig. 6, decline in button production in China has impacted the total global button production.
- Many countries like Taiwan, Korea, and India initially produced button mushroom primarily for export. Production data of Taiwan and Korea in FAOSTAT possibly represents production of button mushroom in these countries. Actual production of mushrooms is much higher in these countries (Singh et al. 2020). Many countries are not included in FAOSTAT like Mexico, Brazil, Bangladesh, and number of other countries from Asia.
- Cultivation technology of other mushrooms in some cases is simple than that of button and has been fully automated. Even casing material for cultivation of button mushroom that was peat moss is going to be issue in coming years (Noble et al. 2023)
- Cultivation of Button is not only more labor intensive, but button mushroom also has less shelf life as compared to many other mushrooms and cannot be dehydrated
- Availability of limited germplasm restricts achievements of breeding programs. Breeding is not economical for private players in many countries as cultures can be easily propagated vegetatively.

3 Environmental Footprint of Button Mushroom Cultivation

In the nineteenth century, we could produce food just enough to fill our bellies. There were floods, droughts, and famines. However, the food produced was organic. In the twentieth century, we came with a number of innovations like fertilizers, pesticides, weedicides, other chemicals, mechanization, and food processing and preservation. This approach to agriculture helped us to produce enough food. The focus was on production, convenience, ready-to-eat food, and there was enough food available to fill the belly. The negative impacts of this modern agriculture were multifold, and it affected not only soil health and human health but also climate.

In the present century, there is emphasis on the quality of food. People are increasingly curious about health benefits of the food they eat and also where and how their food is produced and what impact it has on the environment, that is, what is its environmental footprint. Genetically improved materials have thus a role to play in producing more quality food by using less resources and thus having positive impact on the environment. The environmental footprint refers to the impact that human activities, products, or services have on the environment. It encompasses various aspects, including the following:

(i) *Carbon footprint*, that is, the amount of greenhouse gas emissions, primarily carbon dioxide, associated with human activities, such as energy consumption, transportation, and industrial processes. Measuring the carbon footprint involves calculating the total amount of greenhouse gas (GHG) emissions produced directly or indirectly like CO_2, methane (CH_4), and nitrous oxide (N_2O). We convert gases (other than CO_2) into a CO_2 equivalent based on their global warming potential and express it in terms of carbon dioxide equivalents (CO_2e).
(ii) *Water footprint*, that is, the amount of water used, consumed, or polluted during the production, processing, and consumption of goods and services.
(iii) *Waste footprint*, that is, the amount of waste generated, including solid waste, wastewater, and hazardous waste, during the production, consumption, and disposal of goods and services.

Reducing our environmental footprint requires adopting sustainable practices, increasing efficiency, and promoting eco-friendly technologies. If we look at mushrooms, the environmental impacts are mainly related to energy usage for temperature and humidity control during cultivation.

Comprehensive studies from cradle to gate aim to quantify the environmental impacts across the entire life cycle, from substrate production to harvesting and post-harvest handling and are commonly referred as Life Cycle (impact) Assessment (LCA).

Cultivation of mushrooms offers number of environmental benefits like use of agro-wastes, pollution mitigation, nutrient recycling, bio-remediation of pollutants, and improvement of soil health. Mushrooms are a sustainable food choice as the environmental footprint of mushrooms is low as compared to many food items, particularly animal-based products, some plant-based options like tofu and even some vegetables. Mushrooms require relatively little land and water to grow compared to many other crops, contributing to their lower environmental impact. While mushroom cultivation itself is relatively efficient, factors like the energy used in controlled environments for growing and transportation can contribute to the overall carbon footprint. However, the carbon footprint can be reduced further by eating organic, reducing food and plastic waste, and purchasing local produce (Howarth 2025), and most importantly using genetically improved strains as we get more by using fewer resources.

Mushroom cultivation involves the use of various agro- and forest wastes and is a good example of economically viable circular economy. Various agro-wastes like cottonseed hulls, corncob meal, wheat straw, rice straw, and bagasse are produced during cultivation of different crops and much of it is either burnt or remains unutilized. Several studies show that instead of burning agro-wastes, mushroom production using agro-wastes can be a profitable business (Pavlík and Halaj 2019; Odediran et al. 2020; Nanje Gowda and Chennappa 2021). Further, mushroom cultivation has low environmental impact due to miniscule or no use of pesticides and chemical fertilizers.

We know that plants sequester carbon by converting carbon dioxide to glucose, whereas animals and fungi produce carbon dioxide. Even though fungi produce

CO_2, the fact is that fungi in nature actually help in carbon sequestration (review by Hannula and Morien 2022). As per reports, CO_2 has increased from 250 to about 420 ppm in last 150 years (>80% of green house gases is CO_2). We need to sequester this, that is, fix it as carbon in soil. Even though fungi, like us produce CO_2, but these play important role in carbon sequestration. Fungi degrade cellulose, hemicellulose, and lignin. Without this, carbon cycle will stop and after few years, all we may see will be forests of dead trees. A lot of carbon stays in soil in the form of fungal mycelium. More than 90% of trees depend on some form of fungal associations for growth and survival. Hence, in the absence of fungi, trees in forest will not be able to grow properly and thus sequester carbon. Further, CO_2 fixed by trees as sugar is shared with fungi associated (up to 40%) and thus gets sequestered in soil.

Promoting afforestation and agriculture is considered important to reduce CO_2 and thus combat climate change. Successful afforestation will require proper fungal associations. And regarding agriculture, many a times, the energy balance in modern intensive agriculture may be negative. That is, we consume more energy to grow food and on net basis, we have actually produced more carbon dioxide that has been fixed by the vegetable we have produced (Fig. 18). When we go for intensive agriculture there are various energy inputs like fossil fuels, tractors, and combines; energy used to produce, transport, and apply various chemical fertilizers and pesticides; and processing and transport of end product. Irrigation systems also use energy to pump water for irrigation and the output of all these activities is the energy embodied in the caloric content of the food. Modern agricultural practices often prioritize yield maximization over energy efficiency, and as stated earlier, we may be consuming more energy to produce less energy as food.

Carbon Foot Print In Fig. 18, we can see that mushrooms relatively produce much less CO2e than many other vegetables. The values will vary for vegetables and mushrooms cultivated in poly-houses and under natural conditions. The values reported for mushroom here are lower than reported for CO2e produced from cradle to gate or what we call life cycle impact assessment studies by other workers as in Table 10. But comparative analysis highlights the lower environmental impact of mushroom cultivation.

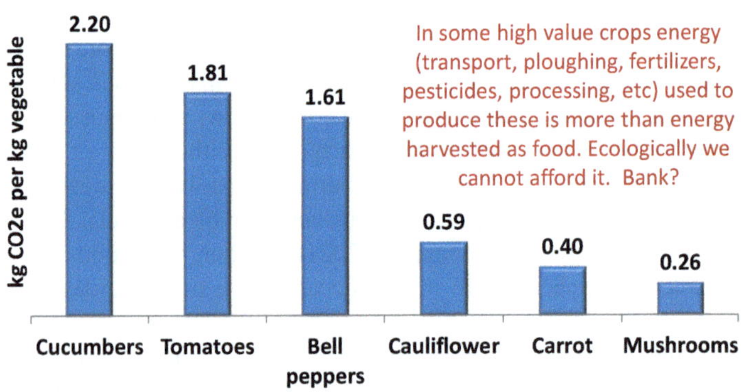

Fig. 18 CO2e balance in different vegetables and mushrooms based on https://impactful.ninja/what-is-the-carbon-footprint-of-mushrooms/

Table 10 Five studies on CO2e in button mushroom

Country	CO2e kg/kg mushroom	Major contributor	References	Remarks
Australia	2.76	Transportation	Gunady et al. (2012)	LCA Button.
Spain	4.41	On-farm energy	Leiva et al. (2015)	LCA Button
USA	2.13–2.95	On-farm energy	Robinson et al. (2019)	LCA Button
Spain	0.52	Compost Electricity/ Diesel	Goglio et al. (2024)	LCA Button
Poland	0.93			
Serbia	1.55			
Italy	2.28		Vinci et al. (2023)	LCA Button
			Jayaraman et al. (2024)	Review

There are only few studies on LCA in button mushroom. These are Gunady et al. (2012) from Australia, Leiva et al. (2015) from Spain, Robinson et al. (2019) from the USA, Vinci et al. (2023) from Italy, and Goglio et al. (2024) from Spain, Poland, and Serbia. All these have reported variable CO2e values. This is expected as resources, environmental conditions, energy sources and requirements, methods of cultivation, etc., may vary (Table 10).

Water Use Mushroom being an indoor crop, the water requirement is much less than other crops unlike wheat and rice where water required to produce one kg is in thousands of liters. In mushrooms, we need 15–25 liters to produce one kg mushroom (modified from SureHarvest 2017). This variation is due to difference in biological efficiency (BE) (Fig. 19).

Energy Consumption Major amount of energy is consumed during the cultivation in cropping rooms followed by the energy required to pasteurize the compost during phase 2 (SureHarvest 2017) (Fig. 20).

As general approach, we need to focus on reducing environmental impact of mushroom cultivation. Some of the issues and approaches to do so are as follows:

Issue	Approach
Energy in transport of raw materials/ choice of raw material	Use locally available raw materials, proper material. Completely recycle agro-wastes—circular economy
Energy (diesel) used by machinery during compositing and cropping	Better efficiency of machinery, boilers, etc.; use of clean energy
Electricity for cooling during cropping	Shift to solar energy (PV), promote seasonal cultivation
Energy use in packing/canning and transport	Promote local fresh consumption

There is a need to focus on complete recycling of agro-wastes, improve soil carbon/health, and thus promote cyclic agriculture. "*A nation that destroys its soils destroys itself*" (Franklin Roosevelt 1937 just after the Dust Bowl). During the green revolution, we forgot this. Wide application of fertilizers and other chemicals

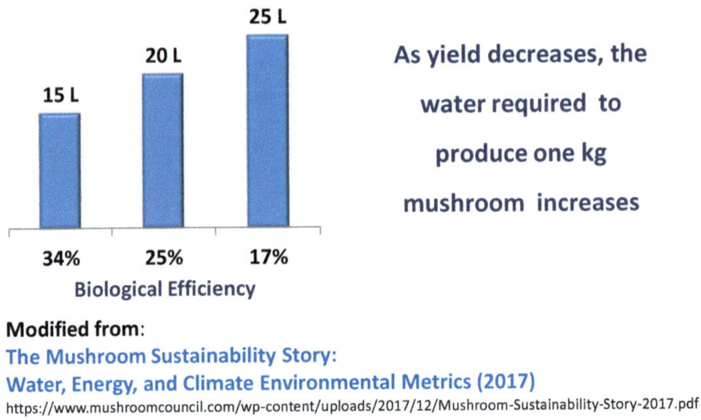

Fig. 19 Water required to produce one kg button mushroom

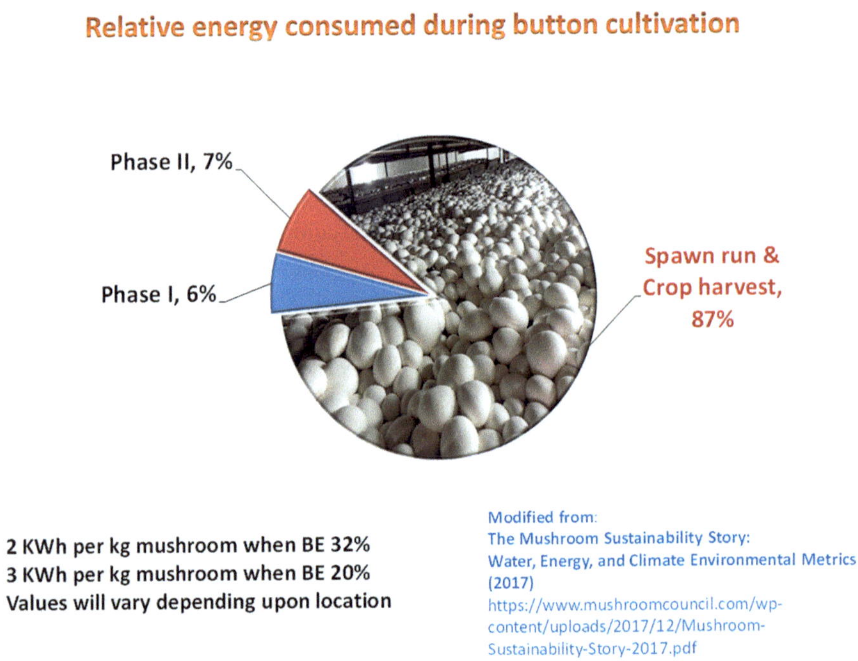

Fig. 20 Energy consumed at different stages of button mushroom cultivation

has reduced carbon content of soil. SMS can help to restore it. Ecologically we cannot afford to spend more energy to produce less on the excuse of feeding increasing population. We need to fix our food systems. *Managing agro-wastes is a challenge.* As mentioned earlier, the next revolution in agriculture will depend on our ability to effectively recycle agro-wastes.

We have seen that mushroom cultivation is an economically and ecologically sustainable option. Despite the benefits, there are few problems associated with mushroom cultivation such as spore production during oyster mushroom cultivation that are health hazard as these may cause respiratory and allergic problems (Nishida and Yatera 2022). Genome editing techniques are likely to play important role in the introduction of sporulation deficiency in major mushroom species (e.g., shiitake mushroom and other oyster mushrooms (Okuda 2022).

Disposal and proper use of spent mushroom substrate is another problem. Proper use of SMS is helpful for promoting soil health and environment. Polypropylene (HDPE) bags used for spawn production and polythene bags/sheets used during mushroom cultivation are other environmental issues that need to be addressed for sustainable mushroom production. According to Okuda 2022, ~17 g of HDPE bags become waste for every one kg of shiitake produced. The current annual shiitake production is over 13 million tonnes, which means production of 220,000 tonne of HDPE waste. Genetically improved materials can mean overall use of less plastics, but it is not a solution of the problem. Similarly, producing one kg of oyster mushrooms generates 40 g of waste (Panjikkaran and Mathew 2012). Hence, the production of 8.28 million tonnes per annum will produce 330,000 tonnes of plastic. We need high-yielding varieties as well as alternatives to plastics like biodegradable plastic bags made of polylactic acid (PLA).

In button mushroom alone, about 10 million tonnes of SMS may be produced. Considering all mushrooms more than 100 million tonnes of SMS is generated annually that places a significant burden on the environment (Okuda 2022). However, SMS as named is not a waste but can be used in various ways to make food, fuel, and compost (Rinker 2017) and thus be a component of circular agriculture. Mushroom production not only utilizes various low economic value by-products generated in forestry and agriculture but is also a model of circular agriculture. Genetic improvement to increase yield and address to various other traits like spore production can substantially contribute to sustainable production.

References

Beaulieu-Fortin S (2019) Economic Profile of the Mushroom Industry in Canada. https://www.150.statcan.gc.ca21-004-X201900100001

Bhandal MS, Mehta KB (1989) Evaluation and improvement of strain in Agaricus bisporus. Mushroom Sci 12:25–35

Bretzloff CW, Robbins WA, Curme JH (1962) Observations on Multisporous isolates from the cultivated mushroom, Agaricus bisporus (Lange) Sing. ISMS 5:Article 20

CEFA (2025) Chinese Edible Fungi Association. https://bigdata.cefa.org.cn/output.html

Chang ST, Miles PG (2004) Mushroom: cultivation, nutritional value, medicinal effect, and environmental impact. CRC Press, Boca Raton

Chang ST, Wasser S (2017) The cultivation and environmental impact of mushrooms. Oxford University Press, Oxford

Choi YJ, Jung S, Eom H, Hoang T, Han HG, Kim S, Ro HS (2023) Structural analysis of the A Mating type locus and development of the mating type marker of *Agaricus bisporus* var. *bisporus*. J Fungi 9:284. https://doi.org/10.3390/jof9030284

Elliott TJ (1971) Sex and the single spore. In: Proceedings of the ISMS held at London in 1971. Mushroom science Vol 8 article 2
FAOSTAT (2025). https://www.fao.org/faostat/en/#data/TM. Accessed April 28, 2025
Fritsche G (1981) Some remarks on the breeding, maintenance of strains and spawn of Agaricus bisporus and A. bitorquis. Mush Sci XI:19
Fritsche G (1991a) A personal view on mushroom breeding from 1957-1991. In: Van Griensven LJLD (ed) Genetics and breeding of Agaricus. Proceedings of the first international seminar on mushroom science. Mushroom Experimental Station, Horst, The Netherlands, Pudoc, Wageningen, pp 3–18
Fritsche G (1991b) Maintenance, rejuvenation and improvement of Horst U1. In: Van Griensven LJLD (ed) Genetics and breeding of Agaricus. Proceedings of the first international seminar on mushroom science. Mushroom Experimental Station, Horst, The Netherlands, Pudoc, Wageningen, pp 145–152
Fritsche G, Sonnenberg ASM (1988) Mushroom strains. In: Van Griensven IJLD (ed) The cultivation of mushroom, pp 101–123
Goglio P, Ponsioen T, Carrasco J, Milenkovi I, Kiwala L, Van Mierlo K, Helmes R, Tei F, Oosterkamp E, Perez M (2024) An environmental assessment of *Agaricus bisporus* ((J.E.Lange) Imbach) mushroom production systems across Europe. Eur J Agron 155:127108. ISSN 1161-0301. https://doi.org/10.1016/j.eja.2024.127108
Gordan HR, Farsi M (2005) Mushroom breeding program in Iran. In: Pisabarro AG, Ramírez L (eds) Genetics and cellular biology of Basidiomycetes-VI. Universidad Pública de Navarra, Spain. ISBN 84-9769-107-5
Gunady MGA, Biswas W, Solah VA, James AP (2012) Evaluating the global warming potential of the fresh produce supply chain for strawberries, romaine/cos lettuces (Lactuca sativa), and button mushrooms (Agaricus bisporus) in Western Australia using life cycle assessment (LCA). J Clean Prod Work 28:81–87. https://doi.org/10.1016/j.jclepro.2011.12.031
Gupta M, Kamal S, Barh A, Bairwa RK, Sharma VP (2019) Assessment of different marker systems for diversity analysis in monokaryotic lines of button mushroom strains using RAPDs, ISSRs, IRAPs and REMAPs. Mushroom Res 28(2):89–97. https://doi.org/10.36036/MR.28.2.2020.97603
Hannula SE, Morri¨en E (2022) Will fungi solve the carbon dilemma? Geoderma 413:115767. https://doi.org/10.1016/j.geoderma.2022.115767
Howarth Grace (2025) What is the carbon footprint of mushrooms? a life-cycle analysis. https://impactful.ninja/what-is-the-carbon-footprint-of-mushrooms/
Jayaraman S, Yadav B, Dalal RC, Naorem A, Sinha NK, Rao CS, Dang YP, Patra AK, Datta SP, Subba Rao A (2024) Mushroom farming: a review focusing on soil health, nutritional security and environmental sustainability. Farm Syst 2(2024):100098. https://doi.org/10.1016/j.farsys.2024.100098
Kamal S, Sharma VP, Gupta M, Barh A, Manjit S (2019) Genetics and breeding of white button mushroom, Agaricus bisporus(Lange.) Imbach. - a comprehensive review. Mushroom Res 28(1):1–22. https://doi.org/10.36036/MR.28.1.2019.91938
Kapoor P, Sharma AKN, Sharma PN, Dhancholia S (2016) Evaluation of single spore isolates of Agaricus bisporus and their genetic characterization. Indian Phytopathol 69(3):304
Kligman AM (1943) Some cultural and genetic problems in the cultivation of the mushroom, *Agaricus campestris* Fr. Am J Bot 30:745. https://doi.org/10.2307/2437548
Ku TY, Lee CL (1989) Mushroom Industry in Taiwan. Proceeding of ISMS held at Braunschweig Volume 12 Article 78
Lambert EB (1950) Comments on twenty years of research in mushroom culture. ISMS 1:Article 1
Lambert EB (1960) Improving spawn cultures of cultivated mushrooms. Proceedings of ISMS held at Copenhagen in 1959. Mushroom Sci IV:33–51
Lee HY, Raveendar S, An H, Oh YL, Jang KY, Kong WS et al (2018) Development of polymorphic simple sequence repeat markers using high-throughput sequencing in button mushroom (*Agaricus bisporus*). Mycobiology 46(4):421–428. https://doi.org/10.1080/12298093.2018.1538072

Leiva FJ, Saenz-Díez JC, Martínez E, Jimenez E, Blanco J (2015) Environmental impact of Agaricus bisporus cultivation process. Eur J Agr 71:141–148. https://doi.org/10.1016/j.eja.2015.09.013

MAFF (2025) Ministery of Agriculture, forestry and Fisheries. https://www.e-stat.go.jp/dbview?sid=0003446939

Nanje Gowda, N.E., and Chennappa, G. (2021). "Mushroom cultivation: a sustainable solution for the management of agriculture crop residues," In Recent advances in mushroom cultivation technology and its application, S. Kulshreshtha (Rohini; New Delhi: Bright Sky Publications, pp. 15–26

Nazrul Mahmudul I, Yin Bing B (2011) Differentiation of homokaryons and heterokaryons of Agaricus bisporus with inter-simple sequence repeat markers. Microbiol Res 166(3):226–236. https://doi.org/10.1016/j.micres.2010.03.001

Nishida C, Yatera K (2022) The impact of ambient environmental and occupational pollution on respiratory diseases. Int J Environ Res Public Health 19:2788. https://doi.org/10.3390/ijerph19052788

Noble R, Grogan H, Corbett E, Seymour G (2023) The future of casing - review of casing materials and availability of peat for mushroom cultivation. https://www.isms.biz/ AMGA Casing Review Report, 55p

Odediran FA, Adekunle OA, Olugbire OO, Olarewaju TO, Kolade RI, Obagunsho OE (2020) Contribution of mushroom production to rural income generation in Oyo state, Nigeria. FUTY J Environ 14:47–54

Oh Y-L, Jang K-Y, Kong W-S, Shin P-G, Oh MJ, Choi I-G (2016) Cultural and morphological characteristics of a new white button mushroom cultivar 'Saedo'. Kor J Mycol 44(2):94–99. https://doi.org/10.4489/KJM.2016.44.2.94. pISSN 0253-651X • eISSN 2383-5249

Okuda Y (2022) Sustainability perspectives for future continuity of mushroom production: the bright and dark sides. Front Sustain Food Syst 6:1026508. https://doi.org/10.3389/fsufs.2022.1026508

Panjikkaran ST, Mathew D (2012) An environmentally friendly and cost effective technique for the commercial cultivation of oyster mushroom [*Pleurotus florida* (Mont.) Singer]. J Sci Food Agric 93:973–976. https://doi.org/10.1002/jsfa.5827

Pavlík M, Halaj D (2019) Production and investment evaluation of oyster mushroom cultivation on the waste dendromass: a case study on aspen wood in Slovakia. Scand J Forest Res 34:313–318. https://doi.org/10.1080/02827581.2019.1584639

Pelham J (1967) Techniques for mushroom genetics. In: Proceeding of ISMS held at Wageningen and Amsterdam in 1965. Mushroom science, vol 6 Article 2

Raper CA, Raper JR, Miller RE (1972) Genetic analysis of the life cycle of *Agaricus bisporus*. Mycologia 64(5):1088–1117. https://doi.org/10.2307/3758075

Rinker DL (2017) Spent mushroom substrate uses. Chapter 20. In: Zied DC, Pardo-Giménez A (eds) Edible and medicinal mushrooms: technology and applications, 1st edn. Wiley, pp 427–454

Robinson B, Winans K, Kendall A, Dlott J, Dlott F (2019) A life cycle assessment of *Agaricus bisporus* mushroom production in the USA. Int J Life Cycle Assess 24:456–467. https://doi.org/10.1007/s11367-018-1456-6

Roosevelt Franklin D (1937) https://www.presidency.ucsb.edu/documents/letter-all-state-governors-uniform-soil-conservation-law

Shandilya TR (2000) Study on vermi-compost as a casing substrate for cultivation of *Agaricus bisporus* (Lange) Imbach. J Mycol Plant Pathol 30(2):285

Sharma VP, Bijla S (2024) Mapping the status of mushroom cultivation: global vis-à-vis India. Mushroom Res 33(2):91–101. https://doi.org/10.36036/MR.33.2.2024.163501

Sharma VP, Annepu SK, Gautam Y, Singh M, Kamal S (2017) Status of mushroom production in India. Mushroom Res 26(2):111–120

Sharma VP, Kumar S, Sharma S (eds) (2020) Technologies developed by ICAR-DMR for commercial use, ICAR- Directorate of mushroom research, Chambaghat, Solan, Himachal Pradesh, India, p 78

Sinden JW (1932) Mushroom spawn and method of making same. USPO No. US1869517A. United States
Singh M, Kamal S (2011a) Validity of mycelial growth on malt extract agar and compost as selection criteria for initial screening of genotypes for yield and quality in *Agaricus bisporus*. In: Proceedings of the 7th international conference on mushroom biology and mushroom products (ICMBMP7) at Arcachon, France 4–7 Oct 2011, pp 71–76
Singh M, Kamal S (2011b) Conventional and molecular approaches for breeding button mushroom. In: Proceedings of the 7th international conference on mushroom biology and mushroom products (ICMBMP7) at Arcachon, France 4–7 Oct 2011, pp 35–42
Singh M, Kamal S (2012) Simplifying breeding procedures in white button mushroom. Mushroom Science XVIII:273–277
Singh M, Kamal S (2017) Genetic aspects and strategies for obtaining hybrids. In: Zied DC, Pardo-Giménez A (eds) Edible and medicinal mushrooms: technology and applications, 1st edn. Wiley, pp 35–87
Singh M, Kamal S, Gupta M (2016) Development and yield evaluation of non-browning hybrids in button mushroom (*Agaricus bisporus*). In: Baars, Sonnenberg (eds) Science and cultivation of edible fungi. © 2016 International Society for Mushroom Science, pp 309–312. ISBN 978-90-9029771-2
Singh M, Kamal S, Sharma VP (2017) Status and trends in world mushroom production-I. Mushroom Res 26(1):1–20
Singh M, Kamal S, Sharma VP (2018) Status and trends in world mushroom production-II -mushroom production in Japan and China. Mushroom Res 27(1):1–25
Singh M, Kamal S, Sharma VP (2020) Status and trends in world mushroom production-III: world production of different mushroom species in 21st century. Mushroom Res 29(2):75–110
Singh M, Kamal S, Sharma VP (2021) Species and region-wise mushroom production in leading mushroom producing countries - China, Japan, USA, Canada and India. Mushroom Res 30(2):99–108. https://doi.org/10.36036/MR.30.2.2021.119394
Sobieralski K, Siwulski M, Frąszczak B, Jasińska A, Spiżewski T (2014) Comparison of fruiting body yields of mushroom varieties grown in Poland and their dry matter content. Nauka Przyr Technol 8:2, ISSN 1897-7820. http://www.npt.up-poznan.net#18
Sonnenberg Anton SM, Baars JJP, Gao W, Visser RGF (2017) Developments in breeding of Agaricus bisporus var. bisporus: progress made and technical and legal hurdles to take. Appl Microbiol Biotechnol 101:1819–1829. https://doi.org/10.1007/s00253-017-8102-2
Sonnenberg AS, Baars JJ, Mikosch TS, Schaap PJ, Van Griensven LJ (1999) Abr1, a transposon-like element in the genome of the cultivated mushroom Agaricus bisporus (Lange) Imbach. Appl Environ Microbiol 65(8):3347–3353. https://doi.org/10.1128/AEM.65.8.3347-3353.1999. PMID: 10427018; PMCID: PMC91503
Sonnenberg ASM, Baars JP, Hendrickx PM, Lavrijssen B, Gao W, Weijn A, Mes JJ (2011) Breeding and strain protection in the button mushroom Agaricus bisporus. In: Proceedings of the 7th international conference on mushroom biology and mushroom products, pp 7–15
Sonnenberg AS, Gao W, Lavrijssen B, Hendrickx P, Sedaghat-Tellgerd N, Foulongne-Oriol M, Kong WS, Schijlen EG, Baars JJ, Visser RG (2016) A detailed analysis of the recombination landscape of the button mushroom Agaricus bisporus var. bisporus. Fung Genet Biol 93:35–45. https://doi.org/10.1016/j.fgb.2016.06.001
Soylu MK, Kang M (2016) Mushroom cultivation in South Korea. Turkish J Agric Food Sci Technol 4(3):225–229
StatCan (2025) Statistics Canada. Table 32-10-0356-01 Area, production and sales of mushrooms. https://doi.org/10.25318/3210035601-eng. https://www150.statcan.gc.ca/t1/tbl1/en/cv.action?pid=3210035601
SureHarvest (2017) The mushroom sustainability story: water, energy, and climate environmental metrics. https://www.mushroomcouncil.com/wp-content/uploads/2017/12/Mushroom-Sustainability-Story-2017.pdf

Thakur Kiran B, Bhandal MS (1992) Hybridisation in *Agaricus bisporus* using non-fertile isolates. Mushroom Res 1:79–81

UNEP (2009) United Nations Environment Programme- converting waste agriculture waste into a resource: compendium of technologies. p6 https://www.fao.org/sustainable-food-value-chains/library/details/en/c/266093/

United Nations (2024) World population prospects (2024: summary of results. UN DESA/POP/2024/TR/NO. 9). United Nations, New York

USDA (2025) United States Department of Agriculture. https://usda.library.cornell.edu/concern/publications/r781wg03d?locale=en

Van Griensven LJLD, Van Roestel AJJ (2004) The cultivation of the button mushroom, Agaricus bisporus, in The Netherlands: a successful industry. Revista Mexicana de Micologia 19:95–102

Vinci G, Prencipe SA, Pucinischi L, Perrotta F, Ruggeri M (2023) Sustainability assessment of waste and wastewater recovery for edible mushroom production through an integrated nexus. A case study in Lazio. Sci Total Environ 903:166044. https://doi.org/10.1016/j.scitotenv.2023.166044

Yating D, Miao R, Feng R, Wang T, Junjie Yan X, Zhao XH, Gan Y, Lin J, Li Y, Gan B, Zhao J (2022) Edible and medicinal fungi breeding techniques, a review: current status and future prospects. Curr Res Food Sci 5:2070–2080. ISSN 2665-9271. https://doi.org/10.1016/j.crfs.2022.09.002

Yoo YB, Oh MJ, Oh YL, Pyung GS, Jang KY, Kong WS (2016) Development trend of the mushroom industry. J Mushrooms 14(4):142–154. https://doi.org/10.14480/JM.2016.14.4.142

Contribution to Sustainable Agriculture Development by Growing Button Mushroom

Sunil Kumar Prajapati, Shivani Jha, Dinesh Kumar, Sandeep Gawdiya, Sandeep Kumar, Mohammad Hasanain, Nilutpal Saikia, Sk Asraful Ali, and Rishabh Srivastava

Abstract Button mushroom (*Agaricus bisporus*) is a most popular edible fungus and the most common variety of mushroom found in cooking worldwide especially in India. It's characterized by their small, round, and smooth caps which are generally white to light brown. They measure 1–3 inches across and have short stems. It's also a low-calorie food source having high fiber, protein, vitamins (vit. B), and several minerals like selenium, potassium, and phosphorus. They also contain various antioxidants and are coupled to various health benefits such as improved heart-health and immune system support. *Agaricus bisporus* is not only a culinary staple but also plays a vital role in sustainable agriculture due to their efficient use of resources and the beneficial by-products they improved soil health. Farming of button mushrooms contributes to sustainable agricultural development in several ways through waste utilization (crop residues), soil health improvement, water conservation, lowering carbon footprint, biodiversity, economic viability, and nutritional value. Agricultural by-products from agricultural biomass and waste used as a substrate for mushroom cultivation will enhance the soil quality with spent mushroom compost, and also there will be efficient use of water resources in mushroom cultivation or farming. There will be low GHG emissions, supporting soil microorganisms and contributing to ecological balance. Button mushroom cultivation will emphasize more on the recycling of resources in food production by offering a

S. K. Prajapati (✉) · D. Kumar · S. Gawdiya · S. Kumar · N. Saikia · S. A. Ali
Division of Agronomy, ICAR-Indian Agricultural Research Institute (IARI), Pusa, New Delhi, India

S. Jha
Division of Plant Pathology, ICAR-Indian Agricultural Research Institute (IARI), Pusa, New Delhi, India

M. Hasanain
ICAR-Indian Agricultural Research Institute (IARI), Regional Station, Pusa Samastipur, Bihar, India

R. Srivastava
Division of Environmental Science, ICAR-Indian Agricultural Research Institute (IARI), Pusa, New Delhi, India

M. N. Owaid et al. (eds.), *Mushroom Production*, Interdisciplinary Biotechnological Advances, https://doi.org/10.1007/978-981-95-7793-4_7

source of high-quality protein and nutrients. Integrating mushroom farming into ecological agricultural systems and preserving natural resources by using lignin-rich plant material as a base for mushroom growth. Hence, button mushroom cultivation can promote a more environmentally friendly approach to sustainable agriculture.

Keywords Bio-conservation · Agro-ecology · Nutritional security · Biodiversity · Waste utilization

Abbreviations

EDS	Analyses using energy dispersive X-ray spectroscopy
EFBs	Empty fruit bunches
FTIR	Fourier transform infrared spectroscopy
HDPE	High-density polyethylene
HSPS	Heat shock proteins
ICAR	Indian council of agricultural research
JICA	Japan International Cooperation Agency
OPFs	Oil palm fronds
OPTs	Oil palm trunks
PABA	Para-aminobenzoic acid
PBS	Poly-butylene succinate
PHBH	3-hydroxybutyrate-co-3-hydroxyhexanoate
PLA	Polylactic acid
SDGs	Sustainable Development Goals
SMS	Spent mushroom substrate
UNEP	United Nations Environment Programme

1 Introduction

Natural resources are finite; the idea of sustainability becomes crucial to preserving the growing population, environment, and their health (Qamruzzaman 2025). A mushroom is one an example of "saprotrophic fungus." The cultivation of mushrooms is integral to their ecological role in converting solid-industrial and "agricultural-wastes" into valuable biomass (Prajapati et al. 2023; Dou et al. 2025). This process not only aids in waste management but also produces edible biomass that can serve as functional food or be used in pharmaceuticals and medications. By breaking down complex organic materials in waste, mushrooms "recycle" nutrients reduce waste volume and create "nutrient-rich food" sources (Ogwu et al. 2025a). Additionally, many mushrooms contain "bio-active" compounds with health benefits and "medicinal properties," making them a sustainable and versatile resource. Research and development into the basic biology of mushrooms as well as growing techniques are necessary if the mushroom industry is to become "ecologically" friendly (Shankar et al. 2024). This is particularly true for well-known Asian delicacies like *Lentinula edodes*, *Ganoderma lucidum*, and *Volvariella volvacea* that are gaining popularity around the world. The most widely used "edible fungus" species of mushrooms for culinary uses globally are *Agaricus bisporus*, commonly referred to as "white button mushrooms." Button mushrooms are identified by their small, round, and smooth caps which are generally white to light brown (Sommer and Vetter 2025). These mushrooms are cultivated in "controlled environments" on composted substrates. The compost is a critical component typically made from crop straw, manure and, composts and gypsum and undergoes a pasteurization process to eliminate contaminants before being inoculated with mushroom spores (Prajapati et al. 2023). Button mushrooms are "high-protein," fiber, vitamins (particularly B vitamins), and minerals (potassium, phosphorus, and selenium) and "low in calories." They also contain various "antioxidants" and are linked to health benefits like improved "heart health" and "immune system" support (Jayaraman et al. 2024). Due to their mild flavor and "meaty texture," button mushrooms are extremely versatile in cooking (Savoie et al. 1993). They can be eaten raw, salted, stir-fried, grilled, braised, or roasted and are a "common ingredient" in dishes like soups, salads, pizzas, pasta, and curry. Fresh button mushrooms should be stored in the refrigerator, where they can last for up to a week. They should be kept in a paper bag to absorb excess moisture and prevent them from becoming slimy. Button mushroom is not only a culinary staple but also play a significant role in "sustainable agriculture" due to their efficient use of resources and the beneficial by-products they improve soil health (Ogwu et al. 2025a; Dou et al. 2025).

For growing of button mushroom, the ideal temperature varies throughout their growing cycle. The "optimal temperature" range during the mycelium growth phase is 20–27.8 °C (68–82 °F). The temperature should be maintained once the mycelium has fully colonized the substrate at 11.7–17.8 °C (53–64 °F) to encourage the mushrooms to fruit (Eastwood et al. 2013). It's cultivated on compost made from "agricultural waste products" which promotes "recycling and reduces waste" in the environment. The discarded mushroom compost can be applied as an "organic fertilizer" to enhance soil fertility and structure after harvest. Mushroom cultivation requires less water compared to traditional agriculture, thus conserving this vital resources for future. The production of button mushrooms has a "lower carbon footprint" than many other crops, as it does not involve the tilling of soil or the use of heavy machinery (Iglesias et al. 2025; Schilla et al. 2025). The mushroom cultivation process can support a diverse range of microorganisms, which can "enhance soil biodiversity." It offers a profitable alternative for farmers, especially those with limited land, as mushrooms can be grown in smaller spaces with higher yields (Beelman et al. 2003; Prajapati et al. 2023). High-quality protein, vitamins, and minerals can be found in button mushrooms, which improve nutrition and food security. Broadly, mushrooms can be categorized into three groups: "symbionts," which live in mutualistic relationships with plants; "saprophytes," which decompose dead matter to obtain nutrients; and "parasites," which derive their nourishment from living plants and insects (Liu et al. 2024; Waqar et al. 2024). Saprophytes are very essential as decomposers in "ecosystems." The only organisms on Earth that can fully degrade wood components including cellulose, hemicellulose, and lignin are white rot fungus, which include numerous "edible fungi" (Miles and Chang 2004).

In the past, people gathered mushrooms by foraging in the wild. However, the significance of mushrooms as a "food source" has been amplified by the development of several cultivation methods. These methods include using wooden logs, bags, bottles and beds, all of which leverage mushrooms natural ability to decompose organic matter in ecosystems (Dhar 2017). Currently, agricultural "by-products" including crop residue, manure, and compost are used in mushroom production, along with forestry by-products such "hardwood logs" and "sawdust" (Grimm and Wösten 2018). Furthermore, since mushroom growing usually doesn't include the use of chemical-fertilizers or pesticides, the environmental-impact of the practice is minimal, making it a prime example of circular agriculture. Between 1990 and 2020, global mushroom production rose significantly, from 13.8 to 42.8 million tons (FAOSTAT 2022).

According to Royse et al. (2017), shiitake mushrooms (*Lentinula edodes*) are thought to produce the most of the primary mushrooms, followed by oyster mushrooms (*Pleurotus* spp.), wood-ear mushrooms (*Auricularia* spp.), and button mushrooms (*Agaricus bisporus*). China led the world in mushroom production (93%) and was followed by Japan and Poland. After reaching a peak in 2012, and the total mushroom production in countries outside China has significantly decreased and subsequently "stagnated." This decline has occurred despite the continuous growth of the "global population" and the apparent increase in "mushroom production" over the previous 30 years (FAOSTAT 2022). Japan, the second largest producer of mushrooms in the world, has also experienced this stagnation. There has been a 10-year lag in production, reflecting a broader trend seen in countries outside of China (MAFF 2021). These tendencies constitute a threat to the existing mushroom crop, and the author suggests that in view of the mushroom's potential to both meet the world's demand and contribute to "environmental sustainability," there is a need to "globally support the industry's" continued existence (Jayaraman et al. 2024). Furthermore, the author will elucidate the frequently overlooked yet noteworthy "environmental stresses," including the deleterious consequences of spores and "waste handling" that arise in the backdrop of mushroom cultivation. Acknowledging the role that mushroom production plays in "sustainability" and addressing these goals and challenges will help guarantee that mushroom cultivation continues in the future (Khan et al. 2024).

2 Contribution of Button Mushroom to Sustainable Agriculture

Sustainable agriculture aims to meet the increasing "global food-demand" and "nutrition" while minimizing "negative impacts" on the environment and promoting social and "economic equity." One way to contribute to sustainable agriculture development is through the cultivation of button mushrooms (Ogwu et al. 2025b; Iglesias et al. 2025). Button mushrooms are an excellent choice for "sustainable agriculture development" for several reasons. Firstly, they can be grown on a variety of organic substrates, making them a versatile option for farmers looking to "minimize environmental impact" (Chang and Wasser 2017). Additionally, button mushrooms have a relatively low "carbon footprint" compared to other crops, as they require minimal space and can be grown vertically in stacked layers, maximizing "land use efficiency" (Qamruzzaman 2025). Furthermore, the cultivation of button mushrooms can contribute to the recycling of agricultural by-products and organic waste, turning them into valuable resources for mushroom production. This not only helps reduce the "environmental burden" of waste disposal but also adds "economic value" to materials that would otherwise be discarded. In addition to their environmental benefits, button mushrooms are "nutritious and high value crop," providing a source of income for farmers. Their high-protein content and rich nutrient profile make them a valuable addition to diets, contributing to "food security and nutrition" (Fig. 1). Overall, the cultivation of button mushrooms presents a promising avenue for sustainable agriculture development, offering a range of environmental, social, and economic benefits. By embracing this cultivation method, farmers can play a significant role in advancing sustainable agriculture practices and contributing to "global food security" (Beelman et al. 2003; El-Ramady et al. 2022; Jayaraman et al. 2024). Button mushroom cultivation also has the potential to improve "soil health and fertility." The growth of button mushrooms involves the decomposition of "organic matter," which releases nutrients back into the soil. This process enhances "soil structure" and promotes "microbial activity," ultimately leading to increased soil fertility. As a result, farmers who integrate button mushroom cultivation into their agricultural practices can benefit from improved soil quality and reduced dependency on "synthetic fertilizers," further contributing to the "sustainability" of their farming operations (Jasinska 2023; Dou et al. 2025; Jayaraman et al. 2024).

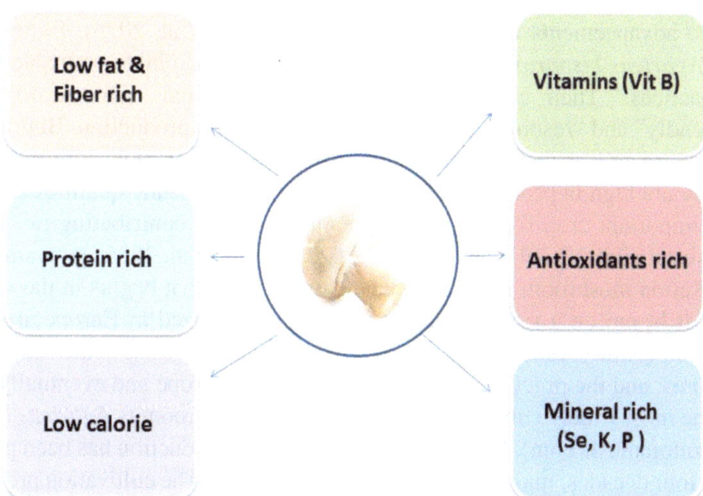

Fig. 1 Health benefits of button mushroom

Moreover, the cultivation of button mushrooms supports "biodiversity" on farms. The "agro-ecosystem" created by integrating mushroom cultivation provides habitats for beneficial microorganisms and insects, contributing to a more balanced and "resilient agro-ecosystem" (Eastwood et al. 2013; Dou et al. 2025). This "ecological diversity" can enhance the overall stability of the farm and reduce the reliance on chemical inputs for pest control, aligning with the principles of "sustainable agriculture." In addition to these "environmental benefits," the cultivation of button mushrooms can also create "employment opportunities" in rural areas, empowering local communities and contributing to inclusive "economic development" (Grimm and Wösten 2018; Schilla et al. 2025). Furthermore, the potential for value addition through the processing and marketing of mushroom-based products can further enhance the economic viability of mushroom cultivation as a sustainable agricultural practice. Hence, the multifaceted benefits of button mushroom cultivation extend beyond "environmental sustainability" to encompass improvements in soil health, "biodiversity conservation," and "socio-economic development." Embracing the cultivation of button mushrooms offers a "holistic approach" to sustainable agriculture, making it a valuable contributor to the pursuit of "global food security" and "resilient agricultural systems." By growing button mushrooms, farmers can contribute to sustainable agriculture development in several ways (Grimm and Wösten 2018; El-Ramady et al. 2022; Schilla et al. 2025; Jayaraman et al. 2024).

3 The Rise of Button Mushroom in Sustainable Farming

The rise of button mushroom cultivation in "sustainable farming" is a significant development in "modern agriculture." Although they were traditionally harvested from the wild mushrooms are now an important part of sustainable agriculture thanks to advancements in growing methods (Schilla et al. 2025). Button mushrooms (*Agaricus bisporus*) have emerged as a cornerstone in "sustainable agricultural practices." Their cultivation represents a paradigm shift toward a more "eco-friendly" and "resource-efficient approach" to food production. Button mushrooms are prized for their "nutritional benefits" as well as their culinary purpose since they are high in proteins, vitamins, and have "therapeutic qualities." They are also an important crop for recycling agricultural waste, contributing to "environmental sustainability" (El-Ramady et al. 2022; Schilla et al. 2025; Jayaraman et al. 2024). Button mushroom cultivation has a long history that begins in the sixteenth century. It began on a small scale but was commercialized in Europe around the seventeenth century. France was one of the first countries to start cultivating button mushrooms, and the practice spread to other parts of Europe and eventually worldwide. The most widely cultivated and eaten type of mushroom is *Agaricus bisporus* (white button mushroom). In India, button mushroom production has been prevalent for over four decades, mainly for the "domestic market." The cultivation process has evolved, with advancements in technology and methods, making it possible to produce mushrooms throughout the year (Sommer and Vetter 2025).

Because button mushroom production "recycles forestry" and "agricultural by-products," it promotes "environmental sustainability" (Grimm and Wösten 2018). This process turns potential waste into valuable food products and reduces the "environmental footprint" of agriculture. The cultivation of button mushrooms is "economically viable" and can be a profitable venture for farmers. It provides an additional income source and can be integrated into existing "farming systems" without the need for extensive additional resources. While the cultivation of button mushrooms has many positive aspects, it also faces challenges such as the disposal of spent mushroom substrates and the environmental impact of spores from cultivated mushrooms (Schilla et al. 2025). Addressing these challenges is essential for the "sustainable future" of mushroom production. Cooperation and education are required to guarantee the growth and continuation of "mushroom farming." These efforts can help spread knowledge about the "ecological," nutritional and cultivation aspects of mushrooms, thereby enhancing their perceived value and production (Sommer and Vetter 2025). Button mushrooms have thus become a symbol of sustainable farming, showcasing how "innovative agricultural practices" can lead to a more sustainable and "environmentally friendly food production system." Growing button mushrooms contributes significantly to "sustainable agricultural development" in several ways. These are the multifaceted role of button mushroom cultivation in promoting "sustainability" in agriculture (Grimm and Wösten 2018; El-Ramady et al. 2022). Button mushrooms are essential (Fig. 2) to the development of sustainable agricultural methods since they may be used to recycle waste, improve soil health, save the environment, and provide food (Schilla et al. 2025; Jayaraman et al. 2024).

- *Waste bioconversion*: Cultivating button mushrooms is a practical way to harvest materials that are left over as recyclable agricultural waste. Agricultural by-products like straw and animal dung are bio converted into the protein-rich biomass that is the mushroom through this process.
- *Soil fertility*: Spent mushroom compost is a by-product of mushroom growing that can be used as a fertilizer for soil. This improves the overall health of agricultural land by increasing soil fertility and functioning as a soil conditioner.

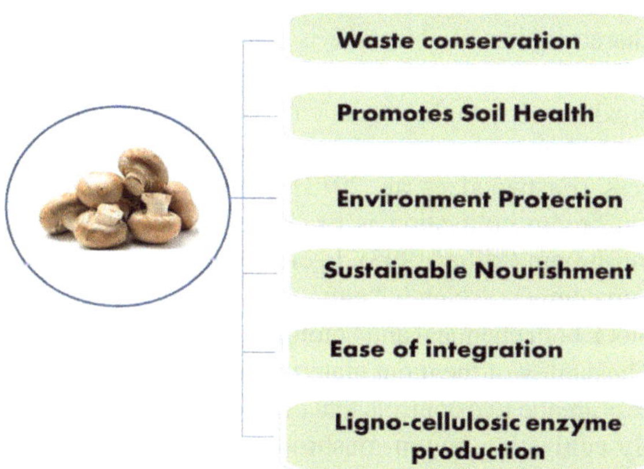

Fig. 2 Mushroom cultivation offers numerous advantages

- *Environmental protection*: Mushroom cultivation aligns with environmental protection strategies. It reduces environmental pollution by processing agro-waste, which would otherwise contribute to landfills or be burned, releasing pollutants into the atmosphere.
- *Lignocellulosic enzyme production*: It produces enzymes that break down lignocellulosic substrates and use the resulting compounds to create their fruiting bodies. This biotechnology method helps to lower and increase the value of agricultural and industrial waste.
- *Sustainable nourishment*: Growing mushrooms is a sustainable way to feed your family. It is now a lucrative area of agriculture, particularly in high-resource environments where farmers may maximize yields and quality while minimizing costs and input resources thanks to technological improvements.

4 Agro-Waste Utilization and Value Addition in Mushroom Production

Both edible and medicinal mushrooms are cultivated on "lignocellulosic residues," and recycling these residues using such fungal species offers significant "economic benefits." At the very least, the wastes can be seen as recyclable materials with favorable effects on the environment and new business prospects (Zotti et al. 2025). *A. bisporus* is a white rot basidiomycete that degrades leaf litter, unlike other mushrooms cultivation like *Lentinula edodes* and *Pleurotus* spp., which function as secondary decomposers. Simple sugars, oligosaccharides, and other "low-molecular-weight" substances are used by a wide population of fungus and bacteria that act as generalist initial colonizers to start the breakdown of forest litter (El-Ramady et al. 2022).

Following this first burst of "microbial activity," more resistant plant polymers like lignocellulose-complexes and humic compounds are created during the first-stage of decomposition break down, providing a pathway for the development of specialized secondary colonizers which are less competitive than "microfungi" in obtaining labile-resources to emerge (Frankland 1992). Because of these ecological characteristics, *A. bisporus* is typically cultivated by inoculating pre-decomposed substrates that are produced through composting. By using microbial populations to break down complex organic matter under conditions of high warmth, high humidity, and oxygenation, composting is a human activity intended to speed up this process (Eastwood et al. 2013). The overall objective of composting for button mushroom cultivation is to create a specialized nutritional media for growth of mushroom (Miller 1993). In order to convert raw agricultural "biomass and wastes" into compost which is "more resistant" to degradation and home to a stable microbiota, both chemical and "biological elements" must be handled (Savoie et al. 1993). The quality of the input materials and "physic-chemical" conditions during the process are the two controls that can be turned on to achieve the composting objectives. By cultivating button mushroom with "sustainable-development" as a goal, it is possible to increase the value of "agro-wastes" and employ locally available resources in energy- and "environmentally friendly" composting processes (El-Ramady et al. 2022; Schilla et al. 2025; Jayaraman et al. 2024).

5 Agricultural and Environmental Impact

Environmental challenges arising from waste disposal and "resource scarcity" have prompted the adoption of sustainable-solutions and the shift to a "circular economy" that prioritizes "waste minimization," recycling, and valorization. Because large amounts of agro-industrial and "agro-forestry leftovers" are rich in organic compounds, growing mushrooms is a "biotechnological technique" that transforms and efficiently uses those wastes that should be recovered. Numerous substrate types have already been studied for their capacity to grow mushrooms; nevertheless, new crop varieties and "cultivation techniques" have been prompted by "climate change," which has resulted in the creation of new potential mushroom-growing media (El-Ramady et al. 2022; Jayaraman et al. 2024; Schilla et al. 2025).

Waste products from mushroom production, such as "wastewater and wasted mushroom" substrate (SMS), must also be included in the loop and used again in order for sustainability to be achieved. Furthermore, as mushrooms are well known for their capacity for "bioaccumulation," culture taking place on "recycled substrates" needs to be closely watched, in addition to sustainable "disease control" (Jasinska 2023). One of the best examples of "sustainable food production" is the button mushroom gardening industry. It reduces pollution in the environment by using selective compost made from recovered "agricultural waste" items. This procedure improves the "fertility and health" of the soil by managing agricultural waste and producing biomass that is high in protein. In general, the cultivation of mushrooms is seen to be "carbon-positive," generating less "greenhouse gas emissions" than other industries including agriculture, fruit production, and the raising of animals for meat. A kilogram of mushrooms is thought to require 9 kg of freshwater to produce, of which part is squandered and part is utilized by the mushroom (Jasinska 2023; Jones et al. 2021; Jayaraman et al. 2024).

There are numerous ways to introduce new substrates into commercial mushroom cultivation. Some of these include trying to use the "problematic waste" from nearby energy production facilities, expanding the production and cultivation of a new crop in the area, or using insufficient amounts of substrate that is typically used. The processing of hazelnuts, for which Italy stands as one of the world's largest-producers, yields substantial waste, particularly "woody pericarps" resulting from the cracking process. These pericarps, typically burned for domestic heating, contribute to air pollution. A potential solution lies in investigating the impact of cultivating various mushroom species on the "high-lignin content" found within these pericarps. This includes edible varieties such as *Agaricus bitorquis*, *Pleurotus djamor*, *Pleurotus ostreatus*, *Lentinula edodes*, *Tuber aestivum*, and *Tuber magnatum*, alongside medicinal mushrooms like *Grifola frondosa*, *Ganoderma lucidum*,

Calvatia gigantean, *Cordyceps sinensis*, *Polyporus umbellatus*, and *Agaricus blazei* (Puliga et al. 2022; Prajapati et al. 2023). Three distinct "hazelnut-shell-based" substrates were used to grow and cultivate *Ganoderma lucidum*, *Lentinula edodes*, and *Pleurotus cornucopiae*: HS (hazelnut shell) and WS (wheat straw) and WSBC (wheat straw mixed with beech chips) as a control. While all these substrates demonstrated effective degradation of the lignocellulosic fraction within the HS, a notable difference emerged in fruiting body production (FBP). Specifically, the presence of HS led to a distinct increase in FBP when compared to the control, even though the "biological efficiency" remained identical across the substrates. This outcome presents promising avenues for both fostering more "sustainable management" practices and enhancing the overall efficiency of mushroom farming (Puliga et al. 2022).

In Norway, anaerobic-digestion of "dairy manure" and "food waste" produces a significant quantity of energy, but water plants account for the majority of the country's energy production (Jasinska et al. 2022; Jasinska 2023). AD produces a "digested effluent." This product, which is rich in nitrogen (N) and organic matter (OM), has been researched as a potential addition to growing substrates for mushroom development. It can be used in combination with barley, oats, or the straw of Norwegian grains. All tested varieties of experimental mushroom compost (EMC) performed admirably during the "composting" process, successfully attaining the required moisture levels of 65–75%, N contents of 1.43–1.93%, and C:N ratio ranging from 21.5 to 29.1 (Jasinska 2023) This process enabled the mycelium to flourish and ultimately produce *Agaricus subrufescens* mushrooms. Although the addition of oat resulted in the lowest productivity and biological efficiency, supplementing with barley straw actually improved the experimental mushroom compost (EMC) structure, leading to both the highest yield and biological efficiency (Jasinska et al. 2022). The yield at the "mid-cycle" of crop growth, or precociousness, was higher for oat substrates (68.9%), although barley EMC had lower earliness (the number of days till harvest from casing). The aforementioned results provide credence to the notion that original digestate can be "utilized straightaway" in mushroom production. A mushroom farm could achieve sustainable asset use if it is located next to an "anaerobic plant digestion" (O'Brien et al. 2019; Jasinska et al. 2022; Jasinska 2023).

Using the prospective new substrate is one technique to address the situation when local mushroom production is prohibited by the restricted availability of the standard cultivation substrate. Large amounts of organic by-products are produced by Malaysia palm oil industry, raising "environmental issues." The potential use of OPFs (oil palm fronds), OPTs (oil-palm trunks), and EFBs (empty fruit bunches) as growing substrates for *Pleurotus ostreatus* was investigated by Aubrey et al. (2022). In general, agronomic performance was higher when 100% EFB was used; the growth of mushroom was 1.9 times faster, and yield was comparable when compared to the control. Mushrooms cultivated on oil palm utilizing substrates treated with the product had a considerably higher crude protein and "beta-glucan content" than mushrooms produced on the control. Furthermore, there was a positive correlation observed between the potassium levels in the substrate and the quantity of fruiting bodies, as well as the crude protein and beta-glucan content of the mushrooms.

As a result, 100% EFB may be utilized in place of RWS in *P. ostreatus* culture and production. There is an immediate need to make use of this waste since SMS (spent mushroom substrate) or WMC (wasted mushroom compost) is created in enormous quantities on average; 5 kg of SMS/WMC are produced for every kg of mushrooms (Zakaria et al. 2022; Jasinska 2023; Martín et al. 2023).

Despite numerous studies highlighting the benefits of SMS (pent mushroom substrate) or SMC (spent mushroom compost) as a viable substitute for plant growth, animal feed, or soil amendment, a significant amount of this valuable by-product still isn't being sustainably reused. This particular technical communication (TC) proposes four innovative methods for recycling SMS/SMC. One key area of examination is the viability of integrating SMSs as a component within the "growing media" for *Pleurotus ostreatus* culture (Zakaria et al. 2022). In an investigation exploring potential supplements for wheat straw substrate, SMSs from *Pholiota nameko* (N-SMS), *Hypsizygus marmoreus* (M-SMS), and *Hericium erinaceus* (E-SMS) were tested at varying supplementation rates by 10%, 20%, and 30%. Among the substrates examined, notable variations emerged in the *P. ostreatus* yield, "protein content," dry matter, and "biological efficiency." The highest yields were observed with 20% E-SMS (254.33 g), 20% N-SMS (253.43 g), and 10% E-SMS (251.67 g). A similar trend to the yield was reflected in the biological efficiency, which ranged from 66.48% (30% M-SMS) to an impressive 72.67% (20% E-SMS). Regarding nutritional content, 30% M-SMS demonstrated the highest protein level, measuring 29.93 g/100 g dry weight. Furthermore, the maximum dry matter for *P. ostreatus* was found in 30% M-SMS (23.74 g) and 10% M-SMS (23.06 g) (Zakaria et al. 2022; Jasinska 2023; Martín et al. 2023).

Past research indicates that spent mushroom substrates (SMSs) could serve as a low cost, "sustainable," and renewable component (at 10 to 30% inclusion rates) in the "conventional growing media" for *Pleurotus ostreatus* culture. Expanding on this, a team led by Kumar et al. (2022) proposed an alternative strategy: investigating the biotransformation of SMSs into biogas after shiitake mushroom cultivation, followed by using the resulting slurry digestate to fertilize tomato (*Solanum lycopersicum* L.) crops. The "biogas" production findings were significant. The 50% SMS treatment yielded the largest reduction in the "physic chemical and proximate" characteristics of the slurry, alongside the highest biogas (8834 mL or 11.93 mL/g of organic carbon) and methane content (61%) among all treatments. Furthermore, when this biogas digestate from the 50% treatment was incorporated at a 10% loading rate, it significantly ($p < 0.05$) enhanced several tomato plant parameters. These included the seed germination rate (93.25%), plant height (53.10 cm), seedling length (9.2 cm), seedling root length (4.19 cm) and total yield (1.86 kg/plant). The nutritional quality of the tomatoes also improved, with higher contents of flavonoid (5.06 mg/g), phenolics (2.78 mg/g), and tannin (3.40 mg/g) (Beelman et al. 2003; Miles and Chang 2004; El-Ramady et al. 2022; Prajapati et al. 2023). The results of this research point to the possibility of reducing the "carbon footprint" of the mushroom production industry by means of the "sustainable up cycling" of SMS motivated by a circular economy approach and achieved through the mutually beneficial bioenergy production and secondary fruit crops (Jasinska 2023).

The production of SMS/SMC will increase due to the unavoidable expansion of the mushroom business, creating an agricultural waste that needs to be properly managed instead of being dumped or burned (Jasinska 2023; Martín et al. 2023). The newest approach to dealing with waste problems is pyrolysis, which yields the useful biochar, which not only absorbs carbon but, in the case of high-quality material, has a variety of applications. Additionally, Chen et al. (2003) demonstrated that SMC/SMS from the cultivation of black fungus (BF) and shiitake fungus (SF) could be transformed into highly porous biochar. Their experiments revealed a significantly increase in biochar yield as the pyrolysis temperature climbed from 400 to 600 °C. The biochar products generated at 800 °C, specifically SMC-SF-BC-800 and SMC-BF-BC-800, exhibited maximal BET (Brunauer-Emmett-Teller) surface areas of 312.5 and 280.9 m^2/g, respectively. Furthermore, analyses using EDS and FTIR confirmed the presence of a substantial number of oxygen-containing functional groups on the surface of these resulting "biochar products." Typically, high-quality biochar like this is most commonly applied for improving "soil quality." According to Deng et al. (2022), biochar made from SMS can be applied as a soil supplement and offer a way to get rid of SMS. They discovered that the addition of biochar made from SMS to the soils of "moso-bamboo" forests reduced the amount of N_2O emissions in the soil.

In their study, Altaf et al. (2022) looked into the use of botanicals and "biocontrol agents" in temperate climates to manage white button mushroom green mold disease (*Trichoderma harzianum*) sustainably. According to their research, adding bacterial and "botanical antagonists" to the white button mushroom pathogen-infested shell lowers green mold infection and increases yield. The least inhibiting botanicals (4.66–7.4%) were *J. regia*, *Curcuma longa*, and *Azadirachta mellea* when tested against the mycelial growth of *A. bisporus*. *J. regia* and *C. longa* had the highest average weight (11.8–11.9 g) of a single fruit body and the combined button yield (11.3–11.9 kg/quintal compost) based on the in vivo evaluation of plant botanicals at 2% concentration. *Pseudomonas fluorescence*, *Azotobacter* sp., and *Bacillus subtilis* were among the "bacterial bioagents" tested in vitro; they showed variable degrees of "stimulatory effects" on the growth of mycelial of *A. bisporus*, but "antagonistic effects" on *T. harzianum*. *P. fluorescens*: 104, and *B. subtilis*: 38. Azotobacter-108 inhibited *T. harzianum's* mycelial growth at the highest rates 97.6%, 97.4%, and 90.3%, respectively.

6 Sustainable Mushroom Production: Opportunities and Benefits

The cultivation of mushroom stands apart from other forms of production crop and nutrient, notably for its ability to utilize low "economic value trees," saw-dust, thinning, and branches all value added and by-products of sustainable forest management as cultivation materials. This approach not only curbs environmental degradation and "health-hazards" linked to incinerations but also stimulates "economic growth" (Chang and Wasser 2017). Consider Japan, for example, forests blanket nearly 70% of the country's land (MAFF 2020), and mushrooms account for a significant 40% of the forestry industry's production value. Beyond timber, mushrooms (edible fungi) contribute over 80% of the value derived from non-wood forest products, a category that includes edible nuts, wild plants, Japanese lacquer, bamboo, charcoal, and firewood. Effective forest management, when focused on mushroom production, offers multiple ecological benefits. It can enhance water storage and "mitigate landslides" by stabilizing tree roots within the soil and "diminishing sediment runoff" (Forest Agency 2019). Furthermore, the improved capacity of soil to retain more water and organic matter bolsters the "ecological and hydrological functions" of diverse natural environments, supplying crucial nutrients to organisms in soil, water, and even marine ecosystems. Therefore, forest management centered on the production of mushroom functions as both an "economic source" and a vital factor in preserving the "health of soil" and "aquatic ecosystems" intrinsically linked to forests. This highlights that both "agricultural and forest by-products" can serve as valuable resources for cultivation of mushroom.

Numerous agricultural activities (farming and operations) across diverse regions generate a wealth of "agricultural by-products," including cottonseed hulls, corncob meal, straw-stover, sugarcane bagasse, rice bran, and wheat bran (Grimm and Wösten 2018). In low-income regions particularly, where "intensive agriculture" is common, these by-products are often incinerated, leading to environmental harm. However, a growing body of research indicates that utilizing these "low-value agricultural and forestry by-products" for mushroom production, rather than burning them, can be a profitable venture (Pavlík and Halaj 2019; Odediran et al. 2020; NanjeGowda and Chennappa 2021). Furthermore, unlike "unsustainable agriculture" which often involves pesticide use and excessively intensive crop production, causing soil pollution, and land-degradation, mushroom cultivation avoids pesticides and chemical fertilizers, thereby exerting a minimal "environmental" impact (Hu et al. 2021). Mushroom cultivation, with its numerous environmental and economic advantages, aligns well with several United Nations Sustainable Development Goals (UN-SDGs), including "11: Sustainable cities and communities," "12: Responsible consumption and production," "13: Climate action," "14: Life below water," and "15: Life on land." It truly serves as a model of "sustainable circular agriculture." Despite these clear benefits, global mushroom production is either declining or stagnating, with China being a notable exception. These trends, however, can be significantly improved by one crucial factor is "education." In developing countries (India) aiming to expand their mushroom production, it's essential to not only accumulate fundamental knowledge and experience but also to establish and maintain "industrial scale" sterilization and cultivation facilities. A successful example of this can be seen in the Japan International Cooperation Agency (JICA) Partnership Program (2016–2019), "Livelihood Improvement of Mushroom Farmers in Western Bhutan," which effectively enhanced "local mushroom production" in Bhutan.

International efforts have also significantly bolstered mushroom cultivation in various regions. For instance, the NUFFIC (Netherlands-University Foundations for International Cooperation) have provided support in Tanzania (Mshandete and Cuff 2008), while the KOICA (Korean International Cooperation Agency) has done so for Sri Lanka. Similarly, the China the CIDCA (International Development Cooperation Agency) has extended comparable assistance to Papua, New Guinea, and Central African Republic (CIDCA 2021). These initiatives have equipped "local communities" with mushroom related-facilities and crucial technical assistance, thereby "strengthening rural economies." Looking ahead, projected declines in China's population after 2040 are anticipated to reduce "domestic food demand," including that for mushrooms (Zhao et al. 2021). This demographic shift will lead to an increase in mushroom exports from China. While it remains uncertain whether China advanced mushroom production technology will be shared with other countries, such technology transfer could profoundly impact the future of "global mushroom production." Given the urgent global demand for food production, international support for mushroom cultivation must expand beyond current partnership programs and not merely await the changes projected for 2040. Currently, training programs largely rely on one-on-one collaborations between public organizations in different countries, a model that may not be efficient enough for expanding mushroom production on a global scale. New efforts and strategies are therefore imperative to enhance the effectiveness and reach of these international support initiatives. Establishing an international forum that incorporates not only "public organizations" but also private companies could foster the creation of higher profitable mushroom production industries. Furthermore, implementing sustainability certification programs can further elevate mushroom cultivation's contribution to sustainability, ensuring that profits are channeled back to the forestry and agriculture stakeholders who supply the raw materials for this valuable industry.

When examining recent production trends, the nutritional value of mushrooms warrants significant attention. Though mushrooms contain lower amounts of three primary macronutrients, proteins, fats, and carbohydrates (excluding dietary fiber), compared to staple crops, they nonetheless represent a valuable food source. For example, while considered relatively high in protein, their protein content, at 3.1 g per 100 g for shiitake, 3.3 g for oyster, and 2.9 g for button mushrooms (MEXT 2015), is less than that found in wheat (10.8 g/100 g), rice (6.8 g/100 g), or corn (8.6 g/100 g). However, mushrooms are notably rich in dietary fiber, which provides numerous "health benefits," and they are also a source of important vitamins, particularly vitamin D (El-Ramady et al. 2022). Consequently, mushrooms serve as excellent ingredients for healthy food and dietary supplements, especially beneficial for individuals with "specific dietary needs" or preferences rooted in religious or philosophical reasons. Despite these advantages, in "developing countries" where the prioritization leans toward food stuffs abundant in the three major nutrients, especially protein, carbohydrates, and minerals (Fe and Zn), the perceived urgency to expand mushroom production might be lower compared to increasing other staple crops. Yet, an exclusive reliance on high-carbohydrate diets could potentially lead to future societal issues like obesity. To mitigate this, the international community should proactively promote the nutritional benefits of mushrooms, highlighting their high dietary fiber content, low-calorie profile, and overall health-promoting properties. Interestingly, the prolonged COVID-19 pandemic might have unintentionally aided dietary education, raising health awareness and boosting consumer interest in immunity-boosting dietary choices, including mushrooms.

7 Challenges and Constraints in Mushroom Production

There are a few, but substantial, "environmental concerns" facing the mushroom industry that are typically disregarded when creating development objectives. The most ancient issue is the damage that mature mushrooms massive spore production causes to the surrounding ecosystem. According to some research, spores from mushrooms that grow naturally outside can create droplets that encourage precipitation or rainfall (Hassett et al. 2015). Modern indoor mushroom cultivation, however, presents several challenges stemming from spore dispersal. These include potential health risks for mushroom growers (Nishida and Yatera 2022), contamination of cultivation sites (Okuda et al. 2013), and even a reduction in genetic diversity within wild populations (Hibbett and Donoghue 1996; Obatake et al. 2003). For certain mushroom species such as button mushrooms, this issue can be managed by harvesting them before the "spore-dispersing pileus" (cap) and hymenium (lamella) fully expand. Harvesting can be timed by altering the cultivation's temperature and humidity levels. However, due to their morphology, several mushroom species, including oyster and shiitake mushrooms, are challenging to manage. It has been observed, for instance, that a significant portion of mushroom cultivators—ranging from 3.5% to 29.0%—who inhale spores from cultivars experience severe respiratory illnesses, including asthma and hypersensitivity pneumonitis (Nishida and Yatera 2022). While spore induced declines in the "quality" of working life (QWL) was first documented as early as the 1970s, they continue to pose a threat to QWL and remain an issue that is not yet fully resolved (Schulz et al. 1974; Nishida and Yatera 2022).

An additional issue includes genetic diversity loss in the natural population as a result of cultivar spores dispersing outside of production facilities and contamination of cultivation facilities (Obatake et al. 2003; Okuda et al. 2013). One promising solution to this issue lies in utilizing cultivars that possess sporulation deficient (spore less) mutations. To the author's knowledge, only four such variations have been developed thus far: one for the oyster mushroom (*Pleurotus ostreatus*) in the Netherlands and three others in Japan for *Agrocybe cylindracea*, *Pholiota adiposa*, and the king trumpet mushroom (*Pleurotus eryngii*). Research has already elucidated that the "MSH4 homolog," a gene involved in meiosis, is responsible for this spore less trait across various species, making it a viable target for molecular breeding (Okuda et al. 2013; Lavrijssen et al. 2020). It is anticipated that introducing sporulation deficiency via genome-editing techniques will enhance both natural and artificial environments for key mushroom species, such as shiitake and oyster mushrooms. However, another pressing global concern in mushroom production is managing postproduction waste from bag cultivation, which is currently the most prevalent method. This waste, known as SMS, is contained within heat-resistant bags, referred to as used culture bags. These bags, typically made of nonbiodegradable polypropylene or high-density polyethylene (HDPE), are disposable and contribute significantly to environmental contamination associated with the mushroom industry. Based on the author's observations, approximately 15 g of HDPE bags are wasted for every 900 g of shiitake mushroom produced. According to previous research (Royse et al. 2017), if the world produces ten million tons of shiitake mushrooms, then more than 160,000 tons of garbage is created every year.

Likewise, 18 g of trash are produced for every 450 g of oyster mushrooms produced (Panjikkaran and Mathew 2012). Because of this, the yearly cultivation of the main mushroom species uses over a million tons of polypropylene and HDPE bags, which has a detrimental environmental effect. While the amount of waste generated by mushroom industry may appear to be less than that of other industries, it is roughly equal to 1000 times weight of the 1 billion "plastic straws" that "Starbucks" used annually prior to the company decision to phase them out in 2020. Bags made from poly-lactic acid (PLA), a biodegradable plastic, have already been developed as a substitute for polypropylene and HDPE. However, PLA is not yet widely adopted, primarily because it is more expensive than its conventional counterparts. Additionally, its decomposition is slow, occurring effectively only in hot and humid environments, such as those found in industrial composting facilities (Poulin et al. 2021). Since bag cultivation preparation involves sterilizing the bags at high temperatures as 120 °C, heat resistance is a crucial component of the bags. Other biodegradable plastics that show promise include polybutylene succinate (PBS) and poly-3-hydroxybutyrate-co-3-hydroxyhexanoate (PHBH), however, the former is not very heat resistant. Therefore, the future focus in creating biodegradable heat-resistant materials will be on reducing costs and improving the biodegradability of materials like PLA and PHBH.

Each year, over 100 million tons of SMS are generated, contributing "negatively to the environment." To mitigate this waste produced by the mushroom industry, considerable research has focused on various management methods, including composting, conversion into animal feed or fuel, and even repurposing SMS as a substrate for subsequent mushroom production (Phan and Sabaratnam 2012; Grimm and Wösten 2018). When processing spent mushroom substrate (SMS), composting and burning are the most straightforward methods, as they don't require specialized techniques. Composting, while promising, faces limitations regarding the time needed and the volume that can be effectively processed. Conversely, burning is environmentally detrimental due to its high ash content and generates additional "environmental problems" (Phan and Sabaratnam 2012; Grimm and Wösten 2018). Mushroom cultivation actually occurs in numerous locations, each possessing a wide array of resources. Consequently, processing the spent mushroom substrate demands a careful consideration of local circumstances, including variations in the chemical composition of SMS components, the specific agricultural environment, and the cost associated with constructing industrial-scale composting systems.

Given this diversity of variables, one-dimensional research findings will not offer a universal solution. It is also important to note that the development of effective waste management procedures for handling both "heat resistant bags" and SMS will likely remain a significant challenge for the eventual disposal of all biodegradable plastics.

Even though a wide range of "biodegradable polymers" are being created, post-consumption treatment techniques are usually overlooked. As with bags used in mushroom cultivation, the best way to dispose of PLA and other biodegradable plastics right now is to compost them in an environment with high temperatures and humidity. Large-scale composting facilities must be built and run in order to handle the massive volume of biodegradable plastics, which will probably be burned like regular garbage. In light of the potential for future regulations pertaining to nonbiodegradable plastics (UNEP 2022), it is evident that incineration will not be a recognized internationally acceptable method of disposal. Therefore, developing more effective and efficient composting systems will be crucial for resolving significant "environmental issues" in the future. These systems will need to be capable of treating waste from mushroom production, as well as nondurable consumer products made from biodegradable plastics.

The impressive strength of fungal mycelium, particularly which derived from mushrooms, has recently garnered notable attention. This has led to the development of innovative technologies that utilize mycelium in various applications, including packaging, building materials, paper, textiles, and as a substitute for leather (Jones et al. 2021). Utilizing these resources improves animal welfare while lowering our reliance on petroleum-based products. Moreover, these materials are "biodegradable and renewable," and their use encourages cycling. Businesses engaged in the creation of leather substitutes have partnered with clothing firms like Adidas and Kering, while companies creating packaging materials have partnered with home furnishings companies like IKEA. The two uses of materials obtained from fungi have already been implemented in real life. These aren't always SMS based, but in the near future, SMS will probably be used for them as a less-expensive material. Therefore, mushroom production holds the potential to transcend its ecological role as decomposers in natural ecosystems, effectively eliminating their "dark side." This transformation would allow them to become a valuable source of specific materials and sustenance.

8 Nutritional and Economic Benefits

Button mushrooms are a great complement to the human diet because they are "low calories" and high in vital nutrients. They take less time and space to cultivate than traditional crops, making them a sustainable option (Prajapati et al. 2023). This component has a great deal of promise to support rural income and food security. Nowadays, mushrooms are widely recognized as wholesome food ingredients that are rich in compounds with multiple health advantages, dietary fiber, and vitamin

D. Originally, the method used to obtain it was gathering wild mushrooms that grew organically in the field, depending on the area (El-Ramady et al. 2022). Modern, established growing practices have significantly contributed to "environmental sustainability" through the successful development of a worldwide industry focused on recycling agricultural and forestry by-products. This expansion in global mushroom production represents the "positive" facet of circular agriculture.

However, the sustainability of mushroom farming has not yet received the widespread recognition it deserves, and the trend of mushroom production worldwide has stalled recently. For the future growth of mushroom production, dissemination initiatives through international cooperation focused on education encompassing the science of mushroom ecology, cultivation, and nutrition are therefore essential. On the other hand, the "dark side," or unfavorable sustainability effects, of mushroom cultivation must also be immediately recognized and addressed. It is common to overlook the harmful impacts of farmed mushroom spores on the "environment and disposal techniques" such used heat-resistant bags and discarded mushroom substrates. Enhancing the global perception of the worth and sustainability of mushroom cultivation can be achieved by outlining the advantages and disadvantages of sustainability in the industry and outlining future possibilities (Miles and Chang 2004; El-Ramady et al. 2022; Prajapati et al. 2023).

9 The Indigenous Biodiversity Reservoir

The genetic diversity present in native fungal populations is vital for the commercial cultivation of mushrooms. These naturally occurring populations are "physiologically" adapted to specific substrates and environmental conditions, offering the essential genetic variation needed to enhance desirable traits in cultivated mushroom varieties/species (Schilla et al. 2025). This diversity plays a pivotal role in advancing the productivity and "resilience" of the mushroom cultivation sector. Regretfully, not many government agencies in developing nations particularly those in tropical and subtropical regions have expressed interest in funding systematic, ongoing research on the uses, preservation, and administration of edible and medicinal fungi as a genetic resource, despite these regions' rich biological diversity. Strain collections are crucial for the advancement of research in edible fungal cultivation. Presently, these collections focus on preserving mycelia and their associated data, with some also offering services for strain identification, verification, and deposit. Instead of studying local germplasm in an isolated and fragmented manner, it would be more beneficial to promote efforts to gather, examine, assess, and integrate the various characteristics of edible mushrooms. This comprehensive approach would provide a true understanding of their advantages, drawbacks, requirements, and current priorities (El-Ramady et al. 2022; Jayaraman et al. 2024; Schilla et al. 2025).

Due to great adaptability, tenacity, and development in the humic-rich leaf litter environment, *A. bisporus* has recently been recognized as a model fungus (Morin et al. 2012). In many different nations, it has been shown that *Cupressus* spp. leaf litter is an excellent home for isolating native strains. Many indigenous strains were also obtained from plant wastes, manures in fields or parks, and the needle litter of *Picea*. On the other hand, *A. bisporus* has been discovered in a variety of environments, including coastal dunes in France and desert areas of the Sonoran Desert in California, USA, under *Prosopis* and *Tamarix* (Kerrigan 1995; Xu et al. 1997; Callac et al. 2002). Given this, it is reasonable to anticipate that geographic populations and/or species variations will adapt to various temperatures and ecosystems (Largeteau et al. 2011). This capacity to employ native isolates as fresh-sources of breeding material to bring back genetic variability in cultivars is linked to "genetic diversity." Maintaining the "natural reservoir" of variation for this species of mushroom should be achieved by protecting the areas in which *A. bisporus* is commonly found. Unfortunately, major weather events and human activity often pose a threat to these kinds of environments. Cypress woods seen along the coasts of North America and Europe are an example of this. For the preservation of *A. bisporus* biodiversity, the creation of a robust germplasm is becoming more and more important.

For the sustainable advancement of button mushroom production, crucial characteristics include disease resistance, substrate adaptation, and especially, the ability to withstand heat. Largeteau et al. (2011) tested a large sample of native strains of *A. bisporus*, representing major geographical species and/or populations, for their capacity to produce mature fruiting bodies at temperatures exceeding the standard 17–19 °C typically used for commercial fruiting induction. While certain *A. bisporus* var. *bisporus* strains fruiting at elevated temperatures, most strains showed noticeably lower yields compared to those at normal temperatures. Interestingly, the frequency of thermo-tolerant strains did not correlate with either climate/environment (Largeteau et al. 2011). In contrast, Navarro and Savoie (2015) found that all *A. bisporus* var. *burnettii* strains originating from the "Sonoran Desert" produced substantial quantities of fruiting bodies at 25 °C, with some strains even fruiting at 30 °C. This suggests that the fruiting ability of this variety is either not influenced by temperature or is considerably less sensitive to it. The small mushrooms produced quickly split open, facilitating rapid spore dispersal. These traits, coupled with an amphiphilic life cycle, enable *A. bisporus* var. *burnettii* to adapt to environments with brief periods of sufficient humidity, allowing the species to complete its life cycle swiftly. Recent research also indicates that *A. bisporus* mushrooms may contain significant concentrations of compounds with potential therapeutic applications. These include tyrosinase, aromatase inhibitors, immune modulating and antitumor polysaccharides, ergosterol, vitamin D_2, and various antioxidants (Beelman et al. 2003).

For the future development of button mushroom as a nutraceuticals, all these ingredients are intriguing (Lelley and Vetter 2005; Gopalakrishnan et al. 2005). *A. bisporus's* genetic background is predicted to have an impact on both the amount and quality of compounds that have health-promoting properties. In comparison to a white hybrid, a brown-cultivated strain had higher quantities of ergosterols and antioxidant activity, according to Shao et al. (2010) in comparison to a white hybrid grown in the same conditions; Savoie et al. (1993) showed better radical scavenging activity correlated with higher phenol concentration in methanol extracts of two native strains. There hasn't been any research on how the genotypes of several *A. bisporus* strains differ in terms of their "potential health advantages," though. A comparison of the agaritine concentrations in 50 *A. bisporus* strains from the native germplasm was the sole published experiment on a significant number of strains (Sabourin et al. 2008). The mean agaritine levels in all American samples were lower than in any European population sample, and the intermediate agaritine levels in cultivars matched the lower-middle range found in strains from Europe. Compared to the cultivars, the *tetrasporic* var. *burnettii* strains exhibited noticeably reduced agaritine levels. This work highlights "valuable resource" for developing *A. bisporus* strains with either high or low concentrations of specific components. For mushroom breeders focused on improving both cultivation qualities and postharvest attributes to satisfy customer expectations and foster sustainable development, the diverse and interesting traits found in native isolates, coupled with an understanding of population aspects, represent a veritable goldmine.

10 Development of Mushroom Strains for Fruiting Under Elevated Temperatures

For two centuries, button mushrooms have been cultivated in cellars and caves across Europe and North America. This practice was driven by the consistently moderate temperatures, typically between 14 and 16 °C, and high relative humidity (RH) found in these environments, which both encouraged the growth of fruiting bodies and allowed for off-season cultivation. During the cultivation process, primordia differentiate when the temperature is lowered from the 25 °C used for spawn runs to 18 °C, subsequently developing into mature fruiting bodies (Eastwood et al. 2013). In order to enhance control over the growth of fruiting bodies and to enhance yield and quality, climatic chambers have taken the place of caves since the 1980s. The organization of mushroom production is "highly efficient," the staff's work environment is microbiologically secure, and managing pests and pathogens is made easier by cleaning and prophylaxis. Raising the temperature 21–25 °C in between flushes will help increase flush synchronization. The fruiting temperature can be adjusted to suit the needs of individual commercial strains, with a range of 16–19 °C. Unfortunately, regulating the climate for modern indoor mushroom production demands costly infrastructure and substantial energy, particularly in hot countries or during warm seasons in "temperate regions." This makes it impractical for small scale, sustainable farming system. Consequently, identifying various sources of thermotolerance species, understanding the genetic basis of this trait and developing thermotolerant strains are critically important for opening up new avenues for sustainable development of button mushroom cultivation. Strains of button mushroom (*A. bisporus*) demonstrating tolerance to relatively high temperatures for both mycelial growth rate and fruiting capacity have been identified within the species germplasm, found in both "European and North American" mushroom populations, as well as in China (Chen et al. 2003; Largeteau et al. 2011). In strains capable of high-temperature fruiting (FHT$^+$), primordial differentiation is not triggered by a temperature drop to 18 °C; rather, it can occur at a consistent 25 °C. *A. bisporus* var. strains that yield substantial quantities of mushrooms at 25 °C have been specifically selected for this characteristic (Navarro and Savoie 2015). Regrettably, none of the existing *A. bisporus* cultivars fully meets growers' expectations regarding rusticity, thermotolerance, disease resistance, high yield, good quality, and extended shelf life. Therefore, achieving the goal of cultivating button mushroom (*A. bisporus*) at temperatures above 20 °C, particularly within the framework of "sustainable agriculture" and "economic development" of emerging nations, necessitates breeding to combine numerous agronomical features into a single cultivar. An understanding of the biological processes underpinning thermotolerance during fruiting will significantly aid this breeding effort.

To investigate gene expression within the vegetative mycelium of a thermotolerant Chinese strain, the initial three gene fragments associated with thermotolerance were isolated using DD-RT-PCR (Chen et al. 2003). A subsequent publication by Chen et al. (2005) of the complete cDNA sequence for one particular gene (028-1, Gen-Bank accession number DQ235473) surprisingly revealed no clear homology in existing databases. Creating a binary expression vector for 028-1 and using *Agrobacterium* spp. to transform a non-thermotolerant strain to transfer the gene, Chen et al. (2005) reported transformants that could grow at 34 °C, while the native type could not thrive at temperatures over 30 °C. Mycelial diameters after 22 days at 34 °C were 15–25% of those at 28 °C, indicating that both the thermotolerant parent and the transformants experienced significant effects on their mycelial development rate. These transformants were not used to investigate the effects on fruiting ability. After studying 028-1 in a number of FHT+ and FHT− strains, Navarro and Savoie (2015) identified the encoded protein as an activating protein Yap-1-like transcription factor, known for its role in yeast thermotolerance. This protein was classified as a member of a novel subfamily of YAP-1 homologs within *Agaricomycotina* (*Basidiomycota*), and *aap1* was proposed as a potential replacement for the previously identified 028-1 gene. The authors further investigated this by examining the deduced polypeptide sequence alignment and sequence polymorphism across 24 button mushroom (*A. bisporus*) strains, which exhibited varying capacities for mycelial growth at temperatures exceeding 30 °C and for producing complete fruiting-bodies at 25 °C. According to Navarro and Savoie (2015), no particular allele was found to be responsible for the mycelium's ability to fruit at 25 °C or for it to survive at 33 °C. However, when nine FHT+ strains' primordia or fruiting bodies were grown at 17 and 25 °C, the relative expression levels of aap1 were measured. This revealed modest levels of regulation and no consistent pattern among the strains. Although the protein that aap1 produces may be engaged as an overall stress-resistance transcription factor, aap-1 is not a dominant contributor to *A. bisporus's* thermotolerance (Navarro and Savoie 2015).

Heat shock proteins (HSPs), a large class of chaperone proteins, are present in nearly all organisms. These highly conserved and widely distributed proteins show increased expression when exposed to environmental stressors like elevated concentrations of "reactive oxygen species" (ROS) or pathogen contact. In button mushroom (*A. bisporus*), the hsp-A-gene (Genbank X98508) was identified through its differential expression in both primordia and fruiting bodies of a commercial strain (Ospina-Giraldo et al. 2000). Using q-PCR, the regulation of hsp-A, a heat-shock-protein from the hsp-70 family, was examined in the sporophores and primordia of ten strains, cultured at both 17 and 25 °C. Unsurprisingly, higher expression levels were observed during the crucial primordia stages compared to sporophores (unpublished data). While expression levels in primordia grown at 25 °C were either lower than or not significantly different from those grown at 17 °C, the hspA gene likely contributed to the primordia's physiological preparedness for rapid differentiation, ultimately leading to the production of mature sporophores. However, when Lu et al. (2014) measured the expression time course of *A. bisporus* heat shock proteins, they found that when vegetative mycelium was exposed to high temperatures, a thermotolerant strain of the bacteria accumulated these defense-related proteins more quickly and robustly than a sensitive strain. While heat stress caused a rapid build-up of H_2O_2 in the strain that was sensitive to it, it had a more moderate effect on the strain that was thermotolerant, where the buildup of H_2O_2 may have been prevented by increased catalase- and superoxide-dismutase activity (Lu et al. 2014).

Para-aminobenzoic acid (PABA) has been identified component involved in the mycelium thermotolerance. This discovery was made through a comprehensive approach combining proteome analysis, gene expression studies; quantify the PABA as an end product, and "thermotolerance analysis" following the addition of PABA or inhibitors of its production. According to Lu et al. (2014), PABA has been demonstrated to mitigate heat-induced oxidative damage by upregulating heat-sensitive genes, catalase and SOD activity. Lastly, the capacity to create mature sporophores at 25 °C is probably unrelated to a "stress response mechanism," even if mycelial thermotolerance incorporates general "stress resistance" pathways. One theory explaining the mechanisms leading to "FHT+ phenotype" (fruiting at high temperatures) is a lack of sensitivity to the temperature signal, which is normally required to induce-sporophore formation in FHT strains. This might help clarify why Foulongne-Oriol et al. (2014) observed no discernible differences in the transcript-profiling of *A. bisporus* var. *brunette* primordia under two different fruiting temperature conditions (17 and 25 °C). It could also explain the correlation found in an analysis of hybrids from an intervarietal *A. bisporus* var. *bisporus* × *A. bisporus* var. *brunette* cross, where the number of fruiting bodies produced at both temperatures was linked.

11 Strain Selection: Health Benefits and Low Safety

The usage of edible and medicinal mushrooms among natural goods is expanding daily due to their affordability and ease of availability in relatively large numbers. The scientific community generally agrees that this resource is still significantly underutilized when it comes to molecules (proteins, enzymes, metabolites, and polymers), some of which we may not even be aware it exists. *A. bisporus* also contains significant concentrations of compounds of potential therapeutic relevance, but some farmed mushrooms such as *Agaricus subrufescens*, *L. edodes*, *Ganoderma lucidum*, and other species are well-known for their medicinal qualities (Beelman et al. 2003). Through breeding programs, one can anticipate to enhance the quality of the mushrooms and their inherent capacity to create certain chemicals that promote health, in addition to the impacts of culture conditions and mushroom maturation. *A. bisporus* and other *Agaricus* species naturally contain a derivative known as phenyl-hydrazine.

The role of agaritine as a potential carcinogen is a subject of ongoing debate, largely depending on "concentrations" used in in vitro testing or feeding studies (Roupas et al. 2010). Interestingly, Endo et al. (2010) demonstrated in vitro "anticancer activity" against leukemic-tumor cells for agaritine derived from *A. subrufescens*, directly contrasting its previously attributed carcinogenic activity. The observed toxicity of agaritine in some studies might stem from its potential transformation into a more "potent carcinogenic derivative" during metabolism (Walton et al. 1997). Despite efforts within the mushroom industry to maintain low agaritine levels, a comprehensive review of feeding studies involving "mushroom carpophores and extracts" suggests that agaritine from cultivated button mushroom (*A. bisporus*) presents no known "toxicological" risk to "healthy humans" (Roupas et al. 2010). There's considerable optimism for developing enhanced intervarietal hybrids with reduced "agaritine content." Notably, tetrasporic var. *brunnea* strains and certain Native American strains of *A. bisporus* var. *bisporus* have exhibited significantly "lower agaritine levels" compared to the *A. bisporus* var. *bisporus* cultivars used in all feeding tests, as well as some European native strains (Sabourin et al. 2008). Regarding the previously discussed FHT (fruiting at high temperatures) trait, phenol-typing the offspring of the intervarietal *A. bisporus* var. *bisporus* × *A. bisporus* var. *burnettii* hybrid—the same hybrid used to construct the species' reference linkage map would allow for the interpretation of "agaritine contents" genetic heritability. This would also facilitate the discovery of markers for marker-assisted selection (MAS) application in breeding low-agaritine strains.

12 Conclusion and Future Perspectives

The cultivation of button mushrooms aligns with sustainable farming principles, enhancing resource efficiency, soil health, and biodiversity. As the industry evolves, balancing its benefits with mitigating adverse effects is crucial. Using low-economic value by-products from forestry and agriculture can prevent burning and support mushroom farming, which is particularly lucrative for underdeveloped nations and promotes circular agriculture without degrading soil quality and health. Despite its sustainability potential, there is limited global awareness and stagnant trends in mushroom cultivation. Increasing international cooperation, sharing knowledge, and implementing sustainability certification schemes can promote the industry's environmental benefits. However, the vast quantity of spores from indoor farms poses health risks, contaminates facilities, and reduces genetic variation. Identifying genes involved in sporulation and improving material biodegradability can mitigate these issues. Economic sustainability and working conditions are also concerns, with intense pricing rivalry affecting profitability. Addressing these challenges can enhance the perceived value of mushrooms globally. Modern cultivation methods, including hybrids, indoor composting, and climate-controlled rooms, have boosted popularity. However, the industry faces environmental concerns, pest control, and food quality issues. The growing demand for localized, environmentally friendly food production in Western countries can be met through small-scale and efficient large-scale mushroom farming. Utilizing genetic information and agro-waste can increase production efficiency. While the industry offers many benefits, addressing challenges like spent mushroom substrate disposal and spore impact is essential for sustainable future production.

References

Altaf S, Jan SK, Ahanger SA, Basu U, Rather RA, Wani OA, Elansary HO (2022) Management of green mold disease in white button mushroom (*Agaricus bisporus*) and its yield improvement. J Fungi 8(6):554. https://doi.org/10.3390/jof8060554

Aubrey MLL, Chin CFS, Seelan JSS, Chye FY, Lee HH, Rakib MRM (2022) Conversion of oil palm by-products into value-added products through oyster mushroom (*Pleurotus ostreatus*) cultivation. Horticulturae 8(11):1040. https://doi.org/10.3390/horticulturae8111040

Beelman RB, Royse D, Chikthimmah N (2003) Bioactive components in button mushroom *Agaricus bisporus* (J. Lge) Imbach (Agaricomycetideae) of nutritional, medicinal, or biological importance. Int J Med Mushrooms 5:321–337. https://doi.org/10.1615/InterJMedicMush.v5.i4.10

Callac P, Theochari I, Kerrigan RW (2002) The germplasm of *Agaricus bisporus*: main results after ten years of collection in France, Greece, and North America. Acta Hortic 579:49–55. https://doi.org/10.17660/ActaHortic.2002.579.4

Chang ST, Wasser S (2017) The cultivation and environmental impact of mushrooms. Oxford University Press, Oxford. https://doi.org/10.1093/acrefore/9780199389414.013.231

Chen R, Chen LF, Song SY (2003) Identification of two thermo-tolerance related genes in *Agaricus bisporus*. Food Technol Biotechnol 41:339–344

Chen MY, Wang ZS, Liao JH et al (2005) Full-length cDNA sequence of a gene related to the thermo-tolerance of *Agaricus bisporus*. Acta Edulis Fungi 12:85–88

CIDCA (2021) Juncao technology promotes international development cooperation. (Ann. date 9/3/2021). Available online at: http://en.cidca.gov.cn/2021-09/03/c_660813.htm

Deng M, Li K, Yan YJ et al (2022) Enhanced cadmium removal by growing Bacillus cereus RC-1 immobilized on different magnetic biochars through simultaneous adsorption and bioaccumulation. Environ Sci Pollut Res 29:18495–18507. https://doi.org/10.1007/s11356-021-17125-x

Dhar BL (2017) Mushrooms and human civilization. In: Zied DC, Pardo-Giménez A (eds) Edible and medicinal mushrooms: technology and applications. John Wiley and Sons Ltd, Hoboken, pp 1–4. https://doi.org/10.1002/9781119149446.ch1

Dou T, Zhang K, Shi X, Liu W, Yu F, Liu D (2025) Crop–mushroom rotation: a comprehensive review of its multifaceted impacts on soil quality, agricultural sustainability, and ecosystem health. Agronomy 15(3):563. https://doi.org/10.3390/agronomy15030563

Eastwood DC, Herman B, Noble R et al (2013) Environmental regulation of reproductive phase change in *Agaricus bisporus* by 1-octen-3-ol, temperature, and CO_2. Fungal Genet Biol 55:54–66. https://doi.org/10.1016/j.fgb.2013.01.001

El-Ramady H, Abdalla N, Badgar K et al (2022) Edible mushrooms for sustainable and healthy human food: nutritional and medicinal attributes. Sustainability 14:4941. https://doi.org/10.3390/su14094941

Eastwood DC, Herman B, Noble R et al (2013) Environmental regulation of reproductive phase change in Agaricus bisporus by 1-octen 3-ol, temperature, and CO2. Fungal Genet Biol 55:54–66. https://doi.org/10.1016/j.fgb.2013.01.001

Endo M, Beppu H, Akiyama H, Wakamatsu K, Ito S, Kawamoto Y, Shimpo K, Sumiya T, Koike T, Matsui T. (2010) Agaritine purified from Agaricus blazei Murrill exerts anti-tumor activity against leukemic cells. Biochim Biophys Acta. 1800(7):669–73. https://doi.org/10.1016/j.bbagen.2010.03.016

FAOSTAT (2022) Food and Agriculture Organization of the United Nations Statistics Database. Available online at: http://www.fao.org/faostat/en/#data. Accessed June 2022

Forest Agency (2019) State of Japan's forests and forest management−3rd country report of Japan to the Montreal process. Available online at: https://www.maff.go.jp/e/policies/forestry/attach/pdf/index-8.pdf

Frankland JC (1992) Mechanisms in fungal succession. In: Carroll GC, Wicklow DT (eds) The fungal community, vol 22, issue 6. Dekker, New York, pp 383–340

Gopalakrishnan C, Pawar RS, Bhutani KK (2005) Development of *Agaricus bisporus* as a nutraceutical of tomorrow. Acta Hortic 680:45–47. https://doi.org/10.17660/ActaHortic.2005.680.5

Grimm D, Wösten HAB (2018) Mushroom cultivation in the circular economy. Appl Microbiol Biotechnol 102:7795–7803. https://doi.org/10.1007/s00253-018-9226-8

Hassett MO, Fischer MWF, Money NP (2015) Mushrooms as rainmakers: how spores act as nuclei for raindrops. PLoS One 10(10):e014040. https://doi.org/10.1371/journal.pone.0140407

Hibbett DS, Donoghue MJ (1996) Implication of phylogenetic studies for conservation of genetic diversity in shiitake mushrooms. Conserv Biol 10:1321–1327. https://doi.org/10.1046/j.1523-1739.1996.10051321.x

Hu Y, Mortimer PE, Hyde KD et al (2021) Mushroom cultivation for soil amendment and bioremediation. Circ Agric Syst 1:11. https://doi.org/10.48130/CAS-2021-0011

Iglesias H, Ortiz AP, Soriano Disla JM, Lara-Guillén AJ (2025) Environmental and economic life cycle impacts of using spent mushroom substrate as a soil improver. Environments 12(1):31. https://doi.org/10.3390/environments12010031

Jasinska A (2023) Sustainability of mushroom cultivation systems. *Horticulturae* 9(11):1191. https://doi.org/10.3390/horticulturae9111191

Jasinska A, Wojciechowska E, Stoknes K, Roszak M (2022) Bioconversion of agricultural wastes into a value-added product: straw of Norwegian grains composted with dairy manure food waste digestate in mushroom cultivation. *Horticulturae* 8(4):331. https://doi.org/10.3390/horticulturae8040331

Jayaraman S, Yadav B, Dalal RC, Naorem A, Sinha NK, Rao CS, Rao AS (2024) Mushroom farming: a review focusing on soil health, nutritional security and environmental sustainability. Farming System 2(3):100098. https://doi.org/10.1016/j.farsys.2024.100098

Jones M, Gandia A, John S, Bismarck A (2021) Leather-like material biofabrication using fungi. Nat Sustainability 4:9–16. https://doi.org/10.1038/s41893-020-00606-1

Kerrigan RW (1995) Global genetic resources for *Agaricus* breeding and cultivation. Can J Bot 73:S973–S979. https://doi.org/10.1139/b95-347

Khan A, Murad W, Ali S, Shah SS, Halim SA, Khalid A, Al-Harrasi A (2024) Contribution of mushroom farming to mitigating food scarcity: current status, challenges and potential future prospects in Pakistan. Heliyon 10(23):e40362. https://doi.org/10.1016/j.heliyon.2024.e40362

Kumar P, Eid EM, Taher AM et al (2022) Biotransforming the spent substrate of shiitake mushroom (*Lentinula edodes* Berk.): a synergistic approach to biogas production and tomato (*Solanum lycopersicum* L.) fertilization. Horticulturae 8(6):479. https://doi.org/10.3390/horticulturae8060479

Largeteau ML, Callac P, Navarro-Rodriguez AM et al (2011) Diversity in the ability of *Agaricus bisporus* wild isolates to fruit at high temperature (25°C). Fungal Biol 115:1186–1195. https://doi.org/10.1016/j.funbio.2011.08.004

Lavrijssen B, Baars JP, Lugones LG et al (2020) Interruption of an MSH4 homolog blocks meiosis in metaphase I and eliminates spore formation in *Pleurotus ostreatus*. PLoS One 15:e0241749. https://doi.org/10.1371/journal.pone.0241749

Lelley JI, Vetter J (2005) The possible role of mushrooms in maintaining good health and preventing diseases. Acta Edulis Fungi 12:412–419

Liu JJ, Yang XQ, Li ZY, Miao JY, Li SB, Zhang WP, Lin LB (2024) The role of symbiotic fungi in the life cycle of Gastrodia elata Blume (Orchidaceae): a comprehensive review. Front Plant Sci 14:1309038. https://doi.org/10.3389/fpls.2023.1309038

Lu Z, Kong X, Xiao M et al (2014) Para-aminobenzoic acid (PABA) synthase enhances thermotolerance of mushroom *Agaricus bisporus*. PLoS One 9:e91298. https://doi.org/10.1371/journal.pone.0091298

MAFF (2020) Ministry of Agriculture, Forestry and Fisheries in Japan. Annual Report on Forest and Forestry in Japan. Available online at: https://www.maff.go.jp/e/data/publish

MAFF (2021) Ministry of Agriculture, Forestry and Fisheries in Japan. Census of Agriculture and Forestry. Available online at: https://www.e-stat.go.jp

Martín C, Zervakis GI, Xiong S, Koutrotsios G, Strætkvern KO (2023) Spent substrate from mushroom cultivation: exploitation potential toward various applications and value-added products. Bioengineered 14(1):2252138. https://doi.org/10.1080/21655979.2023.2252138

MEXT (2015) Ministry of Education, Culture, Sports, Science and Technology in Japan. Standards tables of food composition in Japan–2015, 7th revised edn. Available online at: https://www.mext.go.jp/en/policy/science_technology/policy/title01/detail01/sdetail01/sdetail01/1385122.htm

Miles PG, Chang ST (2004) Mushrooms: cultivation, nutritional value, medicinal effect, and environmental impact, 2nd edn. CRC Press, 480pp. https://doi.org/10.1201/9780203492086

Miller FC (1993) Conventional composting system. In: Tan Nair NG (ed) Proceedings of the second AMGA/ISMS international workshop-seminar on *Agaricus* compost, Sydney, pp 1–18

Morin E, Kohler A, Baker AR, Foulongne-Oriol M, Lombard V, Nagy LG et al (2012) Genome sequence of the button mushroom *Agaricus bisporus* reveals mechanisms governing adaptation to a humic-rich ecological niche. Proc Natl Acad Sci U S A 109:17501–17506. https://doi.org/10.1073/pnas.1206847109

Mshandete AM, Cuff J (2008) Cultivation of three types of indigenous wild edible mushrooms: Coprinus cinereus, Pleurotus flabellatus, and Volvariella volvacea on composted sisal decortications residue in Tanzania. Afr J Biotechnol 7:4551–4562. https://doi.org/10.5897/AJB08.792

NanjeGowda NE, Chennappa G (2021) Mushroom cultivation: a sustainable solution for the management of agriculture crop residues. In: Kulshreshtha S (ed) Recent advances in mushroom cultivation technology and its application. Bright Sky Publications, Rohini, New Delhi, pp 15–26

Nishida C, Yatera K (2022) The impact of ambient environmental and occupational pollution on respiratory diseases. Int J Environ Res Public Health 19:2788. https://doi.org/10.3390/ijerph19052788

Navarro P, Savoie JM (2015) Selected wild strains of Agaricus bisporus produce high yields of mushrooms at 25°C. Rev. Iberoam. Micol. 32:54–58. https://doi.org/10.1016/j.riam.2013.05.014

Obatake Y, Murakami S, Matsumoto T, Fukumasa-Nakai Y (2003) Isolation and characterization of a sporeless mutant in Pleurotus eryngii. Mycoscience 44:33–40. https://doi.org/10.1007/S10267-002-0074-Z

O'Brien BJ, Milligan E, Carver J, Roy ED (2019) Integrating anaerobic co-digestion of dairy manure and food waste with cultivation of edible mushrooms for nutrient recovery. Bioresour Technol 285:121312. https://doi.org/10.1016/j.biortech.2019.121312

Odediran FA, Adekunle OA, Olugbire OO et al (2020) Contribution of mushroom production to rural income generation in Oyo state, Nigeria. FUTY J Environ 14:47–54

Ogwu MC, Chime AO, Sawyer WE, Izah SC, Ovuru KF (2025a) Sustainable agriculture and environmental monitoring through edible mushrooms. In: Izah SC, Ogwu MC, Akram M (eds) Bioactive compounds in edible mushrooms, Reference series in phytochemistry. Springer, Cham. https://doi.org/10.1007/978-3-031-80050-4_20

Ogwu MC, Richard G, Izah SC, Alimba CG, Wangboje OM (2025b) Environmental roles of edible mushrooms. In: Izah SC, Ogwu MC, Akram M (eds) Bioactive compounds in edible mushrooms, Reference series in phytochemistry. Springer, Cham. https://doi.org/10.1007/978-3-031-52642-8_5-1

Okuda Y, Murakami S, Honda Y, Matsumoto T (2013) An MSH4 homolog, stpp1, from Pleurotus pulmonarius is a "silver bullet" for resolving problems caused by spores in cultivated mushrooms. Appl Environ Microbiol 79:4520–4527. https://doi.org/10.1128/AEM.00561-13

Ospina-Giraldo MD, Collopy PD, Romaine CP, Royse DJ (2000) Classification of sequences expressed during the primordial and basidiome stages of the cultivated mushroom *Agaricus bisporus*. Fungal Genet Biol 29:81–94. https://doi.org/10.1006/fgbi.2000.1189

Panjikkaran ST, Mathew D (2012) An environmentally friendly and cost effective technique for the commercial cultivation of oyster mushroom (*Pleurotus florida* (Mont.) Singer). J Sci Food Agric 93:973–976. https://doi.org/10.1002/jsfa.5827

Pavlík M, Halaj D (2019) Production and investment evaluation of oyster mushroom cultivation on the waste dendromass: a case study on aspen wood in Slovakia. Scand J For Res 34:313–318. https://doi.org/10.1080/02827581.2019.1584639

Phan CW, Sabaratnam V (2012) Potential uses of spent mushroom substrate and its associated lignocellulosic enzymes. Appl Microbiol Biotechnol 96:863–873. https://doi.org/10.1007/s00253-012-4446-9

Poulin A, Aeby X, Siqueira G, Nyström G (2021) Versatile carbon-loaded shellac ink for disposable printed electronics. Sci Rep 11:23784. https://doi.org/10.1038/s41598-021-03075-4

Prajapati SK, Kumar V, Rawat DK et al (2023) Mushroom cultivation: a sustainable approach to future agriculture to ensure quality food and nutritional security of current population in India. Int J Multidiscip Res Growth Eval 04(04):697–705

Puliga F, Leonardi P, Minutella F, Zambonelli A, Francioso O (2022) Valorization of hazelnut shells as growing substrate for edible and medicinal mushrooms. Horticulturae 8(3):214. https://doi.org/10.3390/horticulturae8030214

Qamruzzaman M (2025) The effects of natural resources, education, and financial inclusion in achieving environmental sustainability in resources-abundance nations. Discov Sustain 6(1):136. https://doi.org/10.1007/s43621-025-00853-2

Roupas P, Keogh J, Noakes M, Margetts C, Taylor P (2010) Mushrooms and agaritine: a mini-review. J Funct Foods 2(2):91–98. https://doi.org/10.1016/j.jff.2010.04.003

Royse DJ, Baars J, Tan Q (2017) Current overview of mushroom production in the world. In: Zied DC, Pardo-Giménez A (eds) Edible and medicinal mushrooms: technology and applications. John Wiley & Sons, pp 5–13. https://doi.org/10.1002/9781119149446.ch2

Sabourin RE, Brant BL, Wach MP, Kerrigan RW (2008) Variation in agaritine levels among individuals in natural populations of *Agaricus bisporus*. Mushroom Sci 17:184–190

Savoie JM, Chalaux N, Libmond S, Laborde J, Olivier JM (1993) French research progress: quality of raw ingredients and of composts, improvement of microbial activities. In: Tan Nair NG (ed) Proceedings of the second AMGA/ISMS international workshop-seminar on *Agaricus* compost ISMS, Sydney, pp 109–126. https://hal.inrae.fr/hal-02776911

Schilla F, Mumladze G, Soukupova J, Smutka L (2025) Economic analysis of Agaricus bisporus mushrooms production and the perspective of sharing economy. Front Sustain Food Syst 8:1415291. https://doi.org/10.3389/fsufs.2024.1415291

Schulz KH, Felten G, Hausen BM (1974) Allergy to the spores of Pleurotus florida. Lancet 1(6):625. https://doi.org/10.1016/S0140-6736(74)93022-0

Shankar BL, Veena TR, Pawar S, Gowda A, Kadadevaramath AR (2024) A study on value chain of mushroom for value addition: challenges, opportunities and prospects of cultivation of mushroom. Int J Bus Syst Res 18(6):539–554. https://doi.org/10.1504/IJBSR.2024.142079

Shao S, Hernandez M, Kramer JK, Rinker DL, Tsao R (2010) Ergosterol profiles, fatty acid composition, and antioxidant activities of button mushrooms as affected by tissue part and developmental stage. J Agric Food Chem 58(22):11616–11625. https://doi.org/10.1021/jf102285b

Sommer K, Vetter W (2025) Sterols in fresh and preserved button mushroom (Agaricus bisporus) products. Eur Food Res Technol 251(4):613–624. https://doi.org/10.1007/s00217-025-04659-4

UNEP (2022) United Nations Environment Programme. Video: Global plastic pollution agreement: a historic moment. Available online at: https://www.unep.org/news-and-stories/news/unea5-unep50-updates#update168799

Walton K, Coombs MM, Walker R, Ioannides C (1997) Bioactivation of mushroom hydrazines to mutagenic products by mammalian and fungal enzymes. Mutat Res 381:131–139. https://doi.org/10.1016/S0027-5107(97)00160-7

Waqar S, Bhat AA, Khan AA (2024) Endophytic fungi: unravelling plant-endophyte interaction and the multifaceted role of fungal endophytes in stress amelioration. Plant Physiol Biochem 206:108174. https://doi.org/10.1016/j.plaphy.2023.108174

Xu J, Kerrigan RW, Callac P, Horgen PA, Anderson JB (1997) Genetic structure of natural populations of *Agaricus bisporus*, the commercial button mushroom. J Hered 88:482–488. https://doi.org/10.1093/oxfordjournals.jhered.a023141

Zakaria Z, Seng TTC, Zailani SN, Zabidi KAA, Salim SK (2022) Utilization of spent mushroom compost in Grey oyster mushroom cultivation. In: International conference on biomass utilization and sustainable energy. Springer Nature Singapore, Singapore, pp 269–278. https://doi.org/10.1007/978-981-99-1695-5_23

Zhao H, Chang J, Havlík P, Van Dijk M, Valin H, Janssens C et al (2021) China's future food demand and its implications for trade and environment. Nat Sustain 4:1043–1051. https://doi.org/10.1038/s41893-021-00784-6

Zotti M, Cecchi G, Canonica L, Di Piazza S (2025) A review of nature-based solutions for valorizing aromatic plants' lignocellulosic waste through oyster mushroom cultivation. Sustainability 17(10):4410. https://doi.org/10.3390/su17104410

A Recent Database of Desert Truffle Production

Sevda Kırbağ and Mehmet Akyüz

Abstract There has been a growing interest in natural food sources in recent years. These are food sources that have not yet undergone any modifications and that are dependent on the seasons. They are therefore preferred by many as they are perceived to be more natural and healthier. It is known that the consumption of modified products has serious effects on human health, directly or indirectly, as a result of genetic alterations, the use of pesticides and preservatives, out-of-season production, and practices to increase their shelf life. Among natural foods in the world, truffle species are the most expensive, delicious, nutritious, and medicinal due to their organoleptic and flavor characteristics. They have a high level of commercial return and are completely dependent on nature. As unmodified natural products, truffles are the preferred choice of gourmets and the upper classes, who are discerning in their preferences. Truffle species are classified as either true or false truffles and are generally grouped into forest and desert truffles. Significant advances have been made in the industrial production of macrofungi belonging to Basidiomycota. Nevertheless, the fact that truffle species in Ascomycota are entirely reliant on natural conditions, that standard cultivation techniques used for other cultivated mushrooms (Basidiomycota) cannot be applied to truffles, and that there is no sufficient research conducted on the production of ascocarps in culture precludes the possibility of cultivating these species under controlled conditions. In recent years, biotechnological approaches have significantly advanced the cultivation of forest truffle species, enabling their production under controlled conditions to a measurable extent. This is achieved by initially grafting the truffles onto host roots in culture until a certain developmental stage, followed by transferring them to natural environments where they mature over an extended period (5–10 years). This section of the chapter provides a summary of the biotechnological approaches applied to

S. Kırbağ (✉)
Department of Biology, Faculty of Science, Fırat University, Elazığ, Turkey
e-mail: skirbag@firat.edu.tr

M. Akyüz
Department of Biology, Faculty of Arts and Science, Bitlis Eren University, Bitlis, Turkey
e-mail: makyuz@beu.edu.tr

M. N. Owaid et al. (eds.), *Mushroom Production*, Interdisciplinary Biotechnological Advances, https://doi.org/10.1007/978-981-95-7793-4_8

desert truffles, highlighting practices conducted in different countries worldwide. It also encompasses biotechnological studies on the cultivation and reproductive potential of desert truffles and their host plants under controlled conditions up to a specific developmental stage, followed by their transfer to natural environments for subsequent stages over an extended period. These early-stage biotechnological studies, which involve controlled cultivation experiments followed by long-term observations in natural growing areas, will generate valuable baseline data for future research and contribute significantly to the sustainable production of desert truffle species.

Keywords Desert truffle cultivation · *Terfezia* · *Tirmania* · *Picoa* · Host plant · Edible mushroom

1 Introduction

The relationship between humans and macrofungi stretches back to ancient times, with evidence of their consumption dating back to the dawn of civilization (Manzi et al. 1999; Martins 2016). For centuries, humans have gathered and consumed wild mushrooms for commercial, culinary, medicinal, and religious purposes (Dugan 2011; Valverde et al. 2015; Martins 2016; Anusiya et al. 2021; Cateni et al. 2022; Hamza et al. 2024). Edible mushrooms are valued for their highly desirable taste, aroma, texture, and flavor and are a significant source of carbohydrates, proteins, vitamins, and minerals in our diet. Moreover, their bioactive compounds exhibit significant medicinal properties, including antiviral, anti-inflammatory, antitumor, anti-obesity, and antimicrobial activities (Kumar et al. 2021; Mwangi et al. 2022; Zhang et al. 2022; Badalyan and Zambonelli 2023; Badalyan et al. 2023; Devi et al. 2024).

Mushrooms, classified as macrofungi within the phyla Basidiomycota and Ascomycota, are primarily utilized as a food source (Sundari et al. 2018). While it is estimated that there are over 140,000 fungal species in nature—92,724 in Ascomycota and 50,385 in Basidiomycota (Catalogue of Life 2020)—only 3000 of them have been identified as edible. Within the phylum Basidiomycota, approximately 25 or more species—including *Agaricus bisporus*, various *Pleurotus* species, *Lentinula edodes*, different *Auricularia* species, *Hericium erinaceus*, *Flammulina velutipes*, *Volvariella* species, etc.—are extensively cultivated commercially worldwide (Rahi and Malik 2016; Chang and Wasser 2017; Rathore et al. 2017; Royse et al. 2017; Atila 2021; Singh et al. 2021; Xiang et al. 2021; Deshpande and Arya 2022; Khatua et al. 2022; Li and Xu 2022; Ferraro et al. 2022; Shrikhandia et al. 2022; Dedousi et al. 2024; Devi et al. 2024).

For thousands of years, both commercially cultivated edible mushrooms and wild edible fungi have been collected and consumed for their taste, nutritional value, and medicinal properties. While the gathering of wild mushrooms has been a long-standing practice, there is currently a growing interest in cultivating them under controlled environmental conditions. Species within the phylum Basidiomycota have been extensively cultivated worldwide, and numerous studies have reported well-established culture techniques and accumulated expertise for these fungi. In contrast, true truffles (both forest and desert species) belonging to the phylum Ascomycota are naturally occurring, seasonally emerging fungi that remain highly dependent on specific ecological conditions. However, true truffles (forest and desert truffles) within the phylum Ascomycota are special species that grow naturally, emerge seasonally, and remain entirely dependent on nature. Consumers seek access to these species during specific seasons and in particular regions; however, their availability is typically limited due to short fruiting periods and their restricted natural distribution (Kagan–Zur et al. 2014; Zambonelli et al. 2016; Khan et al. 2022). Growing interest in gastronomy, gourmet markets, and evolving consumer preferences has further increased demand for truffle species, which current supply cannot adequately meet. Therefore, greater emphasis should be placed on developing cultivation strategies and enhancing their production in natural habitats.

Desert truffles are edible underground mushrooms that grow in semi-arid and arid regions. Valued for their culinary and medicinal properties, they hold significance across the Mediterranean basin, the Middle East, and the Gulf regions. These truffles usually form mycorrhizal associations with Cistaceae plants, particularly *Helianthemum* species, and are mainly harvested from the wild. However, habitat loss due to urbanization, desertification, intensive agriculture, and climate change, combined with the increasing need for arid-adapted crops, has accelerated efforts toward their domestication (Morte et al. 2021a).

In years of continuous research, it has been observed that the methods utilized in the production of Basidiomycota-class-cultivated mushrooms are not advantageous for the cultivation of truffle mushroom species (forest truffles and desert truffles), prompting the exploration of new methodologies and the development of novel biotechnological approaches. While significant and advanced experience has been gained in the cultivation of Basidiomycota mushrooms, using the same methods for the production of truffle mushroom (forest and desert truffle) species is not appropriate. The experience and methods acquired for mushroom species cultivated within this group may assist in the development of new and diverse biotechnological approaches to truffle mushroom production (Wang 2012; Zambonelli et al. 2015; Bajaj et al. 2021; Leonardi et al. 2021; Morte et al. 2021a, b). However, they cannot be entirely applied to the production of truffle species.

The increased demand for both forest and desert truffle species in some countries not only prompted the search for new strategies to enhance ascocarp production in their natural habitats but also paved the way for the development of production protocols for mycorrhizal plants, as well as new research into the biology, biodiversity, and cultural aspects of truffle fungi through biotechnological approaches (Morte et al. 2017, 2021a, b). Therefore, it is proposed to increase truffle production and preserve the natural habitats of desert truffles by employing such biotechnological methods.

In recent years, a two-stage biotechnological culture approach has been developed to increase the production of forest truffles in their natural habitats. This approach involves associating the mycelia of forest truffles with the root systems of suitable trees (such as oak, pine, beech, and hazelnut) in sterile culture environments. The first stage begins with the establishment of mycorrhizal structures by connecting the roots of tree seedlings cultivated in the laboratory with the mycelia. Subsequently, mycorrhizal seedlings are inoculated into appropriate natural environments in the second stage. Therefore, even though it is not entirely produced in a controlled environment, this method offers a promising strategy: truffle-inoculated seedlings planted in suitable areas enable the creation of truffle forests within approximately 5–10 years. Cultural studies of forest truffles have continued successfully through various methods (Chevalier 1973; Hall et al. 2007; Zambonelli et al. 2015, 2021; Sourzat 2017; Leonardi et al. 2021, etc.). Similarly, in the cultivation of desert truffles, mycorrhizal synthesis has been achieved using both seedlings and micropropagated plants of *Helianthemum* species, together with *Terfezia* spores and mycelium (Morte et al. 2008, 2009; Morte and Andrino 2014). The use of *Helianthemum* seeds and *Terfezia* spores for mycorrhizal plant production is generally preferred as it is more cost-effective than using micropropagated plants and mycelium as inoculum (Morte et al. 2017, 2021a, b).

The successful biotechnological production method derived from forest truffle species (Hall and Haslam 2012; Agueda et al. 2014; Bajaj et al. 2021; Leonardi et al. 2021) has opened new avenues for enhancing the yield and production of desert truffles within their natural habitats. This advancement has streamlined the development of a comparable production technique adapted from the methods employed for forest truffle species, which is particularly beneficial considering the vital role desert truffles play as a food source despite the substantial declines in their quantities in recent years. Although forest truffles cannot be entirely cultivated under controlled conditions, new methods inspired by such approaches can be developed to increase yields and ensure the production of desert truffles in their natural habitats (Morte et al. 2021a, b). In this chapter of the book, recent biotechnological studies on the cultivation of desert truffles worldwide will be summarized. This chapter reviews recent biotechnological studies on the global cultivation of desert truffles and presents our work on mycelium production for these species up to a defined developmental stage, along with projections for their future cultivation.

2 Desert Truffle Cultivation and Biotechnological Stages

In greenhouse conditions where desert truffles are cultivated alongside their host plants (Helianthemum species), studies on mycorrhizal characterization have been limited. These studies have examined the cultivation of desert truffles and their interactions with Helianthemum under simulated desert conditions. Alrawi et al. (1979) emphasized the importance of factors such as soil composition, texture, moisture, temperature, and irrigation for desert truffle production. Bawadekji (2011) showed that drip irrigation systems positively affect production and yield in regions where wild desert truffles are productive. Additionally, environmental conditions, soil characteristics, host plants, and irrigation have been identified as key factors influencing yield (Morte et al. 2008, 2009).

Biotechnological tools and management techniques have been developed to study fungal inocula and mycorrhizal associations with host plants, enabling subsequent plantations, especially of selected *Terfezia* species (Honrubia et al. 2001). Since the first plantation of *T. claveryi* mycorrhizal plants in Spain (Honrubia et al. 2001), the growing demand has driven research toward transitioning from experimental to medium- and large-scale cultivation worldwide. The biotechnological phases of these studies have been extensively detailed by Morte et al. (2008, 2009, 2012). Particularly, comprehensive studies on topics such as biological aspects, soil characteristics, ecological parameters, host and mycorrhizal types, cryptic species and distributions, and nutritional and medicinal properties related to the cultivation of desert truffles have been presented to researchers (Kagan–Zur et al. 2014, and in various chapters of this book).

Roth–Bejerano et al. (2004) found that the mycelia of *T. pfeilii* can be obtained via two methods: from the sterile hyphae of fresh fruit bodies or by germinating ascospores. This marks the initial stage in the subsequent production processes of desert truffle mushrooms, offering researchers a convenient means of acquiring pure mycelium.

Mycelium has been identified as the optimal source of inoculum for infecting plants with ectomycorrhizal fungal strains (Marx and Kenney 1982; Iotti et al. 2012). Furthermore, the utilization of mycelium, rather than spores, as a fungal inoculum has been shown to decrease the time needed to obtain mycorrhizal plants with *Terfezia* species by up to 1–3 months (Morte et al. 2009, 2012). In their study, Lopez–Nicolas et al. (2013) observed the development of stimulated mycelium growth for use as inocula in mycorrhizal plant production.

Two desert truffle species have been successfully cultivated: *T. claveryi* in Spain (Honrubia et al. 2001; Morte et al. 2008, 2009, 2012) and *T. boudieri* in Tunisia and Israel (Slama et al. 2010; Kagan–Zur et al. 2014). Additionally, in Spain, the mycorrhizal plants of *Picoa lefebvrei* and *Tirmania nivea* were planted in 2014 and that of *T. arenaria* in 2015 (Morte et al. 2017), providing valuable insights into cultivation practices despite the absence of ascocarps.

Morte et al. (2017) detailed the production of host plants (*Helianthemum* species), fungal inoculum, mycorrhizal plants, and desert truffle plantations. They highlight two main factors for ascocarp production: the high density of mycorrhizal plants and proper irrigation. The first step in establishing a desert truffle plantation is selecting suitable host plants and fungal species adapted to the environment and soil (Honrubia et al. 2014). Honrubia et al. (2014) have thoroughly documented all cultivation practices for plantation management and sylviculture to enhance wild production. Desert truffle fructification usually occurs 1–3 years after planting, depending on seedling quality, site conditions, season, plantation design, and management practices such as irrigation and weed control. A successful sylviculture example in Abu Dhabi (UAE) demonstrated that sprinkler irrigation and spore inoculation in areas with *H. lippii* plants stimulated the production of *T. boudieri* and *T. nivea* (Gouws et al. 2014).

Slama et al. (2012) reported that inoculating *H. sessiliflorum* with *T. boudieri* enhanced growth and improved the plant's nutritional status. Kovacs et al. (2003) studied the anatomical and ultrastructural root interactions of *T. terfezioides* with *Robinia pseudoacacia* and *H. ovatum* in vitro.

Sourzat (2017) reported that since 1973, with the discovery of mycorrhizal plants as well as extensive research and experiments, various truffle cultivation models have been developed, significantly increasing truffle production. In truffle production, the selection of well-mycorrhized host plants, soil characteristics (calcareous, well-aerated, well-drained soils with good biological activity), temperate climate, and appropriate cultivation techniques should be taken into consideration. Sourzat also emphasized the ongoing uncertainties in truffle cultivation, highlighting the continuous renewal of truffle plantation characteristics and their cultural dynamics.

Investigations by Khanaqa (2006) revealed that olive seedlings, inoculated with *T. nivea* and *T. pinoyi*, were initially cultivated in controlled greenhouse conditions before being transplanted to natural areas. Despite the robust development of the olive plants, the truffle yield remained notably poor even after 4 years. However, the introduction of suitable soil amendments to designated areas proved transformative, leading to a significant improvement in truffle development and subsequent increases in yield. Nonetheless, it was stressed that various factors, such as soil properties, rainfall, weather conditions, ecological characteristics, and host selection, exert a significant influence on yield. Thus, the insights gained from desert truffle production highlight the necessity for the continual exploration of new methodologies and advancements.

Morte et al. (2021b) reviewed cultivation practices for two *Terfezia* species, *T. claveryi* and *T. boudieri*, mycorrhized with *Helianthemum almeriense* and *H. sessiliflorum*, respectively. These practices were tested under various soil and climatic conditions, showing that both symbiotic pairs are suitable for arid environments. They also summarized studies on desert truffle cultivation in sandy and silty soils with high pH and low mineral content, detailing methods for producing mycorrhized seedlings and preferred planting techniques. Additionally, they provided comprehensive protocols for production, inoculation, and field management, highlighting the effects of environmental factors, irrigation, and soil characteristics on truffle yield.

Slama et al. (2010) studied the cultivation and fruiting of *T. boudieri* in arid and semi-arid regions, both under semi-controlled (greenhouse) and natural conditions. The fungus was inoculated with *H. sessiliflorum* in two soil types: gypsy and sandy soil. Their results suggest that desert truffle cultivation is a positive step toward ex situ conservation (Varese et al. 2011), but specific measures are necessary. The study highlights the critical importance of evaluating soil and environmental conditions in cultivation areas, especially in semi-arid regions with challenging climates (Benucci et al. 2012; Donnini et al. 2013).

Morte and Honrubia (1997) developed micropropagation protocols for the effective mycorrhization of *Helianthemum* species under in vitro conditions. Morte et al.

(1994) and Morte and Honrubia (1997) highlighted that inocula grown on modified agar media can be directly used as inoculum for in vitro mycorrhization. These inocula have been applied in mycorrhizal bioassays with *Helianthemum* (Kovacs et al. 2003). A study conducted by Morte et al. (2008) presents various approaches to establishing mycorrhizal associations between desert truffles and *Helianthemum* species, considering both the type of fungal inoculum (spores or mycelium) and the source of the host plant (seedlings or micropropagated plantlets). *T. claveryi* was the first desert truffle species to be cultivated, with successful mycorrhization achieved using both spores and mycelial forms (Arenas et al. 2018).

Especially in desert soils, it has been stated that the natural occurrence of *Terfezia* with host *Helianthemum* species enhances water retention, soil erosion control, and soil productivity (Khanaqa 2006). Therefore, it is emphasized that planting in these areas is feasible, and establishing mycorrhizal associations with desert truffles would be highly beneficial (Lopez–Nicolas et al. 2013; Bouzadi et al. 2017; Hashem et al. 2018). Louro et al. (2021) found that *Cistus. salviifolius* and *C. ladanifer* can host *Terfezia* species, potentially enabling their cultivation in acidic soils.

Morte et al. (2009) described the principal biotechnological procedures employed in Spain for the production of the mycorrhizal plants of key desert truffle species, including a complete micropropagation protocol for the host plant *H. violaceum*. Five distinct methods for achieving mycorrhizal synthesis between desert truffles and *Helianthemum* species have been reported, varying according to the type of fungal inoculum, host plant source, and culture conditions. In addition, a practical management protocol has been developed for established desert truffle plantations. Although biotechnological production has proven successful, strict compliance with these protocols is crucial for achieving optimal yield, as soil characteristics, environmental conditions, host plant selection, and particularly irrigation play critical roles.

Morte et al. (2012) reported that *T. claveryi* has been cultivated in southeastern Spain, as well as in specific regions of Israel, Abu Dhabi, and Argentina, using mycorrhizal plants produced with their technology. They detailed the stages of plant propagation, mycorrhizal inoculation, certification of mycorrhizal plants, and plantation phases in truffle production. Fruiting typically occurs 1–2 years after planting, depending on host and seedling quality, soil suitability, and management practices.

Navarro–Ródenas et al. (2016) observed that the mycorrhizal roots, mycorrhizosphere soil, and peridium of desert truffles are enriched with bacteria that can enhance survival and mycorrhization in desert truffle plant production. This bacterial enrichment has potential for semi-industrial scale application to improve these processes. Furthermore, Arenas et al. (2022) reported that climatic parameters and the presence of microorganisms influence host plant physiology, thereby affecting desert truffle production. Andrino et al. (2019) emphasized that the annual yield of *T. claveryi* in semi-arid climates is generally influenced by key agricultural climate parameters, particularly aridity index and soil water potential. These factors are especially critical during the autumn and spring months, underscoring the importance of understanding and managing climate variables to optimize *T. claveryi* yield in semi-arid regions.

3 Isolation of Pure Mycelium from Desert Truffle Ascocarp Through In Vitro Tissue Culture Technique

Desert truffles are a group of economically and ecologically significant fungi that are widely distributed across Türkiye. Although these species are widespread, numerous studies have examined their taxonomy, distribution, soil characteristics, host plants, nutritional and medicinal properties, as well as their ethnomycological and gastronomical aspects in detail and at the local scale. However, no comprehensive study has yet been conducted on the cultivation of pure mycelium, the micropropagation of host plants, the interaction of mycelium and spores with mycorrhizal plants, propagation through mycelial-inoculated hosts, and the cultivation and preservation of natural environments through appropriate field planting of mycelial-inoculated hosts. Despite their wide distribution in Türkiye, and also sufficient knowledge about the cultivation of desert truffles, the practical applicability of this information is limited, making it challenging to implement cultivation procedures in the field. However, when compared to desert truffle culture studies in other countries around the world, we can say that we are still in the early stages.

3.1 Nutrient Agar Culture Preparation and Pouring Operations

Twenty-four grams of malt extract agar were placed into a 500 mL Erlenmeyer flask and adjusted to a final volume of 500 mL with distilled water. After dissolving the nutrient agar in boiling water, the mouth of the flask was sealed with cotton wool and wrapped with aluminum foil, then sterilized in an autoclave at 121 °C and 1.5 atm pressure for 15 min (Fig. 1a). Additionally, 9.00 mm diameter glass Petri dishes were sterilized in a Pasteur oven at 180°C for 1.5 hours, along with the nutrient agar, before being transferred into a laminar flow cabinet equipped with a HEPA

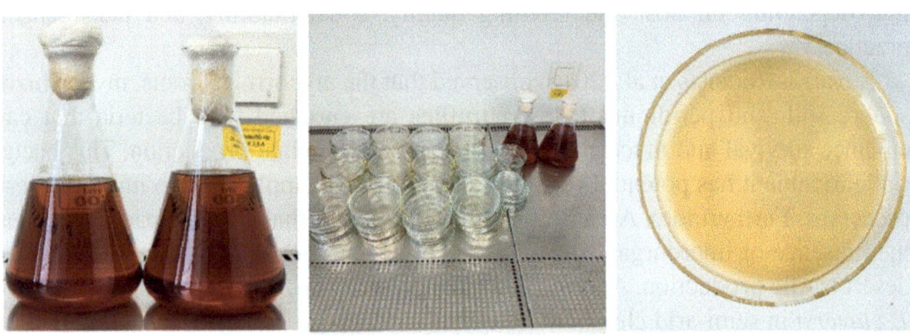

a b c

Fig. 1 Preparation and pouring of nutrient agar culture (**a**): sterile malt extract agar, (**b**): sterile Petri dishes, (**c**): malt extract agar on the Petri

(High-Efficiency Particulate Air) filter, where the inoculation process was carried out (Fig. 1b). During transportation, the UV lamp of the cabinet was operated for 30 min to prevent potential contamination. When the temperature of the agar dropped to 50–55 °C, sterilized glass Petri dishes were opened, and approximately 25 mL of malt extract agar was poured into each one (Fig. 1c). Subsequently, the ultraviolet (UV) lamp of the laminar flow cabinet was kept on for another 30 min to repeat the sterilization process and prevent potential contamination.

3.2 *In Vitro Tissue Culture Technique*

The main method is to obtain pure mycelia by transferring fresh small pieces of truffle tissue, ranging from 1 mm to 0.5 cm, from the inner tissue of fresh truffle to a nutrient agar medium in a sterile environment. Fresh, firm, and elastic *T. boudieri* (Fig. 2a, b) should be taken from their natural environments and brought to the laboratory. The surface of the ascocarp should be wiped with 96% alcohol in a HEPA-filtered laminar flow and then cut with a sterile scalpel, as shown in Fig 2c. A clean

Fig. 2 In vitro tissue culture technique of desert truffles (**a**): desert truffles, (**b**): fresh *T. boudieri*, (**c**): cut *T. boudieri* and its gleba, (**d**): inoculation of the tissue fragment into the nutrient agar medium, (**e**): *T. boudieri* mycelium, (**f**): *P. lefebvrei* mycelium

piece of tissue (1 mm to 0.5 cm in diameter) should be transferred to prepared Petri dishes (Fig. 2d) and wrapped with parafilm.

Subsequently, the Petri dishes should be placed upside down in the incubation device (25 °C). After 6–7 days, mycelial growth should be observed from the tissue pieces, and if there is no risk of infection, the nutrient agar should be incubated with mycelium until completely covered for approximately 25 days. The obtained pure mycelium (Fig. 2e, f) should be stored in the refrigerator (4 °C) for later use.

4 Conclusion

This book chapter provides concise and fundamental information on the cultivation of Terfezia and Tirmania species in the field of desert truffles, while also offering comprehensive and valuable practical knowledge, acknowledging that current practices remain at an early stage. While the production of desert truffle may not entirely occur in a cultured environment, up to a certain stage, such as the inoculation of mycelia into host mycorrhizal root structures, other stages involve the transplantation of mycelium-inoculated mycorrhizal seedlings into suitable natural environments and the monitoring of their growth processes in these natural habitats. Comprehensive and practical studies on this matter have been contributed by Morte et al. (2008, 2009, 2012, 2017, 2021a, b), Honrubia et al. (2001, 2014), and other researchers. These biotechnological approaches will provide significant benefits in increasing the quantities of these species in their natural habitats and multiplying their production potentials.

Although successful biotechnological approaches have been achieved in tissue culture for pure mycelium production, host micropropagation, the inoculation of mycelia or spores with host roots, their transplantation into suitable land, and observation, in many countries where desert truffles naturally occur, these practices either cannot be carried out or are still in their infancy. Despite having sufficient and advanced practical knowledge regarding the culture of desert truffles, comprehensive studies focused on practical implementation are lacking in many countries.

References

Agueda B, Zambonelli A, Molina R (2014) *Tuber* 2013: scientific advances in sustainable truffle culture. Mycorrhiza 24:1–4

Alrawi AM, Aldin MM (1979) New mycorrhizal identification, truffles cultivation, and truffles radiation preservation. Radiat Phys Chem 14:759–767

Andrino A, Navarro–Ródenas A, Marqués–Gálvez JE, Morte A (2019) The crop of desert truffle depends on agroclimatic parameters during two key annual periods. Agron Sustain Dev 39:1–11

Anusiya G, Gowthama Prabu U, Yamini NV, Sivarajasekar N, Rambabu K, Bharath G, Banat F (2021) A review of the therapeutic and biological effects of edible and wild mushrooms. Bioengineered 12:11239–11268

Arenas F, Navarro–Ródenas A, Chávez D, Gutiérrez A, Pérez–Gilabert M, Morte A (2018) Mycelium of *Terfezia claveryi* as inoculum source to produce desert truffle mycorrhizal plants. Mycorrhiza 28:691–701
Arenas F, López–García Á, Berná LM, Morte A, Navarro–Ródenas A (2022) Desert truffle mycorrhizosphere harbors organic acid releasing plant growth–promoting rhizobacteria, essentially during the truffle fruiting season. Mycorrhiza 32:193–202
Atila F (2021) Cultivation and utilization of shiitake mushroom. In: Medicinal plants: sustainable development and biodiversity. Springer, Cham, pp 383–413
Badalyan SM, Zambonelli A (2023) The potential of mushrooms in developing healthy food and biotech products. In: Fungi and fungal products in human welfare and biotechnology. Springer, Singapore, pp 307–344
Badalyan SM, Morel S, Barkhudaryan A, Rapior S (2023) Mushrooms as promising therapeutic resources: review and future perspectives. In: Mushrooms with therapeutic potentials. Springer, Singapore, pp 1–54
Bajaj SR, Marathe SJ, Grebenc T, Zambonelli A, Shamekh S (2021) First report of European truffle ectomycorrhiza in the semi–arid climate of Saudi Arabia. 3 Biotech 11:1–9
Bawadekji A (2011) Success of desert truffles production in muayala natural reserve at Arar city KSA. Al Riyadh Newspaper 15663
Benucci GMN, Bonito G, Baciarelli–Falini L, Bencivenga M, Donnini D (2012) Truffles, timber, food and fuel: sustainable approaches for multi–cropping truffles and economically important plants. In: Edible ectomycorrhizal mushrooms. Springer, Heidelberg, pp 265–280
Bouzadi M, Grebenc T, Turunen O, Kraigher H, Taib H, Alafai A, Sbissi I, Assad MEH, Bedade D, Shamekh S (2017) Characterization of natural habitats and diversity of Libyan desert truffles. 3 Biotech 7:328
Catalogue of Life (2020) Retrieved 15 Jan 2020, from http://www.catalogueoflife.org/col/search/all/key/pleurotus/fossil/1/match/1
Cateni F, Gargano ML, Procida G, Venturella G, Cirlincione F, Ferraro V (2022) Mycochemicals in wild and cultivated mushrooms: nutrition and health. Phytochem Rev 21:339–383
Chang ST, Wasser SP (2017) The cultivation and environmental impact of mushrooms. In: Oxford research encyclopedia of environmental science. Oxford University Press, New York
Chevalier G (1973) Synthese axenique des mycorhizes de *Tuber brumale* Vitt. a partir de cultures pures du champignon. Ann Phytopathol 5:163–182
Dedousi M, Melanouri EM, Karayannis D, Kaminarides EI, Diamantopoulou P (2024) Utilization of spent substrates and waste products of mushroom cultivation to produce new crops of *Pleurotus ostreatus*, *Pleurotus eryngii* and *Agaricus bisporus*. Carbon Resour Convers 7:100196
Deshpande AG, Arya A (2022) Mushroom biotechnology: developing cultivation protocol for four different mushrooms and accessing their potential in pollution management. In: Biology, cultivation and applications of mushrooms. Springer, Singapore, pp 457–485
Devi PV, Islam J, Narzary P, Sharma D, Sultana F (2024) Bioactive compounds, nutraceutical values and its application in food product development of oyster mushroom. J Future Foods 4:335–342
Donnini D, Gargano ML, Perini C, Savino E, Murat C, Di Piazza S, Altobelli E, Salerni E, Rubini A, Rana GL, Bencivenga M, Venanzoni R, Zambonelli A (2013) Wild and cultivated mushrooms as a model of sustainable development. Plant Biosyst 147:226–236
Dugan FM (2011) Conspectus of world ethnomycology: fungi in ceremonies, crafts, diets, medicines and myths. American Phytopathological Society, St. Paul
Ferraro V, Venturella G, Pecoraro L, Gao W, Gargano ML (2022) Cultivated mushrooms: importance of a multipurpose crop, with special focus on Italian fungiculture. Plant Biosyst 156:130–142
Gouws A, De Wet T, Abdullah F, Hassan A, Honrubia M, Morte A (2014) Desert truffle research in UAE Second symposium on hypogeous fungi in Mediterranean basin (HYPOGES2) &

Fifth congress Tuber aestivum/uncinatum European Scientific Group (TAUESG5). Université Mohammed V, Rabat (Morocco), 9–13 April 2014, p 17

Hall IR, Haslam W (2012) Truffle cultivation in the southern hemisphere. In: Edible ectomycorrhizal mushrooms. Springer, Berlin, Heidelberg, pp 191–208

Hall I, Brown G, Zambonelli A (2007) Taming the truffle. The history, lore, and science of the ultimate mushroom. Timber Press, Portland

Hamza A, Mylarapu A, Krishna KV, Kumar DS (2024) An insight into the nutritional and medicinal value of edible mushrooms: a natural treasury for human health. J Biotechnol 381:86–99

Hashem A, Alqarawi AA, Shah MA, Wirth S, Egamberdieva D, Tabassum B, Abduallah EF (2018) Desert truffles in Saudi Arabia: diversity, ecology, and conservation. In: Microbial resource conservation. Springer, Cham, pp 353–369

Honrubia M, Gutierrez A, Morte A (2001) Desert truffle plantation from south–east Spain. Edible Mycorrhizal Mushrooms and Their Cultivation: Proceedings of the Second International Conference on Edible Mycorrhizal Mushrooms. Christchurch, New Zealand, pp. 3–5

Honrubia M, Andrino A, Morte A (2014) Preparation and maintenance of both man–planted and wild plots. In: Desert truffles. Springer, Berlin, Heidelberg, pp 367–387

Iotti M, Piattoni F, Zambonelli A (2012) Techniques for host plant inoculation with truffles and other edible ectomycorrhizal mushrooms. In: Edible ectomycorrhizal mushrooms. Springer, Berlin, Heidelberg, pp 145–161

Kagan–Zur V, Roth–Bejerano N, Sitrit Y, Morte A (2014) Desert truffles. Phylogeny, physiology, distribution and domestication. Springer, Berlin

Khan FS, Hussain I, Akram M, Owaid MN (2022) Truffles: the cultivation and health benefits. In: Biology, cultivation and applications of mushrooms. Springer, Singapore, pp 285–300

Khanaqa A (2006) Truffle production in the Kingdom of Saudi Arabia–potential and limitation. J Appl Bot Food Qual 80:14–18

Khatua S, Sett S, Acharya K (2022) *Auricularia* spp.: from farm to pharmacy. In: Biology, cultivation and applications of mushrooms. Springer, Singapore, pp 301–355

Kovacs GM, Vágvölgyi C, Oberwinkler F (2003) In vitro interaction of the truffle *Terfezia terfezioides* with *Robinia pseudoacacia* and *Helianthemum ovatum*. Folia Microbiol 48:369–378

Kumar K, Mehra R, Guiné RP, Lima MJ, Kumar N, Kaushik R et al (2021) Edible mushrooms: a comprehensive review on bioactive compounds with health benefits and processing aspects. Foods 10:2996

Leonardi M, Iotti M, Pacioni G, Hall IR, Zambonelli A (2021) Truffles: biodiversity, ecological significances, and biotechnological applications. In: Industrially ımportant fungi for sustainable development. Springer, Cham, pp 107–146

Li C, Xu S (2022) Edible mushroom industry in China: current state and perspectives. Appl Microbiol Biotechnol 106:3949–3955

Lopez–Nicolas JM, Perez–Gilabert M, Garcia–Carmona F, Lozano–Carrillo MC, Morte A (2013) Mycelium growth stimulation of the desert truffle *Terfezia claveryi* chatin by β–cyclodextrin. Biotechnol Prog 29:1558–1564

Louro R, Natário B, Santos–Silva C (2021) Morphological characterization of the in vitro mycorrhizae formed between four *Terfezia* species (Pezizaceae) with *Cistus salviifolius* and *Cistus ladanifer*—towards desert truffles production in acid soils. J Fungi 7:35

Manzi P, Gambelli L, Marconi S, Vivanti V, Pizzoferrato L (1999) Nutrients in edible mushrooms: an inter species comparative study. Food Chem 65:477–482

Martins A (2016) The numbers behind mushroom biodiversity. In: Wild plants, mushrooms and nuts: functional food properties and applications. John Wiley & Sons, Ltd, United Kingdom, pp 15–63

Marx DH, Kenney DS (1982) Production of ectomycorrhizal fungus inoculum. In: Methods and principles of mycorrhizal research. American Phytopathology Society, St. Paul

Morte A, Andrino A (2014) Domestication: preparation of mycorrhizal seedlings. In: Desert truffles. Springer, Berlin, pp 343–365

Morte A, Honrubia M (1997) Micropropagation of *Helianthemum almeriense*. In: Biotechnology in agriculture and forestry: high-tech and micropropagation. Springer, Berlin, pp 163–177
Morte A, Cano A, Honrubia M, Torres P (1994) In vitro mycorrhization of micro propagated *Helianthemum almeriense* plantlets with *Terfezia claveryi*. Agric Sci Finl 3:309–314
Morte A, Honrubi AM, Gutierrez A (2008) Biotechnology and cultivation of desert truffles. In: Mycorrhiza: state of the art genetics and molecular biology, eco-function, biotechnology, eco–physiology, structure and systematics. Springer, Berlin, pp 467–483
Morte A, Zamora M, Gutierrez A, Honrubia M (2009) Desert truffle cultivation in semi–arid Mediterranean areas. In: Mycorrhizas–functional processes and ecological impact. Springer, Berlin, pp 221–233
Morte A, Andrino A, Honrubia M, Navarro–Rodenas A (2012) *Terfezia* cultivation in arid and semi-arid soils. In: Edible ectomycorrhizal mushrooms. Springer, Berlin, Heidelberg, pp 241–263
Morte A, Pérez–Gilabert M, Gutiérrez A, Arenas F, Marqués–Gálvez JE, Bordallo JJ et al (2017) Basic and applied research for desert truffle cultivation. In: Mycorrhiza: eco-physiology, secondary metabolites, nanomaterials. Springer, Cham, pp 23–42
Morte A, Kagan–Zur V, Navarro–Ródenas A, Sitrit Y (2021a) Cultivation of desert truffles–a crop suitable for arid and semi–arid zones. Agronomy 11:1462
Morte A, Arenas F, Marqués–Gálvez JE, Andrino A, Guarnizo ÁL, Gutiérrez A et al (2021b) Desert truffles (*Terfezia* spp.) breeding. In: Advances in plant breeding strategies: vegetable crops. Springer, Cham, pp 479–504
Mwangi RW, Macharia JM, Wagara IN, Bence RL (2022) The antioxidant potential of different edible and medicinal mushrooms. Biomed Pharmacother 147:112621
Navarro–Ródenas A, Berná LM, Lozano–Carrillo C, Andrino A, Morte A (2016) Beneficial native bacteria improve survival and mycorrhization of desert truffle mycorrhizal plants in nursery conditions. Mycorrhiza 26:769–779
Rahi DK, Malik D (2016) Diversity of mushrooms and their metabolites of nutraceutical and therapeutic significance. J Mycol 2016:7654123
Rathore H, Prasad S, Sharma S (2017) Mushroom nutraceuticals for improved nutrition and better human health: a review. PharmaNutrition 5:35–46
Roth–Bejerano N, Li YF, Kagan–Zur V (2004) Homokaryotic and heterokaryotic hyphae in *Terfezia*. Antonie Van Leeuwenhoek 85:165–168
Royse DJ, Baars J, Tan Q (2017) Current overview of mushroom production in the world. In: Edible and medicinal mushrooms: technology and applications. John Wiley & Sons, Chichester, pp 5–13
Shrikhandia SPP, Devi S, Sumbali G (2022) Lignocellulosic waste management through cultivation of certain commercially useful and medicinal mushrooms: recent scenario. In: Biology, cultivation and applications of mushrooms. Springer, Singapore, pp 497–534
Singh M, Kamal S, Sharma VP (2021) Species and region–wise mushroom production in leading mushroom producing countries–China Japan, USA, Canada and India. Mushroom Res 30:99–108
Slama A, Fortas Z, Boudabous A, Neffati M (2010) Cultivation of an edible desert truffle (*Terfezia boudieri* Chatin). Afr J Microbiol Res 4:2350–2356
Slama A, Gorai M, Fortas Z, Boudabous A, Neffati M (2012) Growth, root colonization and nutrient status of *Helianthemum sessiliflorum* Desf. inoculated with a desert truffle *Terfezia boudieri* Chatin. Saudi J Biol Sci 19:25–29
Sourzat P (2017) Truffle cultivation in the south of France: technical progress and prospects. Rev Mex Mic 46:63–72
Sundari MT, Anand AAP, Jenifer P, Shenbagarathai R (2018) Bioprospection of basidiomycetes and molecular phylogenetic analysis using internal transcribed spacer (ITS) and 5.8S rRNA gene sequence. Sci Rep 8:10720
Valverde ME, Hernández–Pérez T, Paredes–López O (2015) Edible mushrooms: improving human health and promoting quality life. Int J Microbiol 2015:376–387

Varese GC, Angelini P, Bencivenga M, Buzzini P, Donnini D et al (2011) Ex situ conservation and exploitation of fungi in Italy. Plant Biosyst 145:997–1005
Wang X (2012) Truffle cultivation in China. In: Edible ectomycorrhizal mushrooms. Springer, Berlin, Heidelberg, pp 227–240
Xiang Q, Adil B, Chen Q, Gu Y, Zeng X, Li X (2021) Shiitake mushroom (*Lentinula edodes* (Berk.) Sing.) breeding in China. In: Advances in plant breeding strategies: vegetable crops. Springer, Cham, pp 443–476
Zambonelli A, Iotti M, Hall I (2015) Current status of truffle cultivation: recent results and future perspectives. Ital J Mycol 44:31–40
Zambonelli A, Iotti M, Murat C (2016) True truffle (Tuber spp.) in the world. Springer, Cham
Zambonelli A, Iotti M, Puliga F, Hall IR (2021) Enhancing white truffle (*Tuber magnatum* Picco and *T. borchii* Vittad.) cultivation through biotechnology innovation. In: Advances in plant breeding strategies: vegetable crops. Springer, Cham, pp 505–532
Zhang J, Feng N, Liu Y, Zhang H, Yang Y, Liu L, Feng J (2022) Bioactive compounds from medicinal mushrooms. In: Biochemical engineering and biotechnology of medicinal mushrooms. Advances in biochemical engineering/biotechnology. Springer, Cham, pp 219–268

Biological Improvement of *Volvariella volvacea* Based Spent Mushroom Compost for Plant Substrate

Noor Azrimi Umor and Nurul Anina Ashyqin Amran

Abstract The cultivation of *Volvariella volvacea* generates a high amount of waste in the form of spent mushroom compost (SMC). Utilizing this by-product for sustainable agriculture, particularly as a plant substrate, can contribute to the sustainable management of the environment. This review highlights the application of SMC in vegetable crop production. SMC undergoes various treatment, especially using a biological process to improve the quality and adaptability of plants. Furthermore, an exemplary model using plant-growth-promoting bacteria for the improvement of SMC quality for okra cultivation is also presented.

Keywords SMC · PGPB · Soil amendment · Rhizosphere · Biological treatment

1 Introduction

Sustainable agriculture is becoming increasingly important in order to meet the growing demand for food and, at the same time, preserve the quality of the environment. Conventional farming methods, especially the overuse of chemical fertilizers, have contributed to soil degradation, pollution, and reduced soil biodiversity. To address these issues, sustainable agricultural practices are being explored to maintain crop productivity while minimizing environmental impacts. One such approach involves the use of organic soil amendments and beneficial microbes to improve soil health and plant growth.

Volvariella volvacea, commonly known as straw mushroom and is widely cultivated in Asian countries, produces significant amounts of spent mushroom compost (SMC), which is often discarded as waste (Umor et al. 2021a). However, this compost still contains valuable organic matter and nutrients that can improve soil

N. A. Umor (✉) · N. A. A. Amran
Unit Pembangunan Teknologi Cendawan, Fakulti Sains Gunaan, Universiti Teknologi MARA Cawangan Negeri Sembilan, Kuala Pilah, Negeri Sembilan, Malaysia
e-mail: noorazrimi@uitm.edu.my

M. N. Owaid et al. (eds.), *Mushroom Production*, Interdisciplinary Biotechnological Advances, https://doi.org/10.1007/978-981-95-7793-4_9

structure, increase water retention, and serve as a slow-release fertilizer. According to previous studies, SMC was mostly used in agriculture as a mushroom medium, animal feed, plant compost, and fertilizer, as well as for other applications (Umor et al. 2021b). The use of SMC as a soil conditioner also promotes waste recycling and contributes to environmental sustainability.

In addition to organic amendments, plant-growth-promoting bacteria (PGPB) have gained attention for their role in enhancing plant development through mechanisms such as nitrogen fixation, phosphate solubilization, and the production of phytohormones such as indole-3-acetic acid (IAA). These bacteria also help suppress plant pathogens, improving plant health and resistance to diseases. Many studies have demonstrated that applying PGPB to crops can lead to improved growth, yield, and soil quality while reducing the need for synthetic fertilizers. Microorganisms in the soil play an important role in recycling both major and minor nutrients, thereby helping to improve plant health, soil fertility, and the overall productivity of agricultural systems. The microbial biomass, which holds nutrients, acts like a slow-release fertilizer, gradually providing nutrients to plants (Manjunath et al. 2016). In addition, their metabolic processes help make nutrients more available and easier for plants to absorb. Adding microorganisms to the soil or directly to seeds helps them multiply, which in turn supports nutrient cycling and boosts the growth and yield of crops. To improve both agricultural productivity and ensure environmental sustainability, it is important to combine different methods of plant nutrient management (Manjunath et al. 2016).

2 Agricultural Reuse of *Volvariella volvacea*-SMC

Mushroom cultivation helps transform agricultural and agro-industrial waste into nutritious food, making it an environmentally friendly practice. This transformation occurs through the fungi's natural ability to break down materials; however, for the process to be cost-effective and efficient, the right chemical, physical, environmental, and technical conditions need to be carefully managed (Zied et al. 2020) (Fig. 1).

Spent mushroom compost (SMC), also known as spent mushroom substrate (SMS), is the residual material left after mushroom cultivation. *Volvariella volvacea*, sometimes referred to as the "warm mushroom," is one of the easiest mushrooms to produce and thrives at a relatively high temperature (Bashie Ajikah et al. 2019). *Volvariella volvacea* is also thought to have originated in China in the early eighteenth century. It was once considered a delicacy reserved for royalty because of its rich protein and fiber content, along with its low fat levels (Sharifuddin et al. 2022).

Spent mushroom compost contains high levels of salts and unstable organic matter, so it typically requires around 2 years of aging before it can be safely used. This waiting period allows soluble organic compounds to leach out and break down. The compost itself contains about 12% nitrogen, 0.2% phosphorus, and 1.3% potassium.

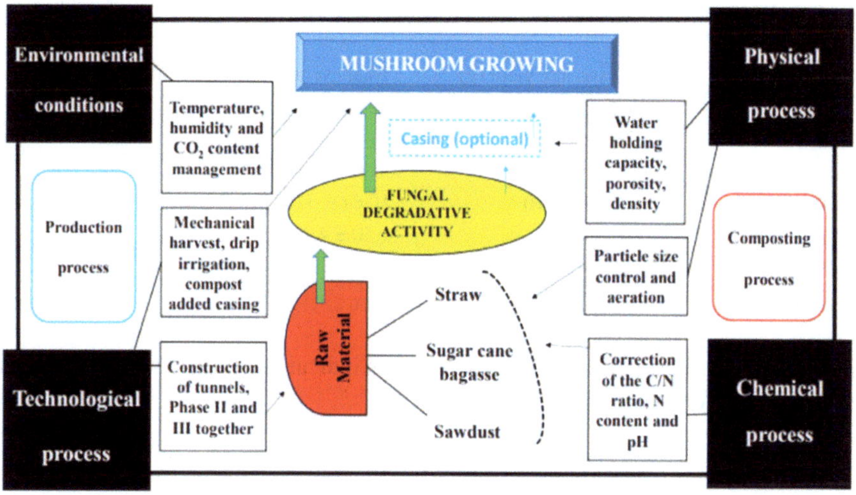

Fig. 1 Diagram of mushroom cultivation outlining the key chemical, physical, environmental, and technological factors that can affect the critical stages of production, as described by Zied et al. (2020)

While nitrogen and phosphorus levels remain stable after about 18 months of storage, the potassium content tends to decline (Olabanjo Olugbenga et al. 2024).

Biological efficiency (BE) is a straightforward measure used in research to determine how effectively a mushroom converts lignocellulosic materials into fruiting bodies. Among edible mushrooms, *Volvariella volvacea* has one of the lowest BEs. For instance, *Pleurotus* species have a BE ranging from 47.0% to 134.5%, while that of *Lentinula edodes* ranges from 22.7% to 85.6%. In comparison, *Volvariella volvacea* achieves only about 0.1% to 26.7% (Sharifuddin et al. 2022). This low BE indicates that *Volvariella volvacea* is less capable of fully using the nutrients in its growth substrate to form fruiting bodies. Sharifuddin et al. (2022) also stated that solid wastes are well suited for cultivating *Volvariella volvacea* because the mushroom produces different types of cellulose-degrading enzymes, allowing it to break down cellulose and access essential nutrients. This is particularly important since *Volvariella volvacea* thrives on substrates that are high in cellulose but low in lignin, using them to develop its fruiting bodies.

Table 1 presents the yield and BE results. The highest values for both yield and BE were obtained from the PS:WP, PS:EFB, and EFB:WP combinations. These three formulations showed better performance than the control (PS alone). Although PS and EFB produced the fastest growth and earliest harvest, their yields were relatively low at 76.67 ± 53.35 g and 70.67 ± 50.14 g, respectively, compared with the other solid waste mixtures. The BE for WP alone was notably lower (3.60 ± 0.00) than when it was supplemented with PS and EFB, which achieved 12.40 ± 3.20 and 9.50 ± 4.10, respectively (Sharifuddin et al. 2022). PS refers to paddy straw, EFB to empty fruit bunch, and WP to waste paper. The values are expressed as a

Table 1 Yield and BE of *Volvariella volvacea* cultivated on different formulations of solid wastes (Sharifuddin et al. 2022)

Formulation	Total average yield (g)	Total average biological efficiency (%)
100PS	76.67 ± 53.35	7.67 ± 5.34
100EFB	70.67 ± 50.14	7.07 ± 5.01
100WP	36.00 ± 0.00	3.60 ± 0.00
50PS:50EFB	106.67 ± 40.70	10.67 ± 4.07
50EFB:50WP	95.00 ± 41.01	9.50 ± 4.10
50PS:50WP	124.00 ± 32.05	12.40 ± 3.20

PS paddy straw, *EFB* empty fruit bunch, *WP* waste paper

mean ± standard deviation, and identical letters indicate that there is no statistically significant difference at $p > 0.05$.

2.1 Recycling Spent Mushroom Compost (SMC) for Sustainable Crop Production

Mushroom production mainly produces used mushroom compost as waste. Some studies (Zied et al. 2020) have explored reusing SMC as a supplement to raw materials in new substrate mixes or as a casing layer for cultivating Agaricaceous mushrooms. In the mushroom circular economy, the only waste produced during the growing cycle comes from the plastic bags or bottles used to cultivate species such as *Lentinula edodes*, *Pleurotus* spp., *Volvariella volvacea*, *Flammulina velutipes*, and others.

Research shows (Mohd Hanafi et al. 2018) that SMS from different mushroom species can be a good feed for ruminants because it contains important nutrients like polysaccharides, vitamins, and minerals, such as iron, calcium, zinc, and magnesium. Feeding SMS to elk has been found to improve their growth and health. Among the types tested, *Pleurotus florida* straw-based SMS was more digestible than *P. sajor-caju* SMS, making it more suitable as feed. *P. ostreatus* corn straw-based SMS also improved straw quality and helped feedlot lambs grow better without affecting meat quality. Table 2 gives examples of agro-based SMS used for ruminant feeding. SMS alone is not the best feed for ruminants because it does not greatly improve the nutritional value of the original crop residues. However, using biological treatments can make it better by increasing protein, fat, and minerals while reducing fiber, making it more nutritious and easier to digest.

Table 2 Agro-industrial spent mushroom substrate as a feed resource for ruminants

Type of mushroom	Substrates types	Finding	Remarks	Feeding trial	References
N/A	Microbially fermented SMS	The use of microbially fermented SMS as feed led to better growth rates and improved overall characteristics.	During the finishing period, animals given microbially fermented SMS weighed 24–40 kg more than those in the control group.	Hanwoo steers	Mohd Hanafi et al. (2018)
Agaricus bisporus	Wheat straw, poultry manure, calcium sulfate, sugar beet molasses, and urea	Pelletized wheat-straw-based SMS can be included in the diet of finishing calves.	Offering it as a mash-type total mixed ration can lower both feed consumption and daily growth rates.	Calves	Mohd Hanafi et al. (2018)
Pleurotus ostreatus	Sawdust-based SMS	SMS can replace wheat bran in broiler production.	The high fiber content in SMS led to an increase in the feed intake of chicks.	Broilers	Mohd Hanafi et al. (2018)
Pleurotus ostreatus	Fresh SMS	With a 15% ration, SMS can be utilized in place of barley without affecting the majority of carcass qualities.	An increase in the proportion of SMS in the diet resulted in a lower final weight of ruminants.	Awassi lambs	Mohd Hanafi et al. (2018)
N/A	Sawdust, beet pulp, corncobs, and cottonseed meal	Sawdust-based SMS was efficiently conserved for animal feed by ensiling with *L. plantarum* KU5 inoculant and molasses.	*L. plantarum* KU5 and molasses improved the aerobic stability of SMS made from sawdust used for animal feed.	N/A	Mohd Hanafi et al. (2018)
Flammulina velutipes	Fresh SMS	SMS-based silage, whether with or without urea, lowers protozoa populations and methane emissions.	Rumen protozoa populations are significantly decreased when Holstein steers are fed SMS-based silage.	Holstein steers	Mohd Hanafi et al. (2018)

(continued)

Table 2 (continued)

Type of mushroom	Substrates types	Finding	Remarks	Feeding trial	References
Pleurotus ostreatus	Fresh SMS	Compared to fresh SMS, the ruminal disappearance of SMS ensiled with *L. brevis* was somewhat higher.	A less expensive option to prepared feeding pellets is fresh SMS treated with *L. brevis*.	Hanwoo steers	Mohd Hanafi et al. (2018)
Volvariella volvacea	PS:EFB, PS:WP, EFB:WP mixtures	In comparison to single substrates, they achieved the greatest yield and biological efficiency (BE).	Productivity and growth are improved by blended substrates.	No direct animal feeding trials	Sharifuddin et al. (2022)
Volvariella volvacea	100% PS (paddy straw) and 100% EFB	Reduced yields were observed despite quicker harvesting and developmental performance.	*V. volvacea* exhibited a quick growth cycle but lower total production; may need enhancement.	Not used directly in feeding trials	Sharifuddin et al. (2022)
Volvariella volvacea	Paddy straw + cotton mill wastes	Yield differences are dependent on strain.	Performance and robustness are impacted by the substrate-strain combination.	No direct animal feeding trials	Ahlawat et al. (2011)

2.2 *Application of SMC Toward Vegetable Plants*

While there is limited research specifically on the application of *Volvariella volvacea*-SMC to okra (*Abelmoschus esculentus*), several studies have examined the application of spent mushroom compost in growing vegetables, including okra. Compost for mushrooms is created from a mixture of hay, poultry manure, gypsum, and water. After the mushrooms are collected, the remaining substrate can be added to soil in organic farming to improve its structure, helping it retain water, allow better airflow, and promote easier water movement (Umor et al. 2021a). As for the SMC from the cultivation of *Volvariella volvacea*, there are no documented cases of its reuse in Malaysia. Reports on the use of SMC from other mushroom species are also limited in the country.

SMC contains high levels of calcium, nitrogen, ash, and protein, making it a valuable fertilizer. Research has shown that SMC can be used effectively in horticulture, either alone or mixed with other materials. Its high nitrogen content makes it useful for improving soil fertility, and it can also enhance soil structure by speeding up the breakdown of pesticides. The use of SMC as a substrate for growing potato seedlings showed remarkable benefits. The substrate was mainly composed of a *Pleurotus ostreatus* medium, supplemented with organic and chemical fertilizers. The study indicates that SMC could effectively replace vermiculite and peat moss to create an optimal growing medium (Mwangi et al. 2024).

The term *vegetable* is not a scientific classification but a general word for many edible plant parts. It can include fruits like garden eggs, roots such as carrots, mushrooms, leafy plants, and other plant parts. In general, vegetables are parts of herbaceous plants, either above or below the ground, that are eaten as savory foods. They may be eaten raw or lightly cooked to preserve their nutrients. Studies have tested the use of spent *Agaricus subrufescens* compost in producing tomato seedlings and plants. Results show that adding this compost in different amounts helps improve seedling quality and growth after transplanting. Tomato seedlings grown with compost from *A. bisporus* and *A. subrufescens* develop different root environments (rhizospheres), which include both helpful microbes like *Chaetomium globosum* and harmful ones like *Olpidium brassicae*, as well as many unidentified species (Mwangi et al. 2024). Mixing spent *Pleurotus pulmonarius* compost with poor garden soil improved all growth traits compared to the control. It significantly increased plant height, stem thickness, number of leaves, flowers, and fruits in the tested vegetables. Using a mix of organic and chemical fertilizers gave better yields than organic fertilizer alone, and tomato yield rose greatly when nitrogen fertilizer was combined with compost (Ashrafi et al. 2016).

Studies on SMC from *Agaricus subrufescens*, *Pleurotus* spp., and *Lentinula edodes* showed that using it as part of the growing medium for lettuce and cucumber improved plant growth. One study showed that using SMC as a soil additive and for disease control in cucumber fields greatly improved germination rates—from 35% up to 92%. As the amount of SMC increased, lettuce plants grew taller with longer roots, although they had fewer leaves. At the same time, leaf phosphorus and potassium levels went up. A mixture containing 50% SMC gave the best results, with higher root length, more leaves, greater shoot weight, and overall enhanced biomass. In addition, *A. subrufescens* SMC has been used successfully as a soil conditioner and organic fertilizer for cabbage. Similar studies on organic fertilization found that SMC can boost lettuce and onion yields, making it a good alternative to chemical fertilizers (Mwangi et al. 2024).

3 Plant-Growth-Promoting Bacteria

Microorganisms are valuable resources for supporting plant growth. They can enhance seed germination, stimulate root and shoot development, increase yield, enlarge leaf size, boost chlorophyll and nitrogen-protein levels, improve drought resistance, and slow down leaf aging. Salehin et al. (2023) found that plant-growth-promoting bacteria (PGPB) are increasingly recognized as effective bio-fertilizers that enhance soil quality, promote sustainable agriculture, and boost crop yields. These beneficial microorganisms can effectively colonize the rhizosphere, the surface of plant roots, or even the inside of the roots. They are known to promote plant development through a number of methods, even in the face of diverse stressors. These include creating compounds that promote growth, balancing plant nutrition and hormones, increasing the availability of soil nutrients, and assisting plants in fending off disease.

PGPB support plant development by producing bioactive compounds, sequestering iron, and solubilizing inorganic phosphate, among other functions. Additionally, PGPB contribute to rhizoremediation by aiding in the breakdown of petroleum hydrocarbons and phenol, and they help minimize the absorption of heavy metals by plants (Alali et al. 2022).

3.1 Endophytic and Ectophytic PGPB

Microorganisms known as endophytes live inside the tissues of plants and develop a mutually beneficial relationship with them without endangering their hosts. Inoculation is a common technique used in sustainable farming to promote plant growth.

Research has shown that the introduction of PGPB can significantly increase the growth and yield of important agricultural crops. PGPB are recognized as an environmentally friendly alternative to chemical fertilizers and insecticides. They present a viable strategy for improving crop productivity, quality, and safety in sustainable agriculture. In addition, PGPB release various compounds that can indirectly enhance plant structure and overall development (Rehan et al. 2023). Aside from their ability to colonize plants, PGPB interact with other bacteria in the environment, influencing their colonization success and plant-growth-promoting activities. However, some of these interactions may be harmful. For example, investigations on sugarcane have revealed that certain coexisting bacteria can inhibit colonization and reduce the amount of injected PGPB, with similar antagonistic effects found in laboratory conditions (Salehin et al. 2023).

Ectophytic, or epiphytic, bacteria exist on the surfaces of plants, such as soils, leaves, or stems, and are also helpful to plant growth, while endophytic, or bacteria that live inside plants, have been the subject of much research. The soil area closest to plant roots, known as the rhizosphere, is formed by the substances that the roots exude. Rhizospheric bacteria can exploit these root exudates, which contain vitamins, amino acids, and other resources, permitting the most adapted bacteria to thrive and proliferate. Rhizobia, which include genera like *Bradyrhizobium*, *Sinorhizobium*/*Ensifer*, *Mesorhizobium*, *Rhizobium*, *Azorhizobium*, *Neorhizobium*, and *Pararhizobium*, are soil bacteria that can develop a symbiotic association with leguminous plants. Through chemical signal exchange, these diazotrophic (nitrogen-fixing) bacteria can infiltrate plant tissues and cause root nodules to grow. They are free-living organisms in the soil (Orozco-Mosqueda et al. 2021).

3.2 Mechanism of PGPB for the Improvement of Plant Traits

Soil microorganisms play a key role in cycling both macro- and micronutrients, contributing to plant health, soil fertility, and overall agroecosystem productivity. Through their growth, reproduction, and metabolic processes, these microorganisms release nutrients that are otherwise locked in unavailable forms and increase the soil's carbon content. Introducing microorganisms to the soil or seeds supports their proliferation, enhances nutrient cycling, and boosts the yield and productivity of agricultural crops (Manjunath et al. 2016). PGPB use both direct and indirect methods to promote the growth and development of plants. By fixing nitrogen, solubilizing vital minerals like phosphate and potassium, and generating plant hormones—capabilities that have been noted in a variety of bacterial genera—PGPB directly promote development in both symbiotic and nonsymbiotic forms (Mustapha et al. 2024).

PGPB can help plants grow indirectly by inhibiting harmful pathogens. They produce various compounds and enzymes that limit or kill these pathogens. For example, *Pseudomonas* bacteria produce siderophores, which bind iron in the environment, making it less available for pathogens. This type of pathogen suppression was one of the first mechanisms discovered in plant-growth-promoting bacteria and has been widely reported. Bacterial endophytes can protect plants from fungi by producing enzymes that break down fungal cell walls. *Bacillus* species, including *B. thuringiensis* UM96, and *Pseudomonas*, use these enzymes to fight pathogens like *Botrytis cinerea*, which causes gray mold in many plants. Bacilli are among the most studied endophytic and rhizospheric bacteria and are considered potential biocontrol agents. They can fight pathogens by producing enzymes and antibiotic compounds (Orozco-Mosqueda et al. 2021). Table 3 lists recent studies highlighting the benefits of plant-growth-promoting bacteria and their use in various crops.

3.3 Nitrogen-Fixing Bacteria

Nitrogen is a key nutrient and is often regarded as one of the most limiting factors for plant growth. Through the natural process of biological nitrogen fixation (BNF), atmospheric nitrogen in its inaccessible form is converted into a form that plants can readily use. Only bacteria and specific actinomycetes naturally carry out BNF's reliance on the nitrogenase enzyme to transform atmospheric nitrogen into ammonia (NH_3), a form that plants may use, through symbiotic or nonsymbiotic relationships with plants (Mustapha et al. 2024). The multienzyme complex known as nitrogenase is responsible for nitrogen fixation. This implies that DPM17 might be a nitrogen-fixing bacterium possessing the nitrogenase enzyme. By helping plants satisfy their nitrogen needs, such nitrogen-fixing PGPB can improve soil quality and increase agricultural output (Alali et al. 2022) (Fig. 2).

Table 3 Examples of studies demonstrating both direct and indirect biocontrol and plant-growth-promoting mechanisms in plant-associated bacteria

Bacterial species and strain	Mechanism and/or benefit in the host plant	Plant host species	References
Agrobacterium rhizogenes K599	Enhanced nitrogen uptake and alteration in the release of organic compounds	*Phaseolus vulgaris* L.	Orozco-Mosqueda et al. (2021)
Pseudomonas stutzeri E25, *Stenotrophomonas maltophilia* CR71	Release of volatile organic chemicals, inhibition of pathogens, and collaborative bacterial species interactions	*Physalis ixocarpa* (Mexican tomato), *Lycopersicon esculentum* (tomato) cv Saladette	Orozco-Mosqueda et al. (2021)
Bacillus megaterium, *Enterobacter* C7	Improved Na, Ca, and Mg acquisition; antioxidant and phytohormone synthesis; and secondary metabolite secretion	*Solanum lycopersicum* (tomato)	Orozco-Mosqueda et al. (2021)
Bacillus subtilis SWR01	Swarm signaling and motility-related genes	*Solanum lycopersicum* (tomato)	Orozco-Mosqueda et al. (2021)
Rhizobium etli G12, *Pseudomonas trivialis*, *Pseudomonas jessenii*, *Serratia plymuthica*, *Bacillus subtilis* Sb4–23, Mc5-Re2, Mc2-Re2	Antibiotic production, biofilm growth, chemotaxis, phytohormone synthesis, nematode toxicity release, and systemic resistance activation	*Solanum lycopersicum* (tomato) cv moneymaker	Orozco-Mosqueda et al. (2021)
Enterobacter cloacae	Helps plants grow under salinity/drought stress	*Solanum lycopersicum* (tomato), *Abelmoschus esculentus* (okra), *Spinacia oleracea* (spinach)	Tran et al. (2017)
Rhizobium sp.	Increased nitrogen supply and plant yield	*Phaseolus vulgaris* (bean), *Pisum sativum* (pea)	Tran et al. (2017)
Endophytic bacteria (e.g., *Bacillus*, *Pseudomonas*, *Azospirillum*, *Serratia*, *Enterobacter*)	Enhanced germination, root/shoot growth, nutrient content, and stress resistance	*Abelmoschus esculentus* (okra)	Salehin et al. (2023)
Serratia marcescens	Produces siderophores and antibiotics, suppresses fungal pathogens, improves vigor	*Lycopersicon esculentum* (tomato)	Phane Bronner et al. (2000)

BNF offers a sustainable way to reduce dependence on chemical nitrogen fertilizers, contributing over 60% of the fixed nitrogen on Earth, making its optimization in agriculture vital to meet the growing food demands of the global population (de Andrade et al. 2023). Rhizobium, known for its strong host specificity in infecting

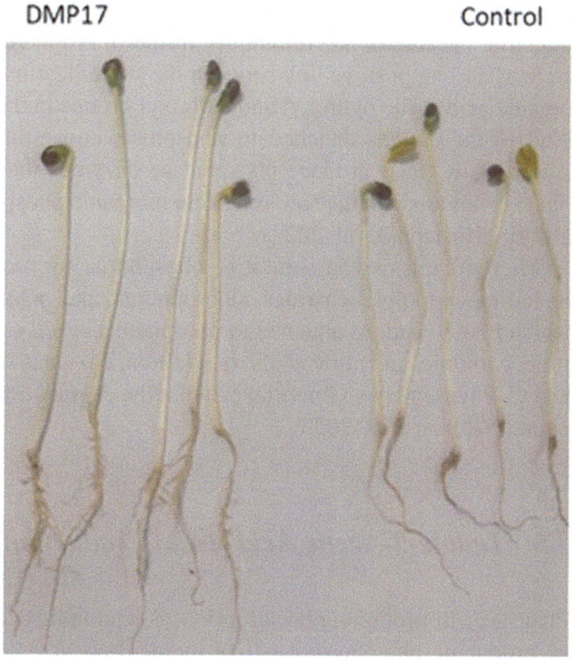

Fig. 2 Effect of DPM17 on okra plant growth (Alali et al. 2022)

the roots of leguminous plants, is a prime example of a symbiotic partnership between nitrogen-fixing microbes and plants, while only certain microorganisms, such as free-living bacteria and blue-green algae, are capable of fixing nitrogen independently (Mustapha et al. 2024).

3.4 Phosphate-Solubilizing Bacteria

Phosphates are vital for agriculture as they support key physiological and biochemical processes, including photosynthesis, root and stem growth, flower and seed formation, crop maturation, nitrogen fixation in legumes, and enhancement of plant disease resistance. Phosphorus (P) is an essential macronutrient for plant growth and metabolism, but in soil, it rapidly binds with metal cations like aluminum (Al), iron (Fe), and calcium (Ca) or attaches to mineral surfaces, reducing its availability for plant absorption (de Andrade et al. 2023). A key factor in enhancing soil phosphorus availability is converting insoluble and fixed forms of phosphorus into forms that plants can absorb.

Many PGPB, particularly from the genera *Bacillus* and *Pseudomonas*, are known as phosphate-solubilizing bacteria (PSB) and have the ability to break down both organic and inorganic phosphorus from insoluble sources (Mustapha et al. 2024). Also, de Andrade et al. (2023), in one of their research articles, stated that phosphorus-solubilizing bacteria have shown to release small organic acids that dissolve inorganic phosphorus, thereby modifying soil properties and indirectly affecting the microbial community in the rhizosphere. P-solubilization tests have shown a strong positive link between the solubilization index and the production of organic acids. The hydroxyl and carboxyl groups in these acids, produced by PSB, bind to the cations attached to phosphate, converting them into soluble forms. Additionally, PSB produce phosphatase enzymes that break down organic phosphate compounds in the soil, releasing inorganic phosphorus that plants can readily absorb (Mustapha et al. 2024).

Microorganisms can solubilize phosphorus by producing various compounds, including exopolysaccharides, siderophores, and organic acids like gluconic acid, which bind to cations attached to phosphate, as well as inorganic acids such as sulfuric, carbonic, and nitric acids. In addition, enzymes such as phosphatase, phytase, and C–P lyase play an important role in the degradation of organic phosphate compounds (Rehan et al. 2023).

3.5 Indole-3-Acetic Acid (IAA) Production

Plant-growth-promoting bacteria (PGPB) can influence a plant's hormonal balance by producing phytohormones like indole-3-acetic acid (IAA), cytokinin, and gibberellin. The impact of bacterial IAA on root growth depends on how much IAA the plant receives and how sensitive the plant is to it, which can lead to either beneficial or harmful effects. IAA is the most common auxin phytohormone, produced in the shoot apical meristem, and distributed throughout the plant. Its production is thought to be a bacterial strategy for colonizing plants, in addition to stimulating the plant's basal defense mechanisms (Mustapha et al. 2024).

Bacterial IAA stimulates the growth of lateral and adventitious roots, enhancing nutrient uptake efficiency and encouraging root exudation. This, in turn, supports further bacterial growth. However, IAA production alone does not fully account for plant growth promotion as it can also indirectly restrict root elongation (de Andrade et al. 2023). Tomato plants inoculated with IAA-producing *Klebsiella* sp. showed noticeable changes in root structure, while IAA-deficient *Bacillus* sp. had no impact on lateral root growth. A similar trend was observed in okra, where inoculation with IAA-producing *Pseudomonas* spp. led to significant improvements in growth parameters (Salehin et al. 2023). PGPB release substantial amounts of IAA, stimulating root growth and improving the plant's ability to absorb water and nutrients efficiently.

3.6 Siderophores Production

Iron (Fe) is a vital nutrient for the metabolism of soil microorganisms and usually exists in the forms Fe^{3+} and Fe^{2+}. In aerobic soils, iron often occurs as insoluble hydroxides and oxyhydroxides, which are not readily accessible for plant uptake.

One of the key traits of PGPB in relation to nutrient acquisition is their ability to produce siderophores, which bind and extract iron from the soil, making it accessible to plants. The bacterial strains in this study are recognized as siderophore producers. This capability not only increases the iron concentration in the plant's root zone but also enhances root activity, improves Fe uptake, and supports its distribution to different parts of the plant (Manjunath et al. 2016).

In soil, iron is mainly present in the ferric form (Fe^{3+}), which tends to form insoluble hydroxides with very low solubility, making it unavailable to plants and rhizosphere bacteria. In contrast, the ferrous form (Fe^{2+}) is much more soluble and accessible to plants, but it quickly oxidizes to Fe^{3+} in the environment, leading to precipitation (de Andrade et al. 2023). Various plant-growth-promoting bacteria, such as *Enterobacter*, *Pseudomonas*, *Azotobacter*, *Bacillus*, *Serratia*, and *Rhizobium*, can produce siderophores that may be extracellular or intracellular, water-soluble, and capable of extracting iron from minerals or organic compounds in iron-deficient conditions. These siderophores can also form stable complexes with heavy metals and radioactive elements. Such PGPR strains not only support plant growth but also help reduce heavy metal toxicity in polluted soils (de Andrade et al. 2023).

An important factor to consider about siderophores in the environment is their vulnerability to abiotic breakdown through processes like hydrolysis and oxidation. For siderophores with hydroxamate groups, hydrolysis can produce hydroxylamine compounds that help convert Fe^{3+} into the more soluble Fe^{2+} form.

4 Conclusion

Although mushroom compost has been shown to enhance plant growth, its application remains relatively limited. The use of mushroom compost contributes to improvements in plant height, fruit development, leaf production, and overall yield performance. When combined with growth-promoting bacteria, soil fertility and plant vigor can be further optimized, as these microorganisms release beneficial nutrients that support plant health. Thus, the integration of mushroom compost with beneficial microbial inoculants offers a sustainable approach to soil restoration and crop productivity enhancement (Fig. 3). In conclusion, the utilization of spent mushroom compost (SMC), in synergy with biological processes, represents a viable waste management strategy that supports the principles of a circular economy and promotes sustainable agriculture.

Fig. 3 PGPB for the improvement of the SMC quality model for okra plant growing

References

Ahlawat OP, Singh R, Kumar S (2011) Evaluation of *Volvariella volvacea* strains for yield and diseases/insect-pests resistance using composted substrate of Paddy straw and cotton mill wastes. Indian J Microbiol 51(2):200–205. https://doi.org/10.1007/s12088-011-0126-1

Alali HA, Khalifa A, Almalki M (2022) Plant growth-promoting bacterium from non-agricultural soil improves okra plant growth. Agriculture (Basel) 12(6). https://doi.org/10.3390/agriculture12060873

Ashrafi R, Rajib, MRR, Sultana R, Rahman MM, Mian MH, Shanta FH (2016) Effect of spent mushroom compost on yield and fruit quality of tomato. Asian J Med Biol Res 1:471. https://doi.org/10.3329/ajmbr.v1i3.26464

Bashie Ajikah L, Markson A-A, Sanni R, John Walter O (2019) Assessing the growth performance of Volvariella volvaceae on local homemade substrate. www.ijriar.com

de Andrade LA, Santos CHB, Frezarin ET, Sales LR, Rigobelo EC (2023) Plant growth-promoting Rhizobacteria for sustainable agricultural production. Microorganisms 11(4) MDPI. https://doi.org/10.3390/microorganisms11041088

Manjunath M, Kanchan A, Ranjan K, Venkatachalam S, Prasanna R, Ramakrishnan B, Hossain F, Nain L, Shivay YS, Rai AB, Singh B (2016) Beneficial cyanobacteria and eubacteria synergistically enhance bioavailability of soil nutrients and yield of okra. Heliyon 2(2). https://doi.org/10.1016/j.heliyon.2016.e00066

Mohd Hanafi FH, Rezania S, Mat Taib S, Md Din MF, Yamauchi M, Sakamoto M, Hara H, Park J, Ebrahimi SS (2018) Environmentally sustainable applications of agro-based spent mushroom substrate (SMS): an overview. J Mater Cycles Waste Manag 20(3):1383–1396). Springer. https://doi.org/10.1007/s10163-018-0739-0

Mustapha Z, Mohd KS, Othman R, Nasir NNNM, Khandaker MM, Juahir H, Bakar MFA (2024) A review of plant growth promoting Rhizobacteria and their characteristics as potential biofertilizer. Pertanika J Trop Agric Sci 47(3):659–676). Universiti Putra Malaysia. https://doi.org/10.47836/pjtas.47.3.05

Mwangi RW, Mustafa M, Kappel N, Csambalik L, Szabó A (2024) Practical applications of spent mushroom compost in cultivation and disease control of selected vegetables species. J Mater Cycles Waste Manag 26(4):1918–1933). Springer. https://doi.org/10.1007/s10163-024-01969-9

Olabanjo Olugbenga I, Gbenga Olorunshola A, Charles AA (2024) Evaluating the impact of spent mushroom substrate on growth and yield of okra (Abelmoschus esculentus (L.) Moench). https://asric.africa/natural-sciences

Orozco-Mosqueda M d C, Flores a, Rojas-Sánchez B, Urtis-Flores CA, Morales-Cedeño LR, Valencia-Marin MF, Chávez-Avila S, Rojas-Solis D, Santoyo G (2021) Plant growth-promoting bacteria as bioinoculants: attributes and challenges for sustainable crop improvement. Agronomy 11(6) MDPI AG. https://doi.org/10.3390/agronomy11061167

Phane Bronner S, Stoessel P, Gravet A, Monteil H, Pré Vost G (2000) Variable expressions of Staphylococcus aureus bicomponent leucotoxins semiquantified by competitive reverse transcription-PCR. *Appl Environ Microbiol* 66(9)

Rehan M, Al-Turki A, Abdelmageed AHA, Abdelhameid NM, Omar AF (2023) Performance of plant-growth-promoting Rhizobacteria (PGPR) isolated from sandy soil on growth of tomato (Solanum lycopersicum L.). Plants 12(8). https://doi.org/10.3390/plants12081588

Salehin A, Yamane S, Ueno M, Hayashi S (2023) Identification, characterization, and growth-promoting effects of bacterial endophytes isolated from okra (Abelmoschus esculentus L.). Agronomy 13(5). https://doi.org/10.3390/agronomy13051226

Sharifuddin AS, Ahmad LWZ, Mohamad A, Sidik NJ, Yunus NM, Yusof SCM, Osman ND, Rusli IS (2022) Growth performance and fruiting body characterisation of the paddy straw mushroom (Volvariella volvacea) cultivated on different types of solid wastes. Asia-Pac J Sci Technol. https://www.tci-thaijo.org/index.php/APST/index

Tran TTT, Mangenot S, Magdelenat G, Payen E, Rouy Z, Belahbib H, Grail BM, Johnson DB, Bonnefoy V, Talla E (2017) Comparative genome analysis provides insights into both the lifestyle of Acidithiobacillus ferrivorans Strain CF27 and the chimeric nature of the iron-oxidizing acidithiobacilli genomes. Front Microbiol 8(JUN). https://doi.org/10.3389/fmicb.2017.01009

Umor NA, Abdullah S, Mohamad A, Ismail SB, Ismail SI, Misran A (2021a) Energy potential of oil palm empty fruit bunch (Efb) fiber from subsequent cultivation of volvariella volvacea (bull.) singer. Sustainability 13(23). https://doi.org/10.3390/su132313008

Umor NA, Ismail S, Abdullah S, Huzaifah MHR, Huzir NM, Mahmood NAN, Zahrim AY (2021b) Zero waste management of spent mushroom compost. J Mater Cycles Waste Manag 23(5):1726–1736). Springer. https://doi.org/10.1007/s10163-021-01250-3

Zied DC, Sánchez JE, Noble R, Pardo-Giménez A (2020) Use of spent mushroom substrate in new mushroom crops to promote the transition towards a circular economy. Agronomy 10(9) MDPI. https://doi.org/10.3390/agronomy10091239

Therapeutic and Immunological Activities of the Cultivated Edible Mushrooms

Ahmed Farhan Shallal, Kazhal Shekh Mohamad Ibrahim, and Mustafa Nadhim Owaid

Abstract This chapter elucidates the considerable health and clinical benefits linked to the intake of edible mushrooms, highlighting their nutritional profile and therapeutic properties. Fungi are rich in protein, dietary fiber, vitamins, and vital minerals while concurrently demonstrating low levels of fat and caloric density. Moreover, they are acknowledged for their bioactive components, such as β-glucans, terpenoids, and phenolic compounds, which bestow antioxidant, anti-inflammatory, anticancer, and immune-modulatory effects. These features position mushrooms not only as a crucial component of the diet but also as a promising therapeutic approach for various ailments, including cancer, diabetes, obesity, and neurodegenerative diseases. A wide variety of fungal species, particularly *Agaricus bisporus* and *Ganoderma lucidum*, have exhibited considerable potential in the therapeutic management of immune-related disorders, the prevention of genomic damage, and the inhibition of neoplastic cell proliferation. Enhanced extraction techniques, including microwave-assisted and ultrasound-assisted methodologies, effectively isolate these pharmacologically active compounds. The growing interest in medicinal fungi underscores their essential role in integrative health practices, aiding both the prevention and treatment of diverse diseases. This chapter calls for a more

A. F. Shallal
Department of Biology, College of Science, University of Raparin, Kurdistan Region, Iraq

College of Medicine, University of Sulaimani, Sulaymaniyah, Iraq
e-mail: ahmed.farhan@uor.edu.krd

K. S. M. Ibrahim
Department of Medical Laboratory Sciences, College of Sciences, University of Raparin, Ranya, Kurdistan Region, Iraq
e-mail: kazhalgallaly@uor.edu.krd

M. N. Owaid (✉)
Department of Environment, College of Applied Sciences-Heet, University of Anbar, Hit, Anbar, Iraq

Department of Heet Education, General Directorate of Education in Anbar, Ministry of Education, Hit, Anbar, Iraq
e-mail: mustafanowaid@uoanbar.edu.iq

M. N. Owaid et al. (eds.), *Mushroom Production*, Interdisciplinary Biotechnological Advances, https://doi.org/10.1007/978-981-95-7793-4_10

extensive exploration of the applications of mushrooms within modern medicinal and nutritional contexts.

Keywords Immunodulatory · Edible mushrooms · Pathogens. Health

1 Introduction

Mushrooms represent a predominantly untapped reservoir of nutrient-dense alimentary resources. Their production currently stands as the most economically viable biotechnological approach for turning lignocellulose waste into high-protein food while also greatly reducing environmental pollution (Kakon et al. 2012). Among the approximately 1600 distinct species of mushrooms, a mere 100 are recognized as edible. Globally, only three varieties of mushrooms are classified as acceptable for human consumption and are cultivated systematically on a regular basis: *Agaricus bisporus L.*, *Pleurotus ostreatus L.*, and *Volvariella volvacea L.* Additionally, around 33 species of edible mushrooms are actively cultivated (Erbiai et al. 2021). Edible mushrooms constitute a highly advantageous source of fiber and essential nutrients, including protein, vitamins, and, minerals, while being low in calories and fat. Beyond their culinary benefits, mushrooms are widely used in medicinal contexts due to their antineoplastic, antioxidant, and antimicrobial properties (Roncero-Ramos et al. 2017). Due to their low lipid, elevated protein, and reduced energy compositions, mushrooms are vital to the human diet, in addition to their notable pharmaceutical qualities. Mushrooms contain all the essential amino acids needed by humans, along with various important nutrients, like B vitamins (such as riboflavin and niacin), phosphorus, iron, vitamin C, and ergosterol (Khatun et al. 2012; Kumar 2015). Mushrooms are valued for their texture, flavor, and potential health benefits. Studies have shown that mushrooms contain functional components that may have antiviral, anticancer, antithrombotic, and immune-modulating effects. Among these, mushroom carbohydrates, specifically β-glucans, are garnering attention from researchers and professionals in the food industry on account of their antioxidant, antidiabetic, anticarcinogenic, and immune-boosting properties. To harness these bioactive compounds, a range of extraction techniques are employed, from traditional methods to advanced approaches like enzyme-assisted, ultrasound-assisted, and microwave-assisted extraction, as well as subcritical water and pulsed-electric-field-assisted processes. These innovative methods help effectively extract valuable components from mushrooms (Perera and Li 2011; Leong et al. 2021). This review covers edible mushrooms, their types, nutrition, and potential medical uses. It discusses their benefits, risks, and how they can be part of a healthy diet and therapy.

2 Common Types of Consumable Mushrooms

Commonly encountered edible mushrooms represent the prominent reproductive structures of several macrofungi that are suitable for human consumption like *Ganoderma* (Farhan et al. 2025). The absence of poisonous or harmful compounds characterizes an edible mushroom. Mushrooms are commonly consumed for their nutritional value, therapeutic properties, and pleasant flavor. They contain high concentrations of proteins, fiber, vitamins, minerals, and bioactive substances, such as sterols, terpenoids, polysaccharides, lactones, sesquiterpenes, and polyphenolic compounds (Kumar 2015). They offer many health advantages, including antibacterial, immune-stimulating, anticancer, and cholesterol-lowering properties (Fig. 1). Several important elements and secondary metabolic products drive their biochemical roles. A diverse range of edible fungi, such as *Phaseolus vulgaris*, *Agaricus campestris*, *Agaricus bisporus*, *Grifola frondosa*, *Boletus satanus*, *Flammulina velutipes*, *Tricholoma mongolicum*, *Ganoderma lucidum*, and *Volvariella volvacea*, contain lectins, which are a class of carbohydrate-binding proteins. Lectins enhance the release of insulin, activate the immune response in fighting infections, and demonstrate characteristics that may inhibit cancer progression (EL-maradny et al. 2021). Lectins, found in edible mushrooms like *Clitocybe nebularis*, are essential for physiological processes like dormancy, developmental growth, morphogenesis, and molecular recognition mechanisms during mycorrhization (Pohleven et al. 2011). Glucan, a unique component of mushrooms, has anti-inflammatory, anticancer, immune-stimulatory, and antioxidant properties. Several edible mushroom species, including *Auricularia auricular*, *Lentinula edodes*, *Ganoderma lucidum*, and oysters, contain glucans (Ganeshpurkar et al. 2015). For instance, the beta-glucan from *Pleurotus pulmonarius* exhibits strong anti-inflammatory and pain-relief qualities, which prevent pro-inflammatory processes by modulating cellular

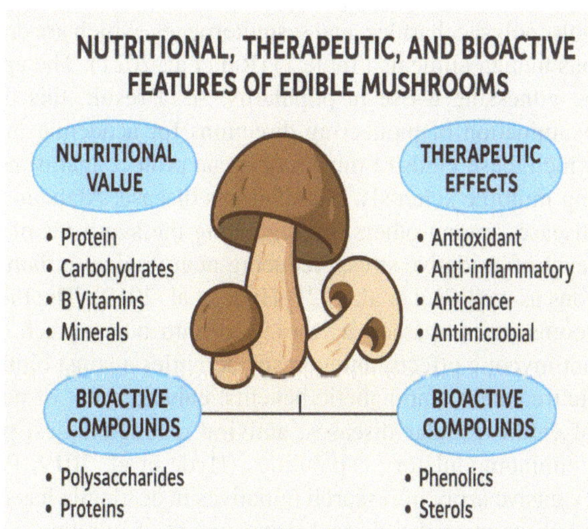

Fig. 1 A detailed assessment of the nutritional, therapeutic, and bioactive features of edible mushrooms

proliferation, initiating programmed cell death, and mitigating inflammatory responses; glucans extracted from *Pleurotus pulmonarius* have also demonstrated a capacity to impede colon carcinogenesis associated with colon inflammation (Murphy et al. 2022; Rathee et al. 2012). Consumable mushrooms contain phenolic acids, which have been shown to prevent various health issues, including p-hydroxybenzoic acid, cinnamic acid, gallic acid, salicylic acid, p-coumaric acid, syringic acid, caffeic acid, ferulic acid, chlorogenic acid, and various flavonoids (Taofiq et al. 2016). Phenolic substances exhibit strong antioxidant properties. Edible mushrooms with high antioxidant capacity include *Meripilus giganteus, Agaricus hydnum*, and *Rufescens silvaticus* (Kała et al. 2021). Flavonoids constitute an important class of polyphenolic compounds known for their antiproliferative activity (Saltarelli et al. 2019). Fungi comprise a total of 285 distinct terpenoids with promising therapeutic implications, notably including ganoderic acids and ergosterol, as they act as natural antioxidants and precursors to vitamin D, found across diverse species (Papoutsis et al. 2020). Terpenoids, which encompass monoterpenoids, sesquiterpenoids, diterpenoids, and triterpenoids present in fungi, have shown significant antibacterial, anticancer, anticholinesterase, anti-inflammatory, antimalarial, and antioxidant activities (Surup et al. 2019). Ergosterols, particularly ergosterol peroxide, exhibit a greater effectiveness than indomethacin in reducing inflammation, with *Agaricus bisporus* showing a remarkably high concentration of ergosterol (Nowak et al. 2022).

3 Therapeutic Applications of Edible Mushrooms

Medicinal mushrooms are categorized based on their ability to provide health benefits, including antibacterial, anticancer, and antioxidant properties. These fungi contain beneficial components such as metal chelating agents, alkaloids, polyphenols, polysaccharides, and sesquiterpenes, which are crucial in the treatment of various human ailments (Table 1) (Rai et al. 2021). The application of medicinal fungi is witnessing a rise in popularity. As a result, this development is fostering the examination of pioneering directions for academic investigation into the ways in which these kinds of mushrooms can protect against neurological diseases, including multiple sclerosis, Huntington's disease, Alzheimer's disease, and Parkinson's disease, among others, by regulating the synthesis of neurotrophins, lowering the level of oxidative stress, reducing neurological inflammation, and preserving neurons as well (Rai et al. 2021; Hyde et al. 2019). The biomedical relevance of mushrooms encompasses a variety of attributes, such as antibacterial properties, antimycotic effects, antagonistic activities against biofilms, modalities for neoplastic treatment, antidiabetic benefits, enhancement of neural functions, management of cardiovascular diseases, antiviral characteristics, prebiotic functionalities, and immunomodulatory capabilities (Hyde et al. 2019; Panda and Luyten 2022). An extensive array of research initiatives in academia has targeted the potential roles of mushrooms in the clinical management of immune system disorders related to the

Table 1 The pharmacological properties of various edible mushrooms, highlighting their medicinal activities

Mushroom species	Medicinal activity	References
Agaricus blazei	Antitumoricidal, anticarcinogenic, antimetastatic (cancer) properties	Bertollo et al. (2022)
Russula griseocarnosa	Immune control, antioxidants, hypoglycemic/hypolipidemic (diabetic) action	Liu et al. (2022)
Macrolepiota procera, *Gymnopus dryophilus*	Anti-inflammatory, anticancer, decreasing cardiovascular disorders	Nowak et al. (2022)
Lentinus edodes and *Hypsizygus tessellatus*	Immune-suppressive states (inflammatory pulmonary diseases)	Murphy et al. (2022)
Sanghuangprous vaninii	Anticancer, particularly gynecological cancers (cervical cancer)	He et al. (2021)
Russula virescens	Hypoglycemia, immunological responses, and chemotherapeutic	Li et al. (2021)
Ganoderma lucidum, *G. artum*, and *G. tsugae*	Actions against inflammation, cancer, oxidation, hyperglycemia, or hyperlipidemia (insulin resistance)	Wang et al. (2020); Ren et al. (2021)
Sanghuangporus sanghuang	Anticoronavirus (SARS-CoV-2) and antioxidant activities	Zhang et al. (2021)
Ganoderma lucidum	Tumor-causing diseases: arthritis, bronchitis, hypertension, gastric ulcer, and nephritis	Kurd-Anjaraki et al. (2022)
Agaricus spp., *Pleurotus* spp., *Lentinus* spp.	Chemicals that fight cancer and tumors	Hyder and Dutta (2021)
Ganoderma applanatum, *Pleurotus eryngii*, *Lentinula edodes*	Effects that are anti-inflammatory, antibacterial, antitumor, and antiproliferative (associated with human colorectal cancer and Alzheimer's disease)	Venturella et al. (2019)
Species of Agaricus, *Ganoderma*, *Pleurotus*	Mitigating effects of osteoarthritis and antineoplastics (with relation to osteoarthritis and cancer)	Yongabi (2019)
Lignosus rhinocerus, *L. tigris*, *L. cameronensis*	Anticancer, anti-inflammatory, and antioxidative activity of cancer	Fung and Tan (2019)
Antrodia salmonea	Antioxidant, anti-inflammatory, antitumor, and antiatherosclerosis activity	Karuppaiya and Akbar (2019)
Agaricus bisporus, *Ganoderma frondosa*, *Lentinula edodes*	Antimalignancy	Neergheen et al. (2022)
Calvatia gigantea, *Fomes fomentarius*, *Phallus impudicus*	Anti-inflammatory, antitumor, antioxidative, and antibacterial	Gründemann et al. (2020)
Armillaria mellea, *Antrodia camphorata*, *Hericium erinaceus*, *Ganoderma* spp.	Anti-inflammatory properties of neurons (Alzheimer's disease)	Kushairi et al. (2020)
Pleurotus citrinopileatus, *P. djamor*, *P. eryngii*, and *P. ostreatus*	Increased lovastatin concentrations and anti-atherosclerotic qualities (associated with cardiovascular and cancer illnesses)	Krakowska et al. (2020)
Ganoderma lingzhi, *G. applanatum*, *G. lucidum*, and *G. tsugae*	Regarding gastrointestinal (immune) disorders	Mirmazloum et al. (2021)

composition and types of mushrooms and their effects (Chan et al. 2020). These are divided into the following groups.

3.1 Anti-DNA Damaging Group

This category contains mushrooms with compounds, such as polysaccharides, that may protect against deoxyribonucleic acid (DNA) damage, like *Agaricus blazei* and *Ganoderma lucidum* (Živković et al. 2019; Collin-Hansen et al. 2005).

3.2 Anticipates Group

Among these are *Tricholoma lobayense* and *Ganoderma lucidum*, which may provide the antiaging polysaccharide (TLH-3), ergosterol, and ganodermasides (Chaturvedi et al. 2018; Zhang et al. 2019a, b).

3.3 Anti-obesity Group

Mushroom species like *Pleurotus eryngii*, *P. sajor-caju*, and *Ophiocordyceps sinensis* contain bioactive substances like *Cordyceps sinensis* polysaccharide and β-glucan water, which can help combat obesity (Ganesan and Xu 2018; Khatun et al. 2020; Wunjuntuk et al. 2022; Mustafa et al. 2022).

3.4 Antiviral Group

Mushrooms contain biologically active compounds with antiviral and immunomodulatory effects, especially relevant during the COVID-19 pandemic. Fungal extracts, polysaccharides, and melanins from fungal cultures and fruit bodies can be employed to assess efficacy against pathogenic viruses sourced from the SRC VB Vector repository. Melanin extracted from the *Inonotus obliquus* strain F-1244 exhibits properties that inhibit the replication of pandemic viruses in MDCK cell cultures (Seo and Choi 2021; Teplyakova et al. 2021). This category includes fungi such as *Phellinus baumii*, *Inonotus hispidus*, and *Phellinus igniarius*, which contain bioactive compounds that effectively hinder the influenza virus (H5N1) (Teplyakova et al. 2021; Shahzad et al. 2020). Furthermore, compounds obtained from mushrooms demonstrate significant therapeutic potential against a range of viral pathogens, including herpes simplex virus, influenza virus, and human immunodeficiency virus (HIV). The underlying mechanisms of these compounds involve the inhibition of critical phases of the viral life cycle, such as viral entry, replication, and immune system modulation, thereby exhibiting antiviral characteristics that are comparable to those of traditional antiviral therapeutics (Seo and Choi 2021). Prominent instances include *Agaricus blazei* (Murill extract utilized against foot-and-mouth disease virus), *Hypsizygus marmoreus* (sterols effective against Epstein-Barr virus), and Auricularia auricula-judae (polysaccharides targeting Newcastle disease virus), thereby underscoring the extensive array of bioactive compounds contained within these fungi that possess therapeutic potential (Stamets 2006). Moreover, mushrooms present notable antiviral effects against diverse viral pathogens, such as HIV, influenza, herpes simplex virus, and hepatitis B and C viruses (HBV/HCV), accentuating the essential requirement for further exploration into their capacity as effective therapeutic interventions (Teplyakova et al. 2021; Ahmad et al. 2021). Based on their pharmacological uses, this collection encompasses fungi such as *Pleurotus ostreatus* (which possesses lectin HBV), *Ganoderma lucidum* (recognized for its elevated levels of ganoderic acid), *Lentinula edodes* (demonstrating a mycelial solid culture extract that shows effectiveness against HCV), and *Antrodia camphorata* (which contains polysaccharides that specifically target HBV). These fungi are characterized by bioactive components that exhibit efficacy against hepatitis viruses B and C (HBV/HCV). Recently, an abundance of academic literature has explored the biological importance of bioactive compounds present in mushrooms, particularly focusing on polysaccharides, in relation to addressing COVID-19 (Zhang et al. 2022; Azam et al. 2023).

3.5 Neuroprotective Group

Dementia represents a persistent neurological condition characterized by impairments in memory and cognitive learning capabilities. The pathophysiology of this disorder is significantly influenced by oxidative stress and inflammatory processes. Natural medicinal fungi, such as *Hericium erinaceus* and *Coriolus versicolor*, have demonstrated efficacy in enhancing cognitive performance and in the prophylaxis of Alzheimer's and Parkinson's diseases. Their bioactive constituents, particularly erinacine A, have been shown to effectively mitigate neurodegenerative processes and dementia manifestations, with no recognized adverse effects. These mushrooms are considered safe for ingestion and may be incorporated into dietary regimens (Mccain et al. 2023; Tong et al. 2023). Although dementia poses a notable challenge, mycological compounds indicate promise in reducing symptoms related to Alzheimer's disease (AD) among the aging demographic.

Current investigations have revealed a range of constituents in fungi, encompassing carbohydrates, peptides, and phenolic compounds, that reveal anti-AD attributes, including antioxidative and antineuroinflammatory properties. However, continuous research is imperative to uncover novel metabolites and investigate their underlying molecular mechanisms (Tong et al. 2023). Certain medicinal fungi, including *Sarcodon scabrosus*, *Paxillus panuoides*, *Hericium erinaceus*, *Lignosus rhinocerotis*, and *Grifola frondosa*, have exhibited the capacity to augment human cognitive functions. This group of fungi harbors a wide range of bioactive compounds that may be utilized in the treatment of various diseases, such as ganoderic acids, scabronines, hericenones, as well as ethanolic and aqueous extracts (Sahu et al. 2023; Rangsinth et al. 2023).

3.6 Anticancer Group

Cancer is acknowledged as a significant element in the global mortality rates. For instance, *Piptoporus betulinus* has been conventionally employed in the Czech Republic for addressing colorectal cancer, whereas the fruiting bodies of *Inonotus obliquus* have been used in Eastern European traditional medicine since the sixteenth or seventeenth century for oncological interventions. Dietary sources naturally containing compounds endowed with immunostimulatory, antioxidant, and rejuvenating properties, such as edible mushrooms, are well-recognized for their potential role in cancer prevention. However, the periodic manifestation of mushroom reproductive structures in their natural habitat poses challenges to their accessibility (Muszyńska et al. 2017). In addition to leukemia, this group comprises several other malignancies, including breast, hepatic, pulmonary, and prostatic carcinomas, as well as colorectal, oral, renal, and cervical malignancies. The primary bioactive mushroom extracts that have the potential to treat cancer are ethanol extracts, ganoderic acid, polysaccharides, and methanol (Liuzzi et al. 2023; Buranrat et al. 2020).

3.7 Immunostimulatory Group

Proteins extracted from nine common and 13 precious edible mushrooms exhibit diverse biological properties. These proteins have been found to increase macrophage numbers and phagocytosis abilities in mice and to also influence the liberation of reactive oxygen species. Rare mushrooms have greater gastronomic and therapeutic value (Xu et al. 2022; Hyde et al. 2019). The immunostimulatory group encompasses various species, such as *Trametes pubescens* and *Pleurotus ostreatus*. Extracts from these mushrooms, particularly their lectins, have been suggested to bolster immune responses (Hyde et al. 2019; Li et al. 2021; Panda and Luyten 2022). Additionally, mushrooms contain anti-inflammatory ingredients like polysaccharides, phenolics, and biometals. These agents exhibit antioxidant and anticancer activities. Edible mushroom extracts have positive health benefits, making them a beneficial "superfood" for inflammation-related illnesses (Muszyńska et al. 2017). This includes several species of mushrooms, such as *Psilocybin* spp. and Grifola frondosa. Extracts of polysaccharides or the active metabolite psilocin may be beneficial for rheumatoid arthritis (Prescott et al. 2023; Chaturvedi et al. 2018).

4 Basic Elements of the Immune System in Humans

The immune system includes two basic categories, referred to as innate immunity and acquired immunity (Abbas et al. 2016; Abbas et al. 2021). Furthermore, each primary branch is further categorized into secondary branches, namely, cellular and humoral. Consequently, dietary changes can influence immune response mechanisms (Shallal et al. 2023). The main elements of different immunity types are shown in Fig. 2.

The development of lymphocytes begins with hematopoietic stem cells residing in the bone marrow and proceeds within specialized primary lymphoid organs, where T lymphocytes mature in the thymus and B lymphocytes mature in the bone marrow.

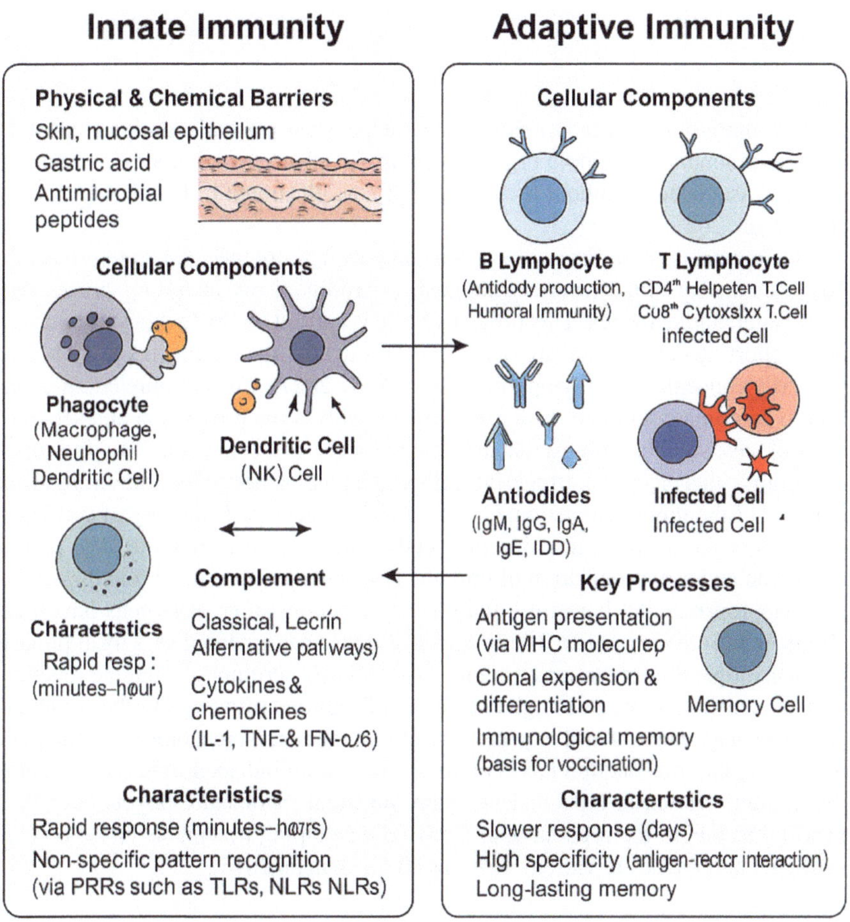

Fig. 2 Basic elements of innate and adaptive immunity

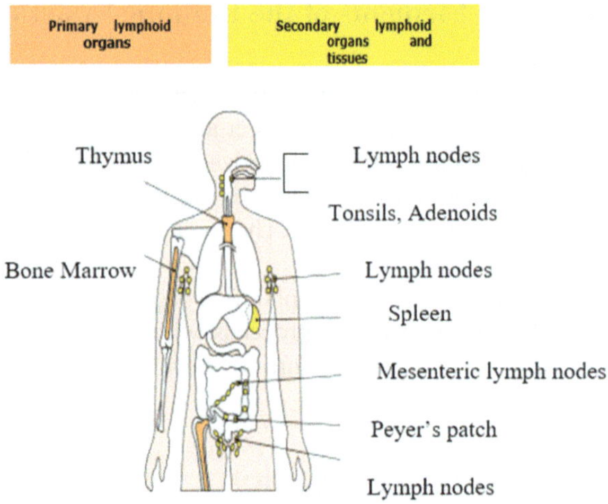

Fig. 3 Primary and secondary lymphoid organs

After maturation, these lymphocytes, which express particular antigen receptors, move to secondary lymphoid organs, such as the spleen, lymph nodes, and various mucosa-associated lymphoid tissues (MALT), as illustrated in Fig. 3 (Owen et al. 2013; Male 2021).

The immune system has evolved as a complex defense mechanism to protect the host from the harmful effects of pathogenic microorganisms, including viruses, bacteria, fungi, and parasites. This protective role is carried out by leukocytes and various supporting cells, which are distributed throughout the organism's body but are primarily concentrated in lymphoid tissues (Male 2021). Cells require communication to coordinate immune responses, which involves the production and secretion of cytokines, soluble proteins that are crucial for the process. These cytokines interact with various receptors, triggering cellular responses that govern the strength and duration of the immune reaction. Chemokines, a specialized category of cytokines, can prompt the movement of immune cells toward an infection site. The precise regulation and synchronization of immune responses play a critical role in preventing inappropriate and harmful inflammatory reactions after pathogens have been eliminated (Sohlberg 2013). The immune response has emerged as a focal point in pharmacological interventions, ranging from a basic scrutiny of immune reactions to pathogens to a more thorough evaluation of immune responses in the context of cancer management. Over the years, examining mediators, receptors, and the pathways of signal transduction has been recognized as an indispensable component of the immune response. Nonetheless, many potential therapeutic targets have been overlooked due to the predominant focus on developing treatments based on conventional approaches to various diseases (B De Sanctis 2016).

5 Immune Modulations

Several natural materials have shown impressive immunomodulatory and health-enhancing effects in humans, exhibiting minimal to negligible toxicity (Bryła 2020). Pharmaceutical compounds of biological origin have demonstrated themselves to be the safest and least harmful when explored as potential alternatives. Mushrooms, owing to their significant nutritional profile and wide variety of compounds, have attracted the interest of both the food and biopharmaceutical sectors (Taofiq et al. 2016). An immunomodulator is a natural or synthetic agent that regulates or modifies the activity of the immune system. In a healthy organism, the immune system produces a diverse array of molecules to maintain physiological homeostasis. An immunomodulator is a substance with the capacity to stimulate, inhibit, or regulate elements of the immune system, encompassing both the innate and adaptive segments of the immune response. These compounds, which may be of biological or artificial origin, are divided into three primary categories: immunosuppressants, immunostimulators, and immunoadjuvants (Enshasy 2011) (Table 2).

While the majority of immunomodulators are synthetic or semi-synthetic compounds, growing scientific interest has recently focused on the exploration of natural immunomodulators. Several bioactive substances derived from natural sources have exhibited significant immunomodulatory effects and have been beneficial to human health, with little to no harmful effects (Zhao et al. 2020; Mannino et al. 2020; Bryła 2020).

6 Immunomodulatory Impacts of Mushroom

Medicinal fungi, known as medicinal mushrooms (MMs), serve as vital components and natural sources of immunomodulating agents. Belonging to a specific group of fungi, medicinal mushrooms (MMs) are macroscopic fungi commonly

Table 2 Types of immunomodulators

No.	Types of immunomodulators	Functions
1	Immunosuppressants	Substances that impede the functioning of the immune system and are commonly employed to manage aberrant immune responses following organ transplants, as well as in the treatment of autoimmune conditions, hypersensitivity reactions, and immune-related pathologies linked to infections
2	Immunostimulators	Compounds that serve to activate the immune system by either triggering the engagement or amplifying the function of its various components; they augment the body's ability to combat allergy, infection, cancer, and autoimmunity
3	Immunoadjuvants	Agents utilized to amplify the effectiveness of vaccines, reflecting a targeted immune-stimulating action

used in the form of extracts or whole mushroom powder to promote human health (Wasser 2010; Zhao et al. 2020). Historically, the dietary value of food has been assessed by examining its nutrient content and its capacity for digestion, absorption, and utilization. Consuming food or food-related products constitutes a crucial and customary element of daily living. Health, nutritional intake, and dietary behaviors are closely interrelated and collectively influence food selection within a social framework, which is further shaped by the nutritional characteristics of food products (Khezri and Shallal 2024). The gastrointestinal tract and the hepatic organ are crucial anatomical systems responsible for the provision of vital nutrients and the implementation of self-protective mechanisms, and these systems are equipped with specialized immunological architectures. Within the gastrointestinal environment, a significant cohort of immunoglobulin A (IgA)-producing cells and intraepithelial T lymphocytes (IELs), each possessing distinct physiological roles, is present, thus greatly enhancing the host's protective mechanisms against pathogenic threats. Additionally, the liver harbors a notable concentration of natural killer (NK) T cells, which demonstrate remarkable efficacy in the elimination of neoplastic cells. Antigens, including pathogens, toxins, and allergens, are recognized by the immune system and trigger immune responses, in which both the liver and the intestine play critical roles (Kaminogawa and Nanno 2004). An extensive range of scientific research and clinical applications has demonstrated the role of fungi as immunomodulators possessing anticancer, anti-inflammatory, antibacterial, antiallergic, and antidiabetic functionalities, which are ascribed to their rich assortment of bioactive metabolites and various phytochemicals (Jiang et al. 2020; Mihich 2000; Zhang et al. 2019a, b). The Food and Drug Administration (FDA) affirms the classification of mushrooms as appropriate health foods due to their notable medicinal and nutritional benefits (Jeitler et al. 2020). Recent investigations have highlighted the effectiveness of extracts obtained from edible mushrooms in the treatment of immunological disorders. Ongoing advances in understanding the physiological mechanisms of the immune system enable the development of more effective and safer pharmaceuticals by deconstructing the complex mechanisms that regulate immune responses (Kała et al. 2024). The proteins derived from edible mushrooms significantly enhance the proliferation and phagocytic activity of murine macrophages, thereby influencing the synthesis of reactive oxygen species and nitric oxide. Further studies on the cytokine profile in murine macrophages demonstrated a significant increase in pro-inflammatory cytokines, suggesting that proteins from edible mushrooms promote differentiation toward the M1 macrophage phenotype. This finding highlights the immunomodulatory characteristics of edible mushroom proteins (Xu et al. 2022).

7 Glucans and Immune Response

Glucans are predominant polysaccharides found in fungi, contributing significantly to the cellular composition of mushrooms. These components are discharged into environments that promote cellular expansion, thereby facilitating their prospective extraction, purification, and analysis (Motta et al. 2021). Immune receptors known as pattern recognition receptors (PRRs) link both innate and adaptive immunity by identifying molecular patterns on pathogens, injured cells, or dying cells. They elicit nonspecific immune responses against infections, cancer, and other dangers by binding particular ligands. Toll-like receptors (TLRs), NOD-like receptors (NLRs), RIG-I-like receptors (RLRs), C-type lectin receptors (CLRs), and AIM2-like receptors (ALRs) are the five primary categories into which PRRs in vertebrates are divided (Li and Wu 2021). Glucans, marked by the presence of pathogen-associated molecular patterns (PAMPs), lack the inherent ability to be synthesized but possess the capacity to trigger an immune response. Detection of these molecular patterns is augmented by a variety of pattern-recognition receptors (PRRs) situated on immune cells, such as macrophages, monocytes, neutrophils, dendritic cells (DCs), and natural killer (NK) cells, including receptors like Dectin-1 and Toll-like receptors (TLRs) (Vetvicka et al. 2019; Plato et al. 2015), which interact with glucans. The engagement of these receptors initiates signaling cascades through either direct activation or the activation of cellular pathways. Studies conducted with animal models indicate a credible mechanism through which the glucan-mediated aggregation of Dectin-1 aids in the recruitment of downstream proteins that ultimately activate the nuclear factor kappa-light-chain-enhancer of activated B cells (NF-κB), thus orchestrating transcriptional regulation (Elcombe et al. 2013). Glucans have been recognized as crucial mediators in the activation of dendritic cells and promote the proliferation and polarization of T lymphocytes, thus playing a significant role in immune responses. Moreover, they are known to stimulate B-cell activation, resulting in increased production of immunoglobulin (Ig) following glucan exposure (Motta et al. 2021).

References

Abbas AK, Lichtman AH, Pillai S (2016) Hypersensitivity. Basic immunology: functions and disorders of the immune system. Cell. Mol. Immunol, 5th edn. Elsevier Health Sciences, pp 231–247

Abbas AK, Lichtman AH, Pillai S (2021) Cellular and molecular immunology, 10e, South Asia edition-e-book. Elsevier Health Sciences

Ahmad MF, Ahmad FA, khan MI, Alsayegh AA, Wahab S, Alam MI, Ahmed F (2021) Ganoderma lucidum: a potential source to surmount viral infections through β-glucans immunomodulatory and triterpenoids antiviral properties. Int J Biol Macromol 187:769–779

Azam MS, Islam MN, Wahiduzzaman M, Alam M, Dhrubo AAK (2023) Antiviral foods in the battle against viral infections: understanding the molecular mechanism. Food Sci Nutr 11:4444–4459

B De Sanctis J (2016) Editorial (thematic issue: immune modulation). Curr Pharm Des 22(41):6207–6208

Bertollo AG, Mingoti MED, Plissari ME, Betti G, Roman Junior WA, Luzardo AR, Ignácio ZM (2022) Agaricus blazei murrill mushroom: a review on the prevention and treatment of cancer. Pharmacol Res Modern Chinese Med 2:100032

Bryła P (2020) Who reads food labels? Selected predictors of consumer interest in front-of-package and back-of-package labels during and after the purchase. Nutrients 12(9):2605

Buranrat B, Sangdee K, Thammawat S, Sangdee A (2020) Mechanisms of crude protein from medicinal mushroom ophiocordyceps sobolifera against human breast mcf-7 cancer cells. Biologia 75:1759–1768
Chan XH, Sabaratnam V, Abdullah N, Phan CW (2020) A 53-year bibliometric and scientometric analysis of research in culinary and medicinal mushrooms. Int J Med Mushrooms 22:521–534
Chaturvedi VK, Agarwal S, Gupta KK, Ramteke PW, Singh MP (2018) Medicinal mushroom: boon for therapeutic applications. 3 Biotech 8:334
Collin-Hansen C, Andersen RA, Steinnes E (2005) Damage to DNA and lipids in boletus edulis exposed to heavy metals. Mycol Res 109:1386–1396
Elcombe SE, Naqvi S, Van Den Bosch MW, MacKenzie KF, Cianfanelli F, Brown GD, Arthur JSC (2013) Dectin-1 regulates il-10 production via a msk1/2 and creb dependent pathway and promotes the induction of regulatory macrophage markers. PLoS One 8(3):e60086
EL-maradny YA, EL-fakharany EM, Abu-serie MM, Hashish MH, Selim HS (2021) Lectins purified from medicinal and edible mushrooms: insights into their antiviral activity against pathogenic viruses. Int J Biol Macromol 179:239–258
Enshasy, H.E. (2011). Immunomodulators. Industrial Applications, pp.165–194
Erbiai EH, DA Silva LP, Saidi R, Lamrani Z, Esteves DA Silva JCG, Maouni A (2021) Chemical composition, bioactive compounds, and antioxidant activity of two wild edible mushrooms armillaria mellea and macrolepiota procera from two countries (Morocco and Portugal). Biomolecules 11:575
Farhan A, Razaq C, Owaid MC University of Anbar (2025) Immunomodulatory effects of ganoderma lucidum on NK cells: from traditional use to modern immunotherapy. J Macrofungi 14. https://doi.org/10.65999/macrofungi.2025.27
Fung SY, Tan CS (2019) Tiger milk mushroom (the lignosus trinity) in Malaysia: a medicinal treasure trove. Medicinal Mushrooms
Ganesan K, Xu B (2018) Anti-obesity effects of medicinal and edible mushrooms. Molecules 23
Ganeshpurkar A, Pardhi P, Bhadoriya SS, Jain N, Rai G, Jain AP (2015) Antioxidant potential of white oyster culinary-medicinal mushroom, pleurotus Florida (higher basidiomycetes). Int J Med Mushrooms 17:491–498
Gründemann C, Reinhardt JK, Lindequist U (2020) European medicinal mushrooms: do they have potential for modern medicine? – an update. Phytomedicine 66:153131
He PY, Hou YH, Yang Y, Li N (2021) The anticancer effect of extract of medicinal mushroom sanghuangprous vaninii against human cervical cancer cell via endoplasmic reticulum stress-mitochondrial apoptotic pathway. J Ethnopharmacol 279:114345
Hyde KD, Xu J, Rapior S, Jeewon R, Lumyong S, Niego AGT, Abeywickrama PD, Aluthmuhandiram JVS, Brahamanage RS, Brooks S, Chaiyasen A, Chethana KWT, Chomnunti P, Chepkirui C, Chuankid B, DE Silva NI, Doilom M, Faulds C, Gentekaki E, Gopalan V, Kakumyan P, Harishchandra D, Hemachandran H, Hongsanan S, Karunarathna A, Karunarathna SC, Khan S, Kumla J, Jayawardena RS, Liu J-K, Liu N, Luangharn T, Macabeo APG, Marasinghe DS, Meeks D, Mortimer PE, Mueller P, Nadir S, Nataraja KN, Nontachaiyapoom S, O'Brien M, Penkhrue W, Phukhamsakda C, Ramanan US, Rathnayaka AR, Sadaba RB, Sandargo B, Samarakoon BC, Tennakoon DS, Siva R, Sriprom W, Suryanarayanan TS, Sujarit K, Suwannarach N, Suwunwong T, Thongbai B, Thongklang N, Wei D, Wijesinghe SN, Winiski J, Yan J, Yasanthika E, Stadler M (2019) The amazing potential of fungi: 50 ways we can exploit fungi industrially. Fungal Divers 97:1–136
Hyder MS, Dutta SD (2021) Mushroom-derived polysaccharides as antitumor and anticancer agent: a concise review. Biocatal Agric Biotechnol 35:102085
Jeitler M, Michalsen A, Frings D, Hübner M, Fischer M, Koppold-Liebscher DA, Murthy V, Kessler CS (2020) Significance of medicinal mushrooms in integrative oncology: a narrative review. Front Pharmacol 11:580656
Jiang X, Meng W, Li L, Meng Z, Wang D (2020) Adjuvant therapy with mushroom polysaccharides for diabetic complications. Front Pharmacol 11:168
Kakon AJ, Choudhury MBK, Saha S (2012) Mushroom is an ideal food supplement. J Dhaka National Medical College Hospital 18:58–62

Kała K, Krakowska A, Szewczyk A, Ostachowicz B, Szczurek K, Fijałkowska A, Muszyńska B (2021) Determining the amount of potentially bioavailable phenolic compounds and bioelements in edible mushroom mycelia of agaricus bisporus, cantharellus cibarius, and lentinula edodes. Food Chem 352:129456

Kała K, Lazur J, Sułkowska-Ziaja K, Muszyńska B (2024) Edible mushrooms substances as natural prevention in autoimmunological diseases. In: Fungi bioactive metabolites: integration of pharmaceutical applications. Springer Nature Singapore, Singapore, pp 339–369

Kaminogawa S, Nanno M (2004) Modulation of immune functions by foods. Evid Based Complement Alternat Med 1:241–250

Karuppaiya P, Akbar A (2019) Mycotherapy of antrodia salmonea: a Taiwanese medicinal mushroom. Medicinal Mushrooms

Khatun S, Islam A, Çakılcıoğlu U, Chatterjee NC (2012) Research on mushroom as a potential source of nutraceuticals: a review on indian perspective. Am J Exp Agric 2:47–73

Khatun MA, Sato S, Konishi T (2020) Obesity preventive function of novel edible mushroom, basidiomycetes-x (echigoshirayukidake): manipulations of insulin resistance and lipid metabolism. J Tradit Complement Med 10:245–251

Khezri S, Shallal AF (2024) Assessment of variations in particular immunological and biochemical markers in rats under a standard and non-standard lifestyle: immunological and biochemical markers in rats under a standard and non-standard lifestyle. Iranian J Pharm Sci 20(1):12–21

Krakowska A, Zięba P, Włodarczyk A, Kała K, Sułkowska-Ziaja K, Bernaś E, Sękara A, Ostachowicz B, Muszyńska B (2020) Selected edible medicinal mushrooms from pleurotus genus as an answer for human civilization diseases. Food Chem 327:127084

Kumar K (2015) Role of edible mushrooms as functional foods-a review 1, 5454–6445

Kurd-Anjaraki S, Ramezan D, Ramezani S, Samzadeh-Kermani A, Pirnia M, Shahi BY (2022) Potential of waste reduction of agro-biomasses through reishi medicinal mushroom (ganoderma lucidum) production using different substrates and techniques. Acta Ecol Sin 42:90–101

Kushairi N, Ahmad Tarmizi NAK, Phan C-W, Macreadie I, Sabaratnam V, Naidu M, David P (2020) Modulation of neuroinflammatory pathways by medicinal mushrooms, with particular relevance to alzheimer's disease. Trends Food Sci Technol 104

Leong YK, Yang F-C, Chang J-S (2021) Extraction of polysaccharides from edible mushrooms: emerging technologies and recent advances. Carbohydr Polym 251:117006

Li, D., Wu, M. (2021). Pattern recognition receptors in health and diseases. Signal ansduction and targeted therapy, 6(1), p.291, Signal Transduct Target Ther

Li YM, Zhong RF, Chen J, Luo ZG (2021) B. Structural characterization, anticancer, hypoglycemia and immune activities of polysaccharides from Russula virescens. Int J Biol Macromol 184:380–392

Liu Y, Hu H, Cai M, Liang X, Wu X, Wang A, Chen X, Li X, Xiao C, Huang L, Xie Y, Wu Q (2022) Whole genome sequencing of an edible and medicinal mushroom, russula griseocarnosa, and its association with mycorrhizal characteristics. Gene 808:145996

Liuzzi GM, Petraglia T, Latronico T, Crescenzi A, Rossano R (2023) Antioxidant compounds from edible mushrooms as potential candidates for treating age-related neurodegenerative diseases. Nutrients 15

Male D (2021) Immunology: an illustrated outline. CRC Press

Mannino G, Di Stefano V, Lauria A, Pitonzo R, Gentile C (2020) Vaccinium macrocarpon (cranberry)-based dietary supplements: variation in mass uniformity, proanthocyanidin dosage and anthocyanin profile demonstrates quality control standard needed. Nutrients 12(4):992

Mccain P, Pathak S, Liu K, Nadar R, Ramesh S, Moore T, Agrawal D, Dhanasekaran M (2023) Cognitive enhancing effects of medicinal mushrooms: a potential neuroprotective implication in dementias

Mihich E (2000) On the immunomodulating effects of anti-cancer drugs and their therapeutic exploitation. Jpn J Clin Oncol 30(11):469–471

Mirmazloum I, Ladányi M, Omran M, Papp V, Ronkainen VP, Pónya Z, Papp I, Némedi E, Kiss A (2021) Co-encapsulation of probiotic lactobacillus acidophilus and reishi medicinal mushroom (ganoderma lingzhi) extract in moist calcium alginate beads. Int J Biol Macromol 192:461–470

Motta F, Gershwin ME, Selmi C (2021) Mushrooms and immunity. J Autoimmun 117:102576
Murphy EJ, Rezoagli E, Pogue R, Simonassi-Paiva B, Abidin IIZ, Fehrenbach GW, O'Neil E, Major I, Laffey JG, Rowan N (2022) Immunomodulatory activity of β-glucan polysaccharides isolated from different species of mushroom – a potential treatment for inflammatory lung conditions. Sci Total Environ 809:152177
Mustafa F, Chopra H, Baig AA, Avula SK, Kumari S, Mohanta TK, Saravanan M, Mishra AK, Sharma N, Mohanta YK (2022) Edible mushrooms as novel myco-therapeutics: effects on lipid level, obesity and bmi. J Fungi (Basel) 8
Muszyńska B, kała K, Sułkowska-Ziaja K (2017) Edible mushrooms and their in vitro culture as a source of anticancer compounds. In: Malik S (ed) Biotechnology and production of anti-cancer compounds. Springer International Publishing, Cham
Neergheen VS, Hip Kam A, Pem Y, Ramsaha S, Bahorun T (2022) Regulation of cancer cell signaling pathways as key events for therapeutic relevance of edible and medicinal mushrooms. Semin Cancer Biol 80:145–156
Nowak R, Nowacka-Jechalke N, Pietrzak W, Gawlik-dziki U (2022) A new look at edible and medicinal mushrooms as a source of ergosterol and ergosterol peroxide – uhplc-ms/ms analysis. Food Chem 369:130927
Owen JA, Punt J, Stranford SA (2013) Kuby immunology, vol 27. WH Freeman, New York, p 109
Panda SK, Luyten W (2022) Medicinal mushrooms: clinical perspective and challenges. Drug Discov Today 27:636–651
Papoutsis K, Grasso S, Menon A, Brunton NP, Lyng JG, Jacquier J-C, Bhuyan DJ (2020) Recovery of ergosterol and vitamin d2 from mushroom waste – potential valorization by food and pharmaceutical industries. Trends Food Sci Technol 99:351–366
Perera PK, Li Y (2011) Mushrooms as a functional food mediator in preventing and ameliorating diabetes. Funct Foods Health Disease 1:161–171
Plato, A., Hardison, S.E., Brown, G.D. (2015). Pattern recognition receptors in antifungal immunity. In Semin Immunopathol (Vol. 37, pp. 97–106). Springer Berlin Heidelberg
Pohleven J, Brzin J, Vrabec L, Leonardi A, Cokl A, Strukelj B, Kos J, Sabotič J (2011) Basidiomycete clitocybe nebularis is rich in lectins with insecticidal activities. Appl Microbiol Biotechnol 91:1141–1148
Prescott TAK, Hill R, Mas-Claret E, Gaya E, Burns E (2023) Fungal drug discovery for chronic disease: history, new discoveries and new approaches. Biomolecules 13
Rai SN, Mishra D, Singh P, Vamanu E, Singh MP (2021) Therapeutic applications of mushrooms and their biomolecules along with a glimpse of in silico approach in neurodegenerative diseases. Biomed Pharmacother 137:111377
Rangsinth P, Sharika R, Pattarachotanant N, Duangjan C, Wongwan C, Sillapachaiyaporn C, Nilkhet S, Wongsirojkul N, Prasansuklab A, Tencomnao T, Leung GP-H, Chuchawankul S (2023) Potential beneficial effects and pharmacological properties of ergosterol, a common bioactive compound in edible mushrooms. Foods 12:2529
Rathee S, Rathee D, Rathee D, Kumar V, Rathee P (2012) Mushrooms as therapeutic agents. Revista Brasileira De Farmacognosia-brazilian J Pharmacogn 22:459–474
Ren L, Zhang J, Zhang T (2021) Immunomodulatory activities of polysaccharides from ganoderma on immune effector cells. Food Chem 340:127933
Roncero-Ramos I, Mendiola-Lanao M, Pérez-Clavijo M, Delgado-Andrade C (2017) Effect of different cooking methods on nutritional value and antioxidant activity of cultivated mushrooms. Int J Food Sci Nutr 68:287–297
Sahu R, Sahu K, Chanda R, Chaturvedi K, Sahu B, Chaursia H (2023) Nutritional and pharmacological potential of edible mushroom. Acta Sci Pharm Sci 7. (ISSN: 2581-5423)
Saltarelli R, Palma F, Gioacchini AM, Calcabrini C, Mancini U, DE Bellis R, Stocchi V, Potenza L (2019) Phytochemical composition, antioxidant and antiproliferative activities and effects on nuclear DNA of ethanolic extract from an Italian mycelial isolate of ganoderma lucidum. J Ethnopharmacol 231:464–473
Seo DJ, Choi C (2021) Antiviral bioactive compounds of mushrooms and their antiviral mechanisms: a review. Viruses 13

Shahzad F, Anderson D, Najafzadeh M (2020) The antiviral, anti-inflammatory effects of natural medicinal herbs and mushrooms and sars-cov-2 infection. Nutrients 12
Shallal AF, Ali DS, Eyüpoğlu V (2023) Ameliorative role of diet in some immunological and metabolic indicators. J Univ Anbar Pure Sci 17(1):22–28
Sohlberg E (2013) Immune maturation in early childhood and the influence of herpesvirus infections (doctoral dissertation, department of molecular biosciences, the Wenner-Gren Institute, Stockholm University)
Stamets P (2006) Can mushrooms help save the world? Explore 2:152–161
Surup F, Hennicke F, Sella N, Stroot M, Bernecker S, Pfütze S, Stadler M, Rühl M (2019) New terpenoids from the fermentation broth of the edible mushroom cyclocybe aegerita. Beilstein J Org Chem 15:1000–1007
Taofiq O, González-Paramás AM, Martins A, Barreiro MF, Ferreira IC (2016) Mushrooms extracts and compounds in cosmetics, cosmeceuticals and nutricosmetics—a review. Ind Crop Prod 90:38–48
Teplyakova TV, Ilyicheva TN, Kosogova TA, Wasser SP (2021) Medicinal mushrooms against influenza viruses. Int J Med Mushrooms 23:1–11
Tong Z, Chu G, Wan C, Wang Q, Yang J, Meng Z, Du L, Yang J, Ma H (2023) Multiple metabolites derived from mushrooms and their beneficial effect on alzheimer's diseases. Nutrients 15
Venturella, G., Saporita, P., Gargano, M. (2019). Current research on medicinal mushrooms in Italy
Vetvicka V, Vannucci L, Sima P, Richter J (2019) Beta glucan: supplement or drug? From laboratory to clinical trials. Molecules 24(7):1251
Wang L, Li J-Q, Zhang J, Li Z-M, Liu H-G, Wang Y (2020) Traditional uses, chemical components and pharmacological activities of the genus ganoderma p. karst.: a review. RSC Adv 10:42084–42097
Wasser SP (2010) Medicinal mushroom science: history, current status, future trends, and unsolved problems. Int J Med Mushrooms 12(1):1
Wunjuntuk K, Ahmad M, Techakriengkrai T, Chunhom R, Jaraspermsuk E, Chaisri A, Kiwwongngam R, Wuttimongkolkul S, Charoenkiatkul S (2022) Proximate composition, dietary fibre, beta-glucan content, and inhibition of key enzymes linked to diabetes and obesity in cultivated and wild mushrooms. J Food Compos Anal 105:104226
Xu J, Xu D, Hu Q, Ma N, Pei F, Su A, Ma G (2022) Immune regulatory functions of biologically active proteins from edible fungi. Front Immunol 13:1034545
Yongabi K (2019) African medicinal mushrooms: source of biopharmaceuticals for the treatment of noncommunicable diseases—a review
Zhang J, Zhang J, Zhao L, Shui X, Wang LA, Wu Y (2019a) Antioxidant and anti-aging activities of ethyl acetate extract of the coral tooth mushroom, hericium coralloides (agaricomycetes). Int J Med Mushrooms 21:561–570
Zhang Y, Pan X, Ran S, Wang K (2019b) Purification, structural elucidation and anti-inflammatory activity in vitro of polysaccharides from smilax China l. Int J Biol Macromol 139:233–243
Zhang J-J, Chen B-S, Dai H-Q, Ren J-W, Zhou L-W, Wu S-H, Liu H-W (2021) Sesquiterpenes and polyphenols with glucose-uptake stimulatory and antioxidant activities from the medicinal mushroom sanghuangporus sanghuang. Chin J Nat Med 19:693–699
Zhang Y, Zhang G, Ling J (2022) Medicinal fungi with antiviral effect. Molecules 27:4457
Zhao S, Gao Q, Rong C, Wang S, Zhao Z, Liu Y, Xu J (2020) Immunomodulatory effects of edible and medicinal mushrooms and their bioactive immunoregulatory products. J Fungi 6(4):269
Živković L, Bajić V, Bruić M, Borozan S, Popić K, Topalović D, Santibanez J, Spremo-Potparević B (2019) Antigenotoxic and antioxidant potential of medicinal mushrooms (immune assist) against DNA damage induced by free radicals-an in vitro study. Mutat Res Genet Toxicol Environ Mutagen 845:403078

Lentinula edodes as a Unique Source of Medicinals

Yashvant Patel, Rasim Faraj Muslim, and Mustafa Nadhim Owaid

Abstract *Lentinula edodes* is one of the most popular edible macrofungi, the first recognized medicinal macrofungus with many therapeutic potentials, like anti-inflammatory, antimicrobial, antifungal, and antiviral activities. Also, this mushroom was used to synthesize eco-friendly nanoparticles to treat microbial infections. The present chapter discusses the availability of different molecules in extracts from *L. edodes*, which could be used to prevent and treat many bacterial infections, including antibiotic-resistant strains in humans. In addition, it is also used as an antiviral agent, for the modulation of the immune system, and for the lowering of the sugar and triglyceride levels in obese and diabetic patients. Taken together, these findings suggest *L. edodes's* usefulness as a natural source of bactericidal candidates for human drug applications. The goal of the current chapter is to discuss some medicinal uses against pathogenic microbes.

Keywords Medicinal mushrooms · Nanoparticles · Human health · Bioactivity · Antibacterial activity

Y. Patel
Department of Biotechnology, Veer Bahadur Singh Purvanchal University, Jaunpur, UP, India

R. F. Muslim
Department of Applied Chemistry Sciences, College of Applied Sciences-Heet, University of Anbar, Hit, Anbar, Iraq

M. N. Owaid (✉)
Department of Environment, College of Applied Sciences-Heet, University of Anbar, Hit, Anbar, Iraq

Department of Heet Education, General Directorate of Education in Anbar, Ministry of Education, Hit, Anbar, Iraq
e-mail: mustafanowaid@uoanbar.edu.iq

M. N. Owaid et al. (eds.), *Mushroom Production*, Interdisciplinary Biotechnological Advances, https://doi.org/10.1007/978-981-95-7793-4_11

1 Introduction

Lentinus edodes, also named shiitake mushroom, is a valuable edible fungus that belongs to the Basidiomycetes family. It has wild populations in Asia, Australia, and America (Hibbett 2001) and, for centuries, has also been widely cultivated in various countries in Asia due to its medicinal, nutritional, and unique aroma attributes. It is a popular edible mushroom first recognized as a medicinal macrofungus with many therapeutic properties, including antimicrobial, anticancer, and antiviral activities; lipid-lowering and immune-stimulating effects; and many more (Sugano et al. 1985; Watanabe et al. 2006; Bisen et al. 2010a) This is due to the various compounds present in its fruiting body and mycelium, such as polysaccharides, polyphenols, flavonoids, terpenoids, alkaloids, steroids, polyglucans, polyketides, and dietary fibers, some of which have pharmaceutical applications (Ng and Yap 2002; Acharya et al. 2015).

Aside from these properties, the fruiting body of this mushroom is a rich source of antimicrobial, antioxidant, and antineoplastic compounds (Ng and Yap 2002; Sasidharan et al. 2010; Vetchinkina et al. 2016). The unique aroma in its fruiting body is due to organosulfur compounds, 1-Octen-3-ol, 3-octanone, dimethyl trisulfide, and dimethyl disulfide, which distinguish it from other mushrooms. This enhances the flavor and provides a sufficient amount of selenium, which attracts people to consume it. (Wu and Wang 2000; Savic et al. 2016); in addition, many tasty components are found at different stages of *L. edodes* (Chen et al. 2015), which make it more delicious and more healthy. Consequently, it may be considered a potential therapeutic agent for various disorders and diseases, owing to its unique aroma and taste.

Generally, plants and macrofungi contain a diverse array of bioactive molecules that may help protect against atherosclerosis and cardiovascular diseases (Rajewska and Balasinska 2004). The anti-inflammatory activity of bioactive compounds from *L. edodes* may help prevent psoriasis by inhibiting cyclooxygenase activity (a marker enzyme of skin inflammation) (Prieto et al. 2003). Nevertheless, Zembron-Lacny et al. (2013) reported another bioactive compound, ergothioneine, isolated from *L. edodes* which has potent antioxidant, anti-inflammatory and regenerative properties throughout the NO concentration regulation and status of thiol redox. Also, Nishitani et al. (2013) reported beta-1,3;1,6-glucan (lentinan) which induce immune suppressive impacts in case of chronic inflammations of gastro-intestinal tracts.

Heteroglycans and heteroglycan-protein fractions, including lentinan derived from mycelial cultures and the fruiting body of *L. edodes*, have shown diverse therapeutic activities (Vetchinkina et al. 2016; Zhang et al. 2005; Vannucci et al. 2017). Lentinan significantly enhances the activity of natural killer (NK) cells and cytotoxic T lymphocytes (CTLs), elevates interleukin-2 (IL-2) and interferon-gamma (INF-γ) levels in splenocytes, and has proven very effective in antitumor biotherapy (Leibundgut-Landmann et al. 2008; Goodridge et al. 2011). Another active compound, the seleno-amino acid Se-methyl-seleno-l-cysteine, synthesized from *L. edodes* (Klimaszewska et al. 2016), is a recently discovered nonprotein

Fig. 1 Fresh and dried *L. edodes* fruiting bodies

seleno-amino acid that exhibits activity in the prevention and treatment of cancers (Rayman 2012).

The present review aims to elucidate the multifarious beneficial properties of *L. edodes*, thereby providing valuable insights for the scientific community, producers, and consumers. In line with this, the chapter highlights its diverse applications, including the synthesis of eco-friendly nanoparticles for the treatment of microbial infections. This review also discusses the different molecules present in *L. edodes* extracts, which could be used to prevent and treat many bacterial infections, including antibiotic-resistant strains in humans. In addition, *L. edodes* may also be used as an antiviral agent, for the modulation of the immune system, and for the lowering of sugar and triglyceride levels in obese and diabetic patients. Taken together, these findings underscore the usefulness of this macrofungus as a natural source of bactericidal candidates for therapeutic applications. The goal of the current chapter is to discuss some medicinal applications of *L. edodes* against pathogenic microbes. The fruiting body of the current mushroom is shown in Fig. 1.

2 Nutritional and Medicinal Value of Polysaccharides Isolated from *L. edodes*

Shiitake mushroom (*L. edodes*), a cosmopolitan cultivated species, is grown not only for meal consumption but also for the extraction of many bioactive compounds, including homo- and hetero-polysaccharides, which may possess homo-linkages or hetero-linkages, depending on configuration and/or linkage position (Wang et al. 2017; Finimundy et al. 2016). Basically, polysaccharides found in cell walls and the composition of these walls varying with types of polysaccharides like the outside layer is hetero-polysaccharide and lentinan having β-D-(1–3)-glucan with branches of beta-(1–6) (Sasaki and Takasuka 1976), the middle layer's lentinan having beta-(1–6)-glucan with a few number of branches of beta-(1–3) but the inner layer is in a form complex of beta-glucan, chitin, and few amounts of acid polymers (Sugui et al. 2003; Shida et al. 1981). Along with these, some other polysaccharides extracted from culture extracts viz LAP, eritadenine, L2 and KS-2 also reported (Sugano et al. 1985; Xiaofei et al. 2017; Wasser 2003). The pileus and stipe contain water-soluble sugars, like mannitol, trehalose, arabitol, and glucose in the 9.0% to 15.0% range (Kim et al. 2013; Chen et al. 2015), which are responsible for the sweet taste of this mushroom (Yang et al. 2001; Kim et al. 2009). The polysaccharides found in this basidiomycete mushroom are known to be the most potent antitumor, antiviral, and immuno-modulating substances (Rincão et al. 2012; Xu et al. 2014).

Also, it is mentioned that the polysaccharide isolated from *L. edodes* lead to the restore age-attenuated immune response by raising levels of cytokine in the blood and reverse age-associated alteration of the composition of microbiota in gut of the mouse (Xu and Zhang 2015). The administration of polysaccharides in experimental mice induced the expression of many antioxidant enzymes and reduced necrotic injuries (Kim et al. 2013).

However, the main immune-modulating impacts of active substances derived from this mushroom involve immune effecter cells activation and mitogenicity, like the macrophage, natural killer cell, and lymphocyte, resulting in the cytokines production, INF (interferon-gamma)-γ, (TNF)-α (tumor necrosis factor-alpha), and interleukins (ILs), (Drori et al. 2017). Lentinan, the beta-glucan obtained from shiitake, enhances cell-mediated immune response by inducing the gene expression of immunomodulatory cytokines and their receptors (Xu and Zhang 2015; Vetvicka and Vetvickova 2014). However, the extracts from mycelia and fruit body positively responded to characteristics of the immuno-stimulatory in terms of co-mitogenic and mitogenic efficacy as support to the thymocytes proliferation of rats directly and acted as co-stimulatory in the case of finding T-mitogen PHA (in vitro) (Vetvicka and Vetvickova 2014). A water-soluble extract, isolated from mycelia culture and fruit body of *L. edodes* stimulated the macrophages functional activation, involving the producing NO, phagocytosis, and cytokine expression (Israilides et al. 2008). Treatment of macrophage in culture response to lipopolysaccharide (LPS)-induced stress, the mixture of extracts of shiitake mushroom, Uncaria, and ribes improve gene expression of cytokines which demonstrated that this mixture could be an important feature to modulate the function of immune (Lee et al. 2009). Furthermore, ethanolic extracts of *L. edodes* have been shown to significantly reduce immunoglobulin in atopic dermatitis (Menghini et al. 2014). The key element of this non-protein is selenium and which shows as a key nutrient in the chemoprevention of cancers (Vinceti et al. 2014). The role of selenium (Se) in cancer chemoprevention was inferred by an inverse the relation between the dietary Se intake and the human cancer mortality (Schrauzer et al. 1977). However, the current vital health impact is connected to the immune system stimulation, inhibiting the tumor cell growth using metabolites of selenium and protecting the cells from the impact of free radicals using Se-dependent antioxidant enzymes (Brozmanova et al. 2010). *L. edodes* demonstrates extensive medicinal effects against many microbes and cancers, as illustrated in Fig. 2.

3 Antimicrobial Properties

3.1 Antibacterial Activity of L. edodes

This mushroom has many important therapeutic properties, especially against microorganisms such as bacteria and fungi, because it contains many biologically active compounds. The mycelia and fruiting bodies of *Lentinula* sp. contain

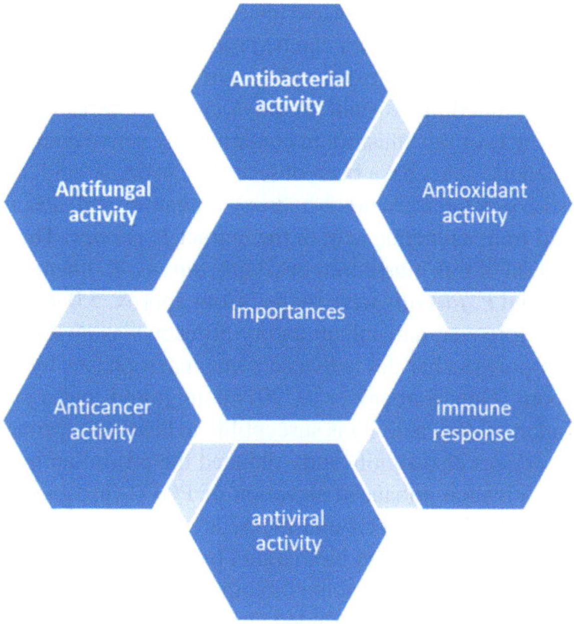

Fig. 2 The medicinal importance of *L. edodes* fruiting bodies

lenthionine, which inhibits the growth of Gram-positive and Gram-negative bacteria, as well as fungi and yeasts, with a stronger inhibitory effect on fungi and yeasts than on bacteria. (Komemushi et al. 1996). Chloroform and ethylacetate extracts of the dried fruit bodies of this mushroom had antibacterial activities against species of *Actinomyces*, *Streptococcus*, *Prevotella*, *Lactobacillus*, and *Porphyromonas* (Oyetayo 2009; Rao et al. 2009); and the most exciting part is its relative heat-stable property.

Antimicrobial activity of this mushroom depends upon the interaction between solvents and the shiitake compounds at very specific conditions of temperature and pressure, and it was observed that extracts of low molecular mass (LMM) of shiitake mushroom showed more antimicrobial activity (Kitzberger et al. 2007), according to the calculation of the microbial population by qPCR (Zaura et al. 2011). This LMM extract inhibited the synthesis of nucleic acids and proteins, which significantly altered the proteome of *Streptococcus mutans* and *Prevotella intermedia* (Signoretto et al. 2014). *L. edodes* has been shown to be active in suppressing symptoms caused by some phytopathogenic bacteria and fungi (Inagaki and Yamaguchi 2009; Ishihara et al. 2018).

Methanolic extracts of *L. edodes* mushroom have also shown inhibitory efficacy against many bacterial groups, including *Escherichia coli*, *Staphylococcus aureus*, *Pseudomonas aeruginosa*, *Bacillus subtilis*, *Klebsiella pneumoniae*, *Salmonella typhi*, and *Candida albicans* (Kuznetsov et al. 2005), as well as the plant pathogenic bacteria *Ralstonia solanacearum*. Microbial growth inhibition activity was observed in *Lentinula edodes*, attributed to its organic acids, including oxalic acid, which exhibited antimicrobial activity against various phytopathogenic bacteria (Kwak et al. 2016). After extensive investigation, it was found that antimicrobial substances from *Lentinula edodes* can be produced under specific pH conditions.

Additionally, it exhibited the strongest antimicrobial activity against the growth of methicillin-resistant *Staphylococcus aureus* (MRSA) (Garcia et al. 2021). Extracts of this mushroom contain many active compounds (Imam et al. 2021). The phenolic acids have been observed as syringic acid, chlorogenic acid, rutin, ferulic acid, p-coumaric acid, 2-hydroxy cinnamic acid, abscisic acid, protocatechuic acid, and trans-cinnamic acid in the extract *L. edodes*. The minimum inhibitory concentration of this mushroom on *Staph. aureus, K. pneumoniae, Acinetobacter baumannii* and *E. faecalis* were ranged from approx. 5.0–6.0 mg/ml (Erdoğan Eliuz 2021).

Moreover, anti-inflammatory efficacy was observed in IFN-γ and (interferon-γ) lipopolysaccharides activated some macrophages by extracts from *L. edodes* mushroom (Gunawardena et al. 2014) suggested that a significant portion of the anti-inflammatory activity is susceptible to heating or prone to evaporation. Extract from *L. edodes* also significantly reduced the pro-inflammatory mediator levels in OVA-sensitization-enhanced elevation of PGD2 (prostaglandin D2)and LTC4 (leukotriene C4). These LTC4 and PGD2 are asthma-associated biomarkers. The reduction of asthma-associated biomarkers by the treatment with extract from *L. edodes* proposed the producing some biological active compounds using the mycelia of mushrooms that maybe included in improving the seen antiasthmatic characteristics (Kim et al. 2016).

The antibacterial activity of *Lentinula edodes* against *Bacillus subtilis* was evaluated using cell-free filtrates obtained after cultivation in 14 different culture media. Filtrates from the growth media, supplemented with molasses, rice bran, or vermiculite, inhibited the growth of the *B. subtilis* bacterium. After 1 month of the development at 25 °C without shaking, the dried mycelial bio-mass in liquid culture with 0.5 percent additional rice bran was 3.2 mg.ml-1, and 4.3 $mg.ml^{-1}$ when shaking was used (150 rpm). The anti-bacterial efficacy was observed in the filtrates of the aerated cultures between 20–24 days of the incubation of the stationary cultures, but not in the filtrates of the aerated cultures. Temperatures between 20 and 25 °C enhanced both antimicrobial efficacy and growth. The optimal pH for the development of mycelial of this mushroom was 3.5–3.0, whereas it was 4.5 for the synthesis of the anti-bacterial substances. The current findings revealed that conditions of the incubation that promote mycelial development differ significantly from those required for *L. edodes* to produce anti-bacterial substances (Hassegawa et al. 2005).

Xanthomonas campestris pv. vesicatoria causes tomato bacterial spot, a severe disease that affects tomato crops all over the world. High humidity and summer temperatures are favorable circumstances for the current disease in the south of United States, resulting in fruits spotting, defoliation, and a substantial loss in fruit output. Because current organic production techniques prohibit the use of synthetic chemicals to prevent and manage disease, active and new bio-pesticides are needed to combat plant infections (Kaur et al. 2016). Since ancient times, this mushroom has been used as a medical treatment. Compared to other species of macrofungi, there have not been many reports on its medicinal applications; however, investigating its impacts may contribute globally to its therapeutic use (Ponugupati 2015).

L. edodes has already been proven to possess antibacterial characteristics, making it one of several biological agents with promise for disease management. To verify the use of shiitake mushroom mycelial culture filtrate foliar treatment to reduce bacterial spot in tomato by the cultivar Agriset761, a controlled-environment research was undertaken. In vitro, the foliar spray of mycelial culture filtrate of the shiitake mushroom dramatically reduced the occurrence of bacterial spots in tomato leaves, but it proved ineffective in vivo. The amount of 423 grams of oxalic acid per 1 ml of the filtrate was blamed for the phyto-toxicity symptoms in the mushroom filtrate-treated tomato leaves (evaluated by HPLC). In the mushroom-filtrate-treated plants, plant height and blooming were normal. Furthermore, while filtrate-seed treatment had no effect on the germination of tomato overall, it greatly improved the germination of marginal tomato seeds that had been exposed to biotic stress. These findings show that after removing oxalic acid from the mushroom filtrate, the product may be used as a bio-pesticide to control tomato bacterial spot. After removing oxalic acid from *L. edodes* culture filtrates, future greenhouse or outdoor studies must be undertaken (Kaur et al. 2016).

The goal of the current chapter is to examine how extracts of *Lentinula edodes* mushroom affect antibacterial activity against *B. subtilis*. When several edible macrofungi extracts were exposed to this pathogenic bacterium, it was predicted that the extract of *L. edodes* would show the highest antibacterial efficacy and the largest clear inhibition zone. Finding out which mushroom extract concentration has the greatest antibacterial impact may lead to the discovery of new chemicals to replace currently ineffective antibacterial treatments. After 2 days of incubation, *L. edodes* exhibited the greatest average inhibition zone. This excellent antibacterial activity of *L. edodes* can be attributed to the presence of lentinan, eritadenine, and lenthionine. On the other hand, antimicrobial efficacy in other mushroom species may be aided by unidentified protein content. The unknown chemicals found in these mushrooms could be extracted and used to produce new antibiotics to address antibiotic resistance in humans, after additional research is done (Ponugupati 2015).

Since the discovery of antibiotics in the twentieth century, they have been used to treat a wide range of illnesses with varying degrees of success. However, multidrug-resistant microorganisms have emerged as a growing global threat to veterinary and human medicine. To reduce the risk of future antimicrobial resistance, it is now commonly accepted that new antimicrobial drugs are required. In order to investigate the antibacterial activities of the aqueous extracts of shiitake mushrooms on a variety of environmental and therapeutically relevant microorganisms, a study was conducted (Hearst et al. 2009). In particular, the antibacterial properties of different mushroom extracts may contribute to a better understanding of the compounds responsible for their efficacy. The field of applied bio-technology identifies edible mushrooms as having a high potential for medical and agricultural advancements (Bisen et al. 2010a; b).

In terms of their potential antibacterial capabilities, edible mushroom extracts have been viewed as rivals to popular synthetic medications (Tehrani et al. 2012). Many recent studies have been investigating the effectiveness of natural antibiotics derived from mushrooms against both Gram-negative and Gram-positive bacteria. Because a growing number of bacteria have become resistant to different commercial antibiotics, society can benefit from recognizing the value of mushrooms as natural sources of antibacterial medicines. However, there is a scarcity of data on the antibacterial and antioxidant properties of wild edible macrofungi. Compounds can be extracted from the fruiting body to establish the origin of the antibacterial efficacy of edible mushrooms by analyzing their antimicrobial effects. The discovery of particular and alternative chemicals to replace existing antibacterial treatments, which were rendered ineffective due to bacterial resistance, may be achieved by determining which species of macrofungi have the highest inhibitory impact on bacteria (Sonawane et al. 2012). This could lead to a better understanding of antibacterial characteristics and could aid in the medicinal application of antibiotics. Proteins and carbohydrates are abundant in wild mushrooms. The proteins present in these fungi have been demonstrated to perform a variety of biological functions that can help prevent diseases.

Edible macrofungi are low in fat, high in polyunsaturated fatty acids, and contain a variety of minerals and vitamins. The role of antibiotics in mushrooms is less well established in terms of their significance in medicine; thus, the identification of novel antimicrobial compounds is still a possibility (Thallaimaharani et al. 2013). For millennia, many mushroom extracts have been used therapeutically and have shown antimicrobial effects against a variety of bacteria. *L. edodes* possesses pharmacological characteristics, which include antitumor activity, antibacterial capabilities, improved liver function, reduction of viremia in chronic hepatitis B patients, and in vitro suppression of HIV infection (Hirasawa et al. 1999). In terms of antibacterial activity, *L. edodes* has been identified as the most active species (Bisen et al. 2010a, b).

Bacillus subtilis is well known for its ability to form endospores, which enable it to survive under harsh environmental conditions. The dependent variable was defined as the diameter of the inhibition zone (mm) formed around each well containing the mushroom extract. The agar diffusion technique can be employed to assess antibacterial activity (Pauliuc and Dorica 2013). Antibacterial properties have been found in all of the mushroom extracts, although to varying degrees (Sonawane et al. 2012). When several edible mushroom extracts were exposed to *B. subtilis*, it was predicted that *L. edodes* would exhibit the highest antibacterial activity and the largest clear zone diameter (in millimeters). Many substances with health-promoting properties have been isolated from *L. edodes* as a result of scientific research. Lentinan, a polymer found in *L. edodes*, is essential for the mushroom's pharmacological and biological activities.

Using the well diffusion method, Pauliuc and Dorica (2013) tested the antimicrobial activity of *L. edodes* against many species of bacteria, including *Bacillus cereus*, *B. subtilis*, *Pseudomonas aeruginosa*, *Streptococcus faecalis*, and *Serratia marcescens*. The results showed that *L. edodes* was effective against pathogenic bacteria, especially against *B. subtilis* (Pauliuc and Dorica 2013).

3.2 *Antifungal Activity of Extracts of* L. **edodes**

In this section the purified extract from *L. edodes* have the anti-fungal compounds from the mushroom extract in a crude form and as purified compounds. This mushroom has many important therapeutic properties, especially against microorganisms such as fungi, because it contains many biologically active compounds (Ishihara et al. 2018). These compounds degrade some organic materials like lignin, because of found the ring of 4-hydroxy-5-methoxy-benzene which derived from the unit of the coniferyl alcohol in the lignin matter. Other compounds have been checked as degradation products for the lignin using the chemical oxidation process with the $KMnO_4$ (de Sousa et al. 1988) or using the action of *Phanerochaete chrysosporium (*Chen et al. 1982). A compound from the current medicinal mushroom is the intermediate one in the path of lignin degradation, while some enzymes included in catabolism of *other compound* (Peng et al. 2002; Vladimirova et al. 2016).

L. edodes has been shown to be active in suppressing symptoms of some phytopathogenic bacteria and fungi (Inagaki and Yamaguchi 2009; Ishihara et al. 2018). These impacts *were* induced the plant resistance to some pathogenic fungi due to the active compounds in this extracts. Given that this mushroom belongs to the kingdom of fungi, it's plausible to believe that some components of macrofungi are identified using phytocells as parts of the pattern of pathogen-related molecular, eliciting resistance responses.

The extract of *L. edodes* demonstrates extensive antifungal efficacy against 50% of yeast and mold species, showing the potential antifungal impacts of shiitake extract. However many works need to isolate and identify different active agents from various extracts of the current mushroom. Once they were discovered, effective pharmaceutical delivery systems must be investigated in order to allow concentrated extracts to be produced and given ideally, rather than crude material ingestion, which might increase fungal resistance (Hearst et al. 2009).

3.3 *Antimicrobial Activity of Crude Nanoparticles Synthesized from* L. **edodes**

Extracellular mycelia of *L. edodes* were used to biosynthesize silver nanoparticles (AgNPs) and study their antibacterial efficacy (Sujatha et al. 2015). These AgNPs exhibited inhibitory activity against *Pseudomonas aeruginosa*, *Escherichia coli*, *Staphylococcus aureus*, and *Bacillus subtilis* due to interactions between the silver NPs and the bacteria (Kaviya et al. 2011). Similarly, extracellular mycelia of *Pleurotus sajor-caju* were employed to mycosynthesize AgNPs (Chan and Don 2013). Likewise, several mushroom species have succeeded in synthesizing silver and gold nanoparticles (AuNPs) (Owaid and Ibraheem 2017; Owaid et al. 2017, 2018).

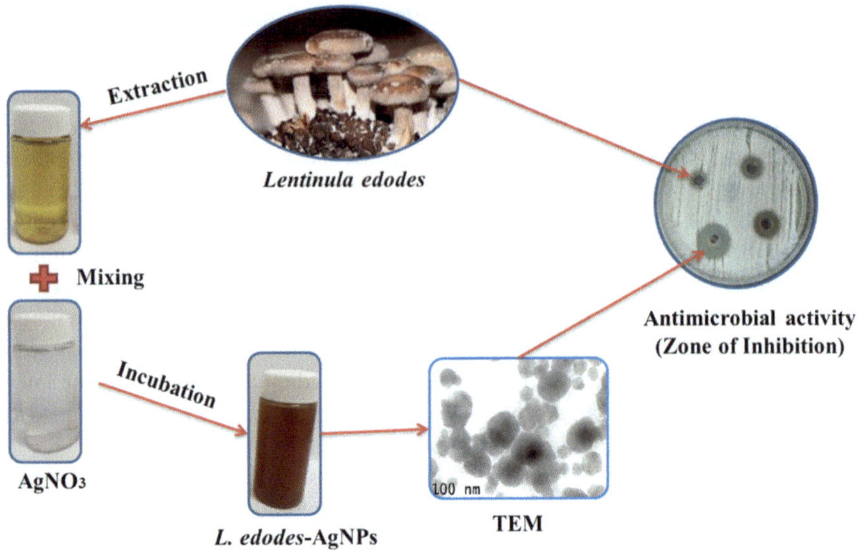

Fig. 3 Mycosynthesis of metallic nanoparticles from *L. edodes* and their applications

Lentinus edodes was used to produce spherical AuNPs from laccase and tyrosinasesVetchinkina et al. (2013) seen the accumulation of gold NPs using the mycelia of *Lentinus edodes* in the form of nano-spheres of 5–50 nm on surfaces and inside cells of fungi (intracellularly) due to laccase and tyrosinases of this mushroom. Besides these beneficial activities, the fruit body of this *L. edodes* is very prone to autolysis system, which was due to change in expression of many genes included in the degradation of cell-wall and other parts of fruiting body, explored by very recent transcriptome analysis of the current mushroom (Sakamoto et al. 2017). This transcriptome analysis would explore a new era to reveal many mysterious but useful products and activities from not only *Lentinula* but many other filamentous/macrofungi. The synthesis of metallic nanoparticles from *L. edodes* and their applications is illustrated in Fig. 3.

4 Conclusion

We reviewed the nutritional, medicinal, and nanoparticle-forming potential of *L. edodes*. It was found that different extracts of this mushroom species possess anticancer, anti-inflammatory, antimicrobial, antifungal, antiviral, antidiabetic, hepatoprotective, and immune-modulator activities due to the presence of various macro- as well as micro-compounds and elements, including polysaccharides, proteins, polypeptides, amino acids, selenium, and others. The numerous therapeutic potentials discussed above, along with lifespan-enhancing studies, suggest that this mushroom has strong potential as a novel organic food.

References

Acharya K, Mukherjee S, Mitra P (2015) Pharmacognostic standardisation of Lentinula Edodes: a widely explored medicinal mushroom. Intl J Pharmacogn Phytochem Res 7(4):866–872

Bisen PS, Baghel RK, Sanodiya BS, Thakur GS, Prasad GBKS (2010a) *Lentinus edodes*: a Macrofungus with Pharmacological Activities. Curr Med Chem 17(22):2419–2430

Bisen PS et al (2010b) Lentinus edodes: a macrofungus with pharmacological activities. Curr Med Chem 17:2419–2430

Brozmanova J, Mániková D, Vlčková V, Chovanec M (2010) Selenium: a double-edged sword for defense and offence in cancer. Arch Toxicol 84:919–938

Chan YS, Don MM (2013) Biosynthesis and structural characterization of ag nanoparticles from white rot fungi. Mater Sci Eng C 33:282–288

Chen C, Chang H, Kirk TK (1982) Holzforschung 36:3–9

Chen W, Li W, Yang Y, Yu H, Zhou S, Feng J, Liu Y (2015) Analysis and evaluation of tasty components in the pileus and stipe of *Lentinula edodes* at different growth stages. J Agric Food Chem 63(3):795–801

de Sousa F, Strömberg LM, Kringstad KP (1988) Water Sci Technol 20:153–160

Drori A, Rotnemer-Golinkin D, Avni S, Drori A, Danay O, Levanon D, Tam J, Zolotarev L, Ilan Y (2017) Attenuating the rate of total body fat accumulation and alleviating liver damage by oral administration of vitamin Denriched edible mushrooms in a dietinduced obesity murine model is mediated by an anti-inflammatory paradigm shift. BMC Gastroenterol 17:130. https://doi.org/10.1186/s12876-017-0688-4

Erdoğan Eliuz EA (2021) Antibacterial activity and antibacterial mechanism of ethanol extracts of Lentinula edodes (Shiitake) and Agaricus bisporus (button mushroom). Int J Environ Health Res:1–14

Finimundy TC, Dillon AJP, Henriques JOANP, Ely MR (2016) A review on general nutritional compounds and pharmacological properties of the mushroom. Food Nutrit Sci 5:1095–1105

Garcia H, Afonso A, Fernandes C, Nunes FM, Marques G, Saavedra MJ (2021) Comparative antioxidant and antimicrobial properties of Lentinula edodes Donko and Koshin varieties against priority multidrug-resistant pathogens. South African J Chem Eng 35:98–106

Goodridge HS, Reyes CN, Becker CA, Katsumoto TR, Ma J (2011) Activation of the innate immune receptor Dectin-1 upon formation of a 'phagocytic synapse'. Nature 472:471–475

Gunawardena D, Bennett L, Shanmugam K, King K, Williams R, Zabaras D, Head R, Ooi L, Gyengesi E, Münch G (2014) Anti-inflammatory effects of five commercially available mushroom species determined in lipopolysaccharide and interferon-c activated murine macrophages. Food Chem 148:92–96

Hassegawa RH, Kasuya MCM, Vanetti MCD (2005) Growth and antibacterial activity of Lentinula edodes in liquid media supplemented with agricultural wastes. Electron J Biotechnol 8(2):212–217

Hearst R, Nelson D, McCollum G, Millar BC, Maeda Y, Goldsmith CE, Rooney PJ, Loughrey A, Rao JR, Moore JE (2009) An examination of antibacterial and antifungal properties of constituents of Shiitake (Lentinula edodes) and oyster (Pleurotus ostreatus) mushrooms. Complement Ther Clin Pract 15(1):5–7. https://doi.org/10.1016/j.ctcp.2008.10.002

Hibbett DS (2001) Shiitake mushrooms and molecular clocks: historical biogeography of Lentinus. J Biogeogr 28:231–241

Hirasawa M et al (1999) Three kinds of antibacterial substances from Lentinus edodes (Berk.) Sing. (Shiitake, an edible mushroom). Int J Antimicrob Agents 11(2):151–157

Imam KMSU, Xie Y, Liu Y, Wang F, Xin F (2021) Extraction, Isolation, and identification of cytotoxic secondary metabolites from Shiitake mushroom 808 Lentinula edodes (Berk.). ACS Food Sci Technol 1:551

Inagaki H, Yamaguchi A (2009) Mushroom Sci. Biotechnol 17:113–115

Ishihara A, Goto N, Kikkawa M, Ube N, Ushijima S, Ueno M, Ueno K, Osaki-Oka K (2018) Identification of antifungal compounds in the spent mushroom substrate of Lentinula edodes. J Pestic Sci 43(2):108–113. https://doi.org/10.1584/jpestics.D17-094

Israilides C, Kletsas D, Arapoglou D, Philippoussis A, Pratsinis B, Ebringerovad A, Hribalovae V, Hardingf SE (2008) *In vitro* cytostatic and immunomodulatory properties of the medicinal mushroom *Lentinus edodes*. Phytomedicine 15(6–7):512–519

Kaur H, Nyochembeng LM, Mentreddy SR, Banerjee P, Cebert E (2016) Assessment of the antimicrobial activity of Lentinula edodes against Xanthomonas campestris pv. vesicatoria. Crop Prot 89:284–288

Kaviya S, Santhanalakshmi J, Viswanathan B, Muthumary J, Srinivasan K (2011) Biosynthesis of silver nanoparticle using *Citrus sinensis* peel extract and its antibacterial activity. Spectrochim Acta A 79:594–598

Kim MY, Chung LM, Lee SJ, Ahn JK, Kim EH, Kim MJ, Kim SL, Moon HI, Ro HM, Kang EY, Seo SH, Song HK (2009) Comparison of free amino acid, carbohydrates concentrations in Korean edible and medicinal mushrooms. Food Chem 113(2):386–393

Kim SP, Park SO, Lee SL, Nam SH, Friedman M (2013) A polysaccharide isolated from the liquid culture of *Lentinus edodes* (shiitake) mushroom mycelia containing black Rice bran protects mice against a salmonella lipopolysaccharide-induced Endotoxemia. J Agric Food Chem 61:10987–10994

Kim SP, Lee SJ, Nam SH, Friedman M (2016) Elm tree (Ulmus parvifolia) bark bioprocessed with mycelia of shiitake (*Lentinus edodes*) mushrooms in liquid culture: composition and mechanism of protection against allergic asthma in mice. J Agric Food Chem 64:773–784

Kitzberger CSG, Smania A Jr, Pedrosa RC, Ferreira SRS (2007) Antioxidant and antimicrobial activities of shiitake (*Lentinus* edodes) extracts obtained by organic solvents and superficial fluids. J Food Eng 80:631–638

Klimaszewska M, Górska S, Dawidowski M, Podsadni P, Turło J (2016) Biosynthesis of Se-methyl-seleno-lcysteine in *Basidiomycetes* fungus *Lentinus edodes* (Berk.) Pegler. SpringerPlus 5:733

Komemushi S, Yamamoto Y, Fujita T (1996) Purification and identification of antimicrobial substances produced by *Lentinus edodes*. J Antibacterial Antifungal Agents 24(1):21–25

Kuznetsov OIU, Mil'kova EV, Sosnina AE, Sotnikova NIU (2005) Antimicrobial action of *Lentinus edodes* juice on human microflora. Zh Mikrobiol Epidemiol Immunobiol 1:80–82

Kwak AM, Lee IK, Lee SY, Yun BS, Kang HW (2016) Oxalic acid from *Lentinus edodes* culture filtrate: antimicrobial activity on Phytopathogenic bacteria and qualitative and quantitative analyses. Mycobiology 44(4):338–342

Lee HH, Lee JS, Cho JY, Kim YE, Hong EK (2009) Study on immunostimulating activity of macrophage treated with purified polysaccharides from liquid culture and fruiting body of *Lentinus edodes*. Microb Biotechnol 19(6):566–572

Leibundgut-Landmann S, Osorio F, Brown GD, Reis e Sousa C (2008) Stimulation of dendritic cells via the dectin-1/Syk pathway allows priming of cytotoxic T-cell responses. Blood 112:4971–4980

Menghini L, Leporini L, Pintore G, Ferrante C, Recinella C, Orlando G, Vacca M, Brunetti L (2014) A natural formulation (imoviral™) increases macrophage resistance to LPS-induced oxidative and inflammatory stress in vitro. J Biol Regul Homeost Agents 28(4):775–782

Ng ML, Yap AT (2002) Inhibition of human colon carcinoma development by lentinan from shiitake mushrooms (*Lentinus edodes*). J Altern Complement Med 8:581–589

Nishitani Y, Zhang L, Yoshida M, Azuma T, Kanazawa K, Hashimoto T, Mizuno M (2013) Intestinal anti-inflammatory activity of lentinan: influence on IL-8 and TNFR1 expression in intestinal epithelial cells. PLoS One 8(4):e62441

Owaid MN, Ibraheem IJ (2017) Mycosynthesis of nanoparticles using edible and medicinal mushrooms. Eur J Nanomed 9(1):5–23

Owaid MN, Al-Saeedi SSS, Abed IA (2017) Biosynthesis of gold nanoparticles using yellow oyster mushroom *Pleurotus cornucopiae* var. *citrinopileatus*. Env Nanotechnol Monit Manag 8:157–162

Owaid MN, Muslim RF, Hamad HA (2018) Mycosynthesis of silver nanoparticles using *Terminia* sp. desert truffle, Pezizaceae, and their antibacterial activity. Jordan. J Biol Sci 11(4):401–405

Oyetayo VO (2009) Free radical scavenging and antimicrobial properties of extracts of wild mushrooms. Brazilian J Microbiol 40:380–386

Pauliuc I, Dorica B (2013) Evaluation of antibacterial activity of Lentinula edodes gemmotherapic extract. J Hortic Fores Biotechnol 17(1):246–249
Peng X, Masai E, Kitayama H, Harada K, Katayama Y, Fukuda M (2002) Characterization of the 5-carboxyvanillate decarboxylase gene and its role in lignin-related biphenyl catabolism in Sphingomonas paucimobilis SYK-6. Appl Environ Microbiol 68(9):4407–4415
Ponugupati A (2015) Macrofungal extracts on the bacteria inhibition of Bacillus subtilis. Eur J Health Biol Educ 4(1):1–8
Prieto JM, Recio MC, Giner RM, Manez S, Giner-Larza EM, Ros JL (2003) Influence of traditional Chinese anti-inflammatory medicinal plants on leukocyte and platelet functions. J Pharm Pharmacol 55(9):1275–1282
Rajewska J, Balasinska B (2004) Biologically active compounds of edible mushrooms and their beneficial impact on health. Postepy Hig Med Dosw 58:352–357
Rao JR, Smyth TJ, Millar BC, Moore JE (2009) Antimicrobial properties of shiitake mushrooms (*Lentinus edodes*). Intl J Antimicrobial Agents 33:591–592
Rayman MP (2012) Selenium and human health. Lancet 379(9822):1256–1268
Rincão VP, Yamamoto KA, Ricardo NMPS, Soares SA, Meirelles LDP, Nozawa C, Linhares REC (2012) Polysaccharide and extracts from *Lentinula edodes*: structural features and antiviral activity. Virol J 2012:9–37
Sakamoto Y, Nakade K, Sato S, Yoshida K, Miyazaki K, Natsume S, Konno N (2017) *Lentinula edodes* genome survey and postharvest transcriptome analysis. Appl Environ Microbiol 83(10):e02990–e02916. https://doi.org/10.1128/AEM.02990-16
Sasaki T, Takasuka N (1976) Further study of the structure of lentinan, an anti-tumor polysaccharide from *Lentinus edodes*. Carbohydr Res 47:99–104
Sasidharan S, Aravindran S, Latha LY, Vijenthi R, Saravanan D, Amutha S (2010) *In vitro* antioxidant activity and Hepatoprotective effects of *Lentinus edodes* against paracetamol-induced hepatotoxicity. Molecules 15:4478–4489
Savic M, Klimaszewska M, Bamburovicz-Klimkowska M, Suchochi P, Niksic M, Szutowski M, Wroczynski P, Turlo J (2016) A search for the optimum selenium source to obtain mushroom-derived chemopreventive preparations. Intl J Med Mushroom 18(4):279–289
Schrauzer GN, White DA, Schneider CJ (1977) Cancer mortality correlation studies–III: statistical associations with dietary selenium intakes. Bioinorg Chem 7(1):23–21
Shida M, Ushioda Y, Nakajima T, Matsuda K (1981) Structure of alkali-insoluble skeletal glucan of *Lentinus edodes*. J Biochem 90(4):1093–1100
Signoretto C, Marchi A, Bertoncelli A, Burlacchini G, Papetti A, Pruzzo C, Zaura E, Lingström P, Ofek I, Pratten J, Spratt DA, Wilson M, Canepari P (2014) The anti-adhesive mode of action of a purified mushroom (*Lentinus edodes*) extract with anticaries and antigingivitis properties in two oral bacterial pathogens. BMC Complement Altern Med 14:75
Sonawane H et al (2012) Antimicrobial activity of some species of Phellinus and Ganoderma sample from Western Ghats of India. IJPSR 3(6):1795–1799
Sugano N, Choji Y, Hibino Y, Yasumura S, Maeda H (1985) Anticarcinogenic action of an alcohol-insoluble fraction (LAP1) from culture medium of *Lentinus edodes* mycelia. Cancer Lett 27:1–6
Sugui MM, De Lima PL, Delmanto RD, Da Eira AF, Salvadori DM, Ribeiro LR (2003) Antimutagenic effect of *Lentinula edodes* (Berk.) Pegler mushroom and possible variation among lineages. Food Chem Toxicol 41:555–560
Sujatha S, Kanimozhi G, Panneerselvam A (2015) Synthesis of silver nanoparticles from *Lentinula edodes* and antibacterial activity. Int J Pharm Sci Rev Res 33(2):189–191
Tehrani MH et al (2012) Search for proteins in the liquid extract of edible mushroom, Agaricus bisporus, and studying their antibacterial effects. Iran J Pharm Res 11:145–150
Thallaimaharani K et al (2013) Studies on antimicrobial and antioxidant properties of oyster mushroom Pleurotus Florida. IJPSR 4:1540–1545
Vannucci L, Šíma P, Vetvicka V, Křižan J (2017) Lentinan properties in anticancer therapy: a review on the last 12-year literature. Am J Immunol 13(1):50–61
Vetchinkina EP, Burov AM, Ageeva MV, Dykman LA, Nikitina VE (2013) Biological synthesis of gold nanoparticles by the xylotrophic basidiomycete *Lentinula edodes*. Prikl Biokhim Mikrobiol 49(4):402–408

Vetchinkina E, Shirokov A, Buckarskaya A, Navolokin N, Prilepskii A, Burov A, Maslyakova G, Nikitina VE (2016) Antitumor activity of extracts from medicinal basidiomycetes mushrooms. Intl J Med Mushroom 18(11):955–964

Vetvicka V, Vetvickova J (2014) Immune-enhancing effects of Maitake (*Grifola frondosa*) and shiitake (*Lentinus edodes*) extracts. Ann Transl Med 2(2):4

Vinceti M, Dennert G, Crespi CM, Zwahlen M, Brinkman M, Zeegers MPA, Horneber M, D'Amico R, Giovane DC (2014) Selenium for preventing cancer. Cochrane Database Syst Rev 30(3):1–193

Vladimirova A, Patskovsky Y, Fedorov AA, Bonanno JB, Fedorov EV, Toro R, Hillerich B, Seidel RD, Richards NG, Almo SC, Raushel FM (2016) Substrate distortion and the catalytic reaction mechanism of 5-Carboxyvanillate decarboxylase. J Am Chem Soc 138(3):826–836

Wang Q, Wang F, Xu Z, Ding Z (2017) Bioactive mushroom polysaccharides: a review on monosaccharide composition, biosynthesis and regulation. Molecules 22(6):955. https://doi.org/10.3390/molecules22060955

Wasser SP (2003) In Shiitake (*Lentinus edodes*). In: Coates PM (ed) Encyclopedia of dietary supplements, 2nd edn. Marcel Dekker, New York, pp 653–664

Watanabe A, Kobayashi M, Hayashi S, Kodama D, Isoda K, Kondoh M, Kawase M, Tamesada M, Yagi K (2006) Protection against D-galactosamine-induced acute liver injury by oral administration of extracts from *Lentinus edodes* mycelia. Biol Pharm Bull 29:1651–1654

Wu C, Wang Z (2000) Volatile compounds in fresh and processed shiitake mushrooms (*Lentinus edodes* sing.). Food Sci Technol Res 6:166–170

Xiaofei X, Jiguo Y, Zhengxiang N, Xuewu Z (2017) ^{1}H NMR-based metabolic investigation of the effect of *Lentinula edodes*-derived polysaccharides on aged mice. J Food Biochem 41:e12371. https://doi.org/10.1111/jfbc/12371. /XIAOFEIET AL.|9of9

Xu X, Zhang X (2015) *Lentinula edodes*-derived polysaccharide alters the spatial structure of gut microbiota in mice. PLoS One 10(1):e0115037. https://doi.org/10.1371/journal.pone.0115037

Xu X, Yan H, Tang J, Chen J, Zhang X (2014) Polysaccharides in *Lentinus edodes*: isolation, structure, immunomodulating activity and future prospective. Crit Rev Food Sci Nutr 54(4):474–487

Yang JH, Lin HC, Mau JL (2001) Non-volatile taste components of several commercial mushrooms. Food Chem 72:465–471

Zaura E, Buijs MJ, Hoogenkamp MA, Ciric L, Papetti A, Signoretto C, Stauder M, Lingstrom P, Pratten J, Spratt DA, Wilson M (2011) The effects of fractions from Shiitake Mushroomon composition and Cariogenicity of dental plaque microcosms in an in vitro caries model. J Biomed Biotechnol. https://doi.org/10.1155/2011/135034

Zembron-Lacny A, Gajewski M, Naczk M, Siatkowski I (2013) Effect of Shiitake (*Lentinus edodes*) extract on antioxidant and inflammatory response to prolonged eccentric exercise. J Physiol Pharmacol 64(2):249–254

Zhang L, Li X, Xu X, Zeng F (2005) Correlation between antitumor activity, molecular weight, and conformation of lentinan. Carbohydr Res 340(8):1515–1521

The Biotechnology of Lignocellulosic Waste Recycling Through *Pleurotus* spp. Cultivation

Mehmet Akyüz and Sevda Kırbağ

Abstract The global agri-food industry produces significant quantities of lignocellulosic residues. It is therefore important to recycle these agro-wastes and investigate alternative uses, including their utilization in food production. *Pleurotus* species are renowned for their ability to decompose agricultural waste, and a number of studies have been conducted to investigate the use of different substrates for their cultivation. The biotechnological valorization of agro-residues through the cultivation of Pleurotus spp. constitutes a sustainable and eco-efficient strategy that enables lignocellulosic bioconversion, resource recovery, and the mitigation of associated environmental impacts. The competitiveness of these species in global production can be attributed to a number of factors, including their diversity of species and varieties, strong enzyme activities, favorable culture conditions, and nutritional and medicinal properties. Although *Pleurotus* species are cultivated worldwide on diverse substrates using different methods, this book chapter will summarize existing studies and provide general practical information on their production in five basic stages: (1) in vitro production of pure mycelia by tissue culture, (2) propagation of the main culture, (3) spawn production, (4) compost preparation and inoculation procedures, and (5) culture conditions. Presenting these methods in stages allows for the delivery of fundamental and practical information on the production of different *Pleurotus* species that yield high quantities in a relatively short period of time. This information is presented in a manner that emphasizes the importance of these unique and distinctive mushroom species and that offers fundamental guidance on how to produce them under culture conditions.

Keywords Recycling of lignocellulosic waste · *Pleurotus* spp. · Mushroom cultivation · Edible mushroom

---

M. Akyüz
Department of Biology, Faculty of Arts and Science, Bitlis Eren University, Bitlis, Türkiye
e-mail: makyuz@beu.edu.tr

S. Kırbağ (✉)
Department of Biology, Faculty of Science, Fırat University, Elazığ, Türkiye
e-mail: skirbag@firat.edu.tr

M. N. Owaid et al. (eds.), *Mushroom Production*, Interdisciplinary Biotechnological Advances, https://doi.org/10.1007/978-981-95-7793-4_12

1 Introduction

A significant number of countries whose economies are heavily dependent on agriculture generate large amounts of lignocellulosic waste, which is rich in organic compounds and thus suitable for recycling (Pilafidis et al. 2022; Munir et al. 2023). Although a small proportion of these wastes are utilized in beneficial applications, such as energy production; biofertilizer, biofuel, biogas, biocomposite, and bioplastic production (Grimm and Wösten 2018; Alves et al. 2021; Yaashikaa et al. 2022; Blasi et al. 2023; Mujtaba et al. 2023); and the production of ethanol, enzymes, volatile oils, additives, etc. (Shirahigue and Ceccato-Antonini 2020; El-Ramady et al. 2022; Soares et al. 2023), a significant proportion still causes serious environmental problems (such as air and soil pollution) due to disposal costs or incineration (Ritota and Manzi 2019; Adejumo and Adebiyi 2020; Kumla et al. 2020; Zárate-Salazar et al. 2020; Shrikhandia et al. 2022; Ajayi and Lateef 2023). Consequently, it is of great importance to transform these agricultural residues into food-based products such as mushrooms by applying fungal biotechnology.

Mushroom cultivation is regarded as a biotechnological process with the potential to transform agricultural waste materials into valuable foodstuffs and reduce environmental pollution through the biological transformation of these wastes (Girmay et al. 2016; Grimm et al. 2021; Ritota and Manzi 2019; Sekan et al. 2019; Kumla et al. 2020; Tesfay et al. 2020; Thakur 2020; Mapayi et al. 2021; Niazi and Ghafoor 2021; Doroški et al. 2022; Pilafidis et al. 2022; Shrikhandia et al. 2022; Wichai et al. 2022). The compost products resulting from this process can also be utilized in a variety of sectors (Grimm and Wösten 2018; Raman et al. 2021; Eliopoulos et al. 2022; Törős et al. 2024), including use as animal feed additives, peat additives in floriculture, and soil fertilizer additives (Umor et al. 2021; Huang et al. 2023).

In recent years, wild edible mushrooms—particularly within the mushroom cultivation sector—have emerged as significant contributors to the agricultural economy (Bellettini et al. 2019; Lu et al. 2020; Thakur 2020; Bakratsas et al. 2021; Mahari et al. 2020; Li and Xu 2022; Llanaj et al. 2023). Economically, nutritionally, medicinally, and commercially valuable species are widely cultivated worldwide, with major production concentrated in America, Europe, and Asia (Raman et al. 2021). Of the 200 mushroom species consumed globally, only about 35 are commercially grown, and ten have achieved industrial production status in many countries (Raman et al. 2021; Niazi and Ghafoor 2021). Naturalists and researchers have extensively studied various mushroom species in their natural habitats, and their findings have been applied by both small businesses and large industries. These studies are conducted in universities, research centers, and research and development units, focusing on cultivation processes, with new research continuously being added. The newly identified and cultured wild species are offered to consumers and enthusiasts with a taste for flavor.

The fungal species of the genus *Pleurotus* (Fr.) P. Kumm. (Pleurotaceae, higher Basidiomycota), commonly known as oyster mushroom, represents the second largest producer of edible mushrooms globally (Sanchez 2010; Royse et al. 2017; Mahari et al. 2020; Magdziak et al. 2021; Raman et al. 2021; Caldas et al. 2022; Melanouri et al. 2022a, b; Amirullah et al. 2023), after *Agaricus bisporus*. The genus *Pleurotus* encompasses more than 200 species found in various tropical and temperate regions worldwide (Correa et al. 2016; Catalogue of Life Search 2020; Kumar 2020). However, only around 30–40 of these species are recognized as commercially edible mushroom species and subspecies (Zervakis and Polemis 2013; Golak-Siwulska et al. 2018). Among these species, only eight to ten are commercially cultivated globally, owing to their superior organoleptic properties, nutritional and medicinal benefits, and significant commercial potential. *P. ostreatus*, *P. pulmonarius*, *P. djamor*, *P. citrinopileatus*, and especially *P. eryngii* are widely cultivated in America, Europe, and Asia due to their high biological efficiency and low-cost cultivation methods (Correa et al. 2016; Zervakis and Koutrotsios 2017; Iossi et al. 2018; Bellettini et al. 2019; Raman et al. 2021; Li and Xu 2022).

Pleurotus species (oyster mushrooms) are widely valued in the mushroom industry for their medicinal, nutritional, and culinary properties (Correa et al. 2016; Carrasco-González et al. 2017; Estrada and Pecchia 2017; Dudekula et al. 2020; Niego et al. 2021; Sharma et al. 2021; Raman et al. 2021; Amirullah et al. 2023). They are also efficient lignocellulosic decomposers, owing to their rapid mycelial growth and strong enzymatic activity (Tsujiyama and Ueno 2013; Bánfi et al. 2015; Salmones 2017; Bumanlag et al. 2018; Hanaf et al. 2018; Jegadeesh et al. 2018; Silva et al. 2018; Bellettini et al. 2019; Ritota and Manzi 2019; Kumla et al. 2020; Araújo et al. 2021; Durán-Aranguren et al. 2021; Leong et al. 2021; Zhang et al. 2021; El-Ramady et al. 2022; Vega et al. 2022). Compared to other edible mushrooms, they exhibit faster growth and greater efficiency under solid-state fermentation.

A significant number of species within the *Pleurotus* genus have garnered commercial interest and are increasingly being cultivated for their edible properties. These include *P. ostreatus* (grey oyster), *P. djamor* (pink oyster), *P. citrinopileatus* (golden oyster), and *P. pulmonarius* (phoenix oyster). Particularly notable in recent years are *P. eryngii* (king oyster) and its variety, *P. nebrodensis* (white oyster), as well as *P. tuoliensis* (Patel et al. 2012; Iossi et al. 2018; Bellettini et al. 2019; Krakowska et al. 2020; Zárate-Salazar et al. 2020; Zou et al. 2020; Espinosa-Páez et al. 2021; Deora et al. 2021; Droki et al. 2022). These species have been cultivated in different experimental groups in various countries around the world. The cultivation practices include pure cultures, 1:1 mixtures, and combinations with varying ratios and additives, comprising different ratios of lignocellulosic substrates, such as wheat/corn/soybean straw, millet/cotton/bean stalk/waste, rice stalk, quinoa stalk, lentil/rice straw, rice husk, corn stalk, maize/straw/stalk/stover, olive mill waste, sugarcane bagasse, beech wood shavings, coffee residue, barley and oats straw, forestry wood wastes, sawdust, oil palm waste, corn cobs and stubble, sorghum stover, groundnut haulms, sugarcane trash, bean/chickpea pods, banana leaves, banana pseudostem, bamboo leaves, teak leaves, grape marc waste, coffee pulp/husk grounds, peanut shells, coconut husk/coir, hazelnut branches/husk, waste paper, and various field-based residues, including waste paper, roadleaf, wood/

hardwood, and other straw-based substrates of plant materials (Ragunathan et al. 1996; Philippoussis et al. 2001; Zervakis and Venturella 2002; Choi et al. 2006; Kırbag and Akyuz 2008; Gargano et al. 2013; Jeznabadi et al. 2016; Atila 2017; Sardar et al. 2017; Zurbano et al. 2017; Liu et al. 2018; Nam et al. 2018; Ritota and Manzi 2019; Zárate-Salazar et al. 2020; Adebayo et al. 2021; Abed et al. 2021; Chai et al. 2021; Grimm et al. 2021; Raman et al. 2021; Koutrotsios et al. 2018, 2022; Akyüz and Kırbağ 2022; Akter et al. 2022; Chilanti et al. 2022; Doroški et al. 2022; Li et al. 2022, 2023; Melanouri et al. 2022a, b; Wachira et al. 2022; Wang et al. 2022; Zhou et al. 2022 2023; Akcay et al. 2023; Dedousi et al. 2023; Kim et al. 2023; Lee et al. 2023; Östbring et al. 2023; Subedi et al. 2023; İnci et al. 2023, 2024; Atila and Çetin 2024; Karpagavalli et al. 2024 etc.).

Pleurotus species are globally renowned for their widespread cultivation and consumption due to their ability to thrive on diverse substrate materials, their shorter cultivation period, and their cost-effective production, as well as their ease of cultivation, high yield potential, and valuable medicinal and nutritional properties (Correa et al. 2016; Ritota and Manzi 2019; De et al. 2023; Devi et al. 2024), which require less environmental control, making them increasingly favored for their flavor and nutritional benefits, alongside their high yield, pest resistance, and versatility in agricultural use (Chang 2006; Sekan et al. 2019; Mahari et al. 2020). Mushroom yield is strongly influenced by substrate properties (composition, pH, moisture), mushroom strain and inoculum level, as well as environmental conditions such as temperature, humidity, and light (Ritota and Manzi 2019; Kumla et al. 2020; Mahari et al. 2020; Yaashikaa et al. 2022; Guragain et al. 2024). Substrate nutritional composition directly affects mushroom quality, and various studies have confirmed the impact of substrate type on both yield and nutritional content (Suwannarach et al. 2022; Zied et al. 2024). Building on these findings, global research on various Pleurotus species has revealed significant variations in cultivation methods, preparation stages, and environmental conditions. These differences primarily arise from variations in substrate materials, resource availability, and production site characteristics. Researchers typically adjust cultivation processes based on species-specific requirements, optimal parameters, and environmental suitability, which can result in variability across cultivation stages and overall production outcomes. The primary reasons for these variations are the differences in the environments and materials used, the diversity of resources obtained, and the variability of the physical environments where the production takes place. It is observed that researchers determine the most suitable process by considering the requirements of the species they use, the appropriate parameters, and the suitability of the physical environment. This situation may lead to differences in cultivation stages.

Although variations may exist in the fundamental production and cultivation stages, at the professional level, the production of cultivated mushrooms is typically carried out in five stages. Before transitioning to commercial and industrial-scale production, it is important to acquire adequate practical knowledge from researchers, universities, and research and development centers. The practical knowledge and experiences obtained from these sources will be beneficial in overcoming the potential challenges encountered during the production stages. This study presents a description of the five stages for the practical cultivation of *Pleurotus* species at the micro and R&D scales. Considered as a valuable resource, it serves both new researchers and producers while providing a useful reference for producers in large-scale and industrial production platforms at educational and operational levels. Furthermore, the knowledge and experience obtained through these five stages can be beneficial for orientation training in universities, small-scale businesses, and industrial production facilities for those interested in working in the field of mushroom cultivation.

2 The Cultivation Stages of *Pleurotus* Species

The remaining stages of the cultivation process—namely, mycelium growth, spawn propagation, compost preparation and spawn inoculation procedures, and culture conditions—were carried out in accordance with the proposed methods summarized in the references (Zadrazil 1978; Delmas and Mamoun 1983; Ragunathan et al. 1996; Philippoussis et al. 2001; Zervakis and Venturella 2002; Choi et al. 2006; Kırbag and Akyuz 2008; Gargano et al. 2013; Jeznabadi et al. 2016; Atila 2017; Sardar et al. 2017; Liu et al. 2018; Nam et al. 2018; Ritota and Manzi 2019; Zárate-Salazar et al. 2020; Adebayo et al. 2021; Chai et al. 2021; Grimm et al. 2021; Raman et al. 2021; Koutrotsios et al. 2018, 2022; Akter et al. 2022; Akyüz and Kırbağ 2022; Chilanti et al. 2022; Doroški et al. 2022; Li et al. 2022, 2023; Melanouri et al. 2022a, b; Wachira et al. 2022; Wang et al. 2022; Zhou et al. 2022, 2023; Akcay et al. 2023; Dedousi et al. 2023; Kim et al. 2023; Lee et al. 2023; Östbring et al. 2023; Subedi et al. 2023; İnci et al. 2023, 2024; Atila and Çetin 2024; Karpagavalli et al. 2024; etc.), with some modifications. The evaluation was conducted taking into consideration the methods outlined in the specified study. Depending on the mushroom species used in the study, the biochemical structure of the compost, and the method of cultivation, differences can be observed among the researchers. The stages for the cultivation of *Pleurotus* species—from the initial production phase to the final stage, including mycelium cultivation and propagation, spawn production, compost preparation, and cultural conditions—have been presented in detail.

2.1 Isolation of Pure Mycelium from Mushroom Basidiocarp Through In Vitro Tissue Culture Techniques

Two methods are generally used to obtain pure mycelia from *Pleurotus* species. The first method involves drying fresh mushrooms on clean A4 paper (Fig. 1a) and shedding the spores (Fig. 1b–c), which are then transferred to a nutrient agar medium. However, this method is less favoured as it carries a high risk of contamination

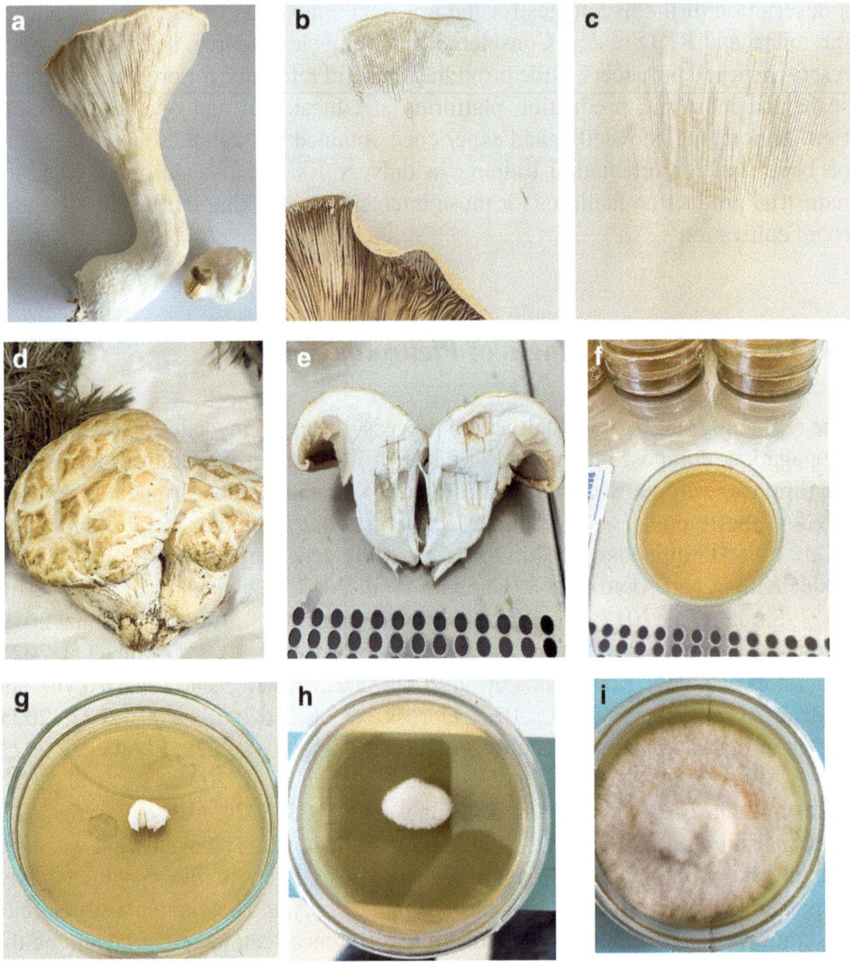

Fig. 1 Isolation of pure mycelium from mushroom basidiocarp (**a**–**c**: based on spores, **d**–**i**: based on a piece of fresh tissue)

during the transfer of spores to nutrient agar media and subsequent development. The other method is to obtain pure mycelia by transferring fresh small pieces of tissue ranging from 1 mm to 0.5 cm from the inner tissue of fresh mushrooms to a nutrient agar medium in a sterile environment. The steps for the second method are as follows: collect fresh, firm, and elastic mushrooms (Fig. 1d) from their natural environment and bring them to the laboratory. In a high-efficiency particulate air (HEPA)-filtered laminar flow, wipe the surface of the pileus with 96% alcohol, and then cut along the pileus and stipe with a sterile scalpel, as shown in Fig. 1e. Excise clean pieces of tissue measuring 1 mm to 0.5 cm in diameter (Fig. 1g) and transfer them to prepared Petri dishes (Fig. 1f). Wrap the dishes with parafilm and place them upside down in the incubation device at 25 °C. After 4–5 days, observe mycelial growth from the tissue pieces (Fig. 1h), and if there is no risk of infection, continue incubating the nutrient agar with the developed mycelium until the surface is completely covered, which requires approximately 2 weeks. Store the obtained pure mycelium (Fig. 1i) in the refrigerator at 4 °C for later use.

2.2 Inoculum Preparation

As mentioned in the first paragraph of this section, some researchers have used different nutrient media for the propagation of the main culture, and it has been noted that malt extract agar is the most effective nutrient medium. However, it is commonly observed that in products provided by different companies, there is a high accumulation of water in Petri dishes after inoculation, causing damage to the mycelial structures. Therefore, researchers should consider using nutrient media obtained from reliable and high-quality sources.

For main culture propagation, 2.0% malt extract agar (MEA) is commonly used (Fig. 2a). MEA plates (90 mm in diameter) are inoculated with a 6 mm mycelium plug taken from the actively growing margin of a young colony (Fig. 2b–c). The plug is placed at the center of a fresh MEA plate (Fig. 2d) and incubated in the dark at 25 °C for 10 days or until the mycelium fully colonizes the plate surface (Fig. 2e–f), after which it is used as an inoculum for spawn.

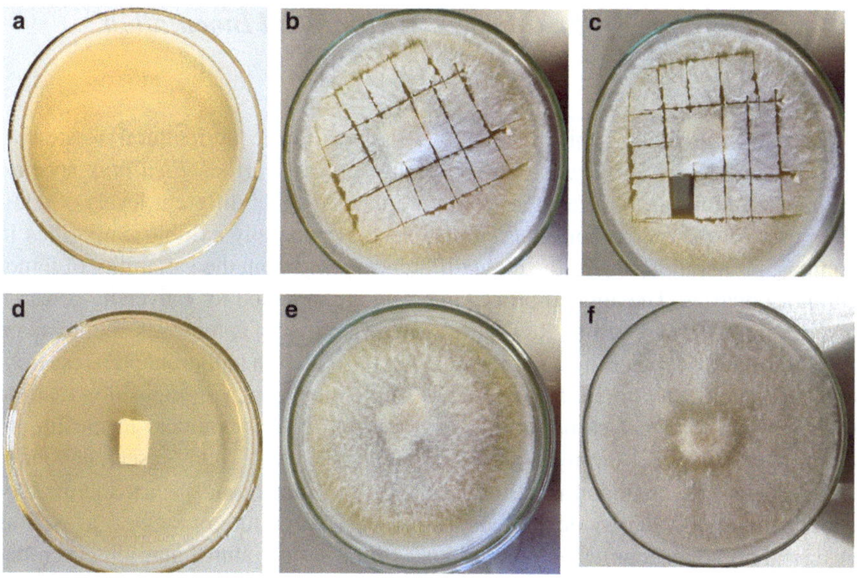

Fig. 2 The stages of main culture propagation

2.3 Spawn Preparation

Again, as mentioned in the first paragraph of this section, researchers have used different agricultural grain media for spawn propagation, and it has been observed that among the different grains tested, barley and wheat are the most effective. Although varying ratios of gypsum, lime, and additives are employed to maintain the pH of the medium at 5.5–6.5 for mycelium development, the stages of spawn propagation generally exhibit similarities. Furthermore, the researchers found that the use of grains is especially beneficial, noting that they can be easily transferred to every part of the compost.

For spawn production, 1 kg of barley or wheat grains is used. The grains are boiled for 40 min (Fig. 3a), washed under running water, and thoroughly drained (Fig. 3b). Subsequently, 2 g of lime and 8 g of gypsum are added and manually mixed to ensure uniform distribution (Fig. 3c). Each mixture is then placed in a 250 mL Erlenmeyer flask containing 120 or 150 g of grains, sterilized in an autoclave at 121 °C for 15 min (Fig. 3d). After cooling, two to three agar discs (1 cm^2) containing mycelium are aseptically transferred to the grains (Fig. 3e). By the end of the fifth day (Fig. 3f), the flasks are shaken to quickly inoculate the mycelium developed on the agar plates onto the grains (Fig. 3g). The prepared material is incubated in complete darkness at 25 °C for 2 weeks until full mycelial colonization is achieved (Fig. 3h–i).

2.4 Preparation of the Compost Medium and Inoculation with Spawn

Mushroom cultivation typically employs lignocellulosic agricultural wastes, as these materials are well suited for the cultivation of various *Pleurotus* species. Although culture media prepared from pure mixtures and different levels of additives used in the cultivation of *Pleurotus* species generally exhibit similarities in compost preparation stages, the main differences arise from the variability of lignocellulosic wastes. The applied procedures here are intended for university research and development, as well as R&D-level enterprises; large enterprises should adapt this information for their own purposes. In the introduction to this chapter, researchers have specifically noted the importance of the biochemical composition (carbon (C), nitrogen (N), and carbon-to-nitrogen ratio (C/N)) of compost in mushroom cultures and its effect on mycelial development and yield. Many researchers have also emphasized the importance of considering the mushroom species and the biochemical composition of the compost (C, N, C/N) in culture processes, as well as ensuring homogeneous culture conditions (light, temperature, humidity, ventilation, etc.).

Experimental groups should consist of various combinations of agricultural residues, including straw, stalk, bran, husk, etc., along with their pure forms, 1:1

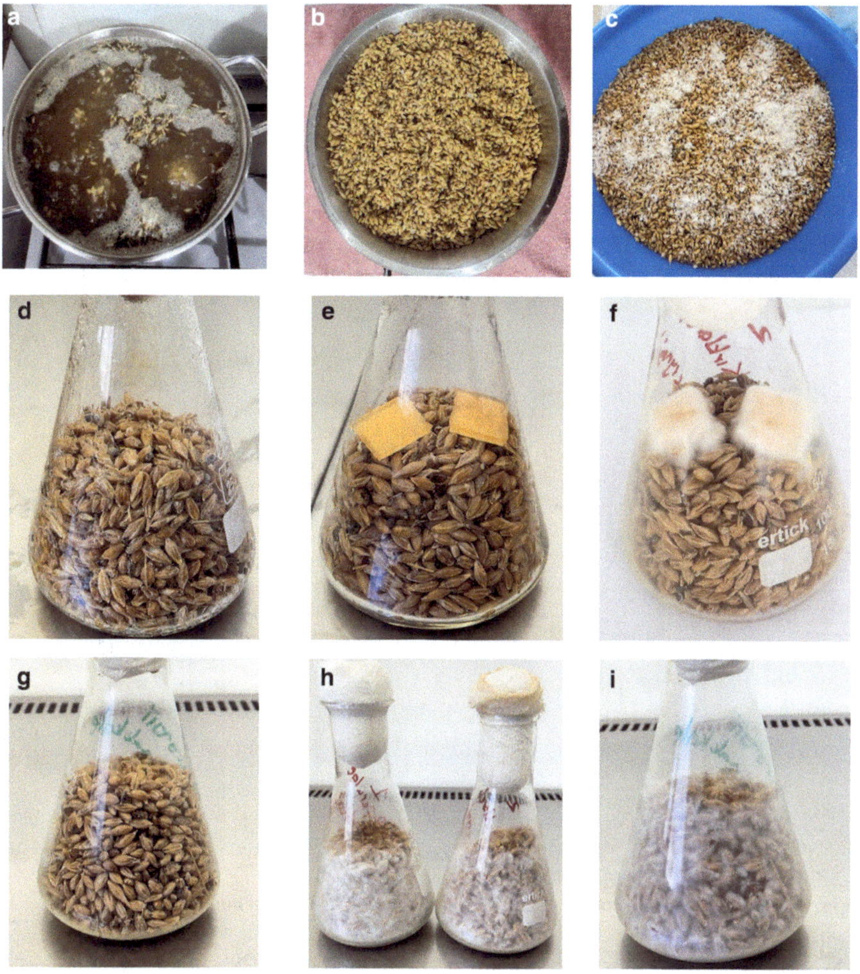

Fig. 3 The stages of spawn propagation

mixtures, and various additives. These groups should be soaked in water for 72 h until they reach a humidity level of 70–75% (Fig. 4a). Subsequently, the drained materials should be spread onto a polyethylene sheet (Fig. 4b); to achieve the desired pH range of 5.5–6.5, 35 g of gypsum and lime per kg of dry compost should be added (Fig. 4c). Following this, the prepared substrates should be placed in cloth bags (Fig. 4d) and subjected to autoclaving at 121 °C for 30 min (or 1.5 h for small-scale and industrial businesses), after which they should be allowed to cool for spawn inoculation. The compost (Fig. 4e) should then be inoculated with 3 or 5% spawn (1000 g compost) using the layering method (Fig. 4f), placed in bags (20 × 30 cm for 1 kg) with closed mouths, and left in a dark culture room (12 m^2) at 25 ± 1 °C for 15 days (Fig. 4g). Once the compost is fully colonized by the spawn

Fig. 4 Preparation of the compost, inoculation of spawn, and cultivation conditions

(Fig. 4h), culture room conditions should be adjusted for primordium induction (Fig. 4i): temperature—18 °C for *P. ostreatus*; 20 °C for *P. pulmonarius*, *P. djamor*, *P. citrinopleatus*, and *P. florida*; and 13–15 °C for *P. nebrodensis*, *P. tuoliensis*, and *P. eryngii* and its varieties); humidity, 80–85%; light exposure, 1000 lux for 12 h daily; and ventilation, 3 h daily. The cultures should also be watered twice daily by spraying.

2.5 *The Harvest of* Pleurotus *spp.*

In the cultivation of *Pleurotus* species, three fundamental factors need to be considered to achieve rapid and high-yield production: (1) the selection of suitable isolates for the desired *Pleurotus* species, (2) the adjustment of substrates composed of pure or 1:1 mixtures of suitable lignocellulosic materials and combinations thereof, and (3) the maintenance of homogeneous culture conditions (such as temperature, humidity, light, ventilation, watering, etc.) for the formation of basidiocarps. For the first factor, rapid composting is crucial for early harvest and to minimize the risk of infection. For the second factor, the biochemical composition of the substrate—particularly parameters such as C, N, C/N ratio, cellulose, hemicellulose, and lignin content—plays a critical role in early and high-yield mushroom production. Researchers must pay attention to the mushroom species and especially the composition of the compost when preparing culture environments for rapid production and nutrient-rich yields. Lastly, for the third factor, maintaining appropriate culture conditions and controlling them within specific parameters are essential for the formation of basidiocarps and for achieving high yields. Primordium growth should be monitored, and fruitbodies should be harvested when they reach the characteristic maturity stage (Fig. 5a–f).

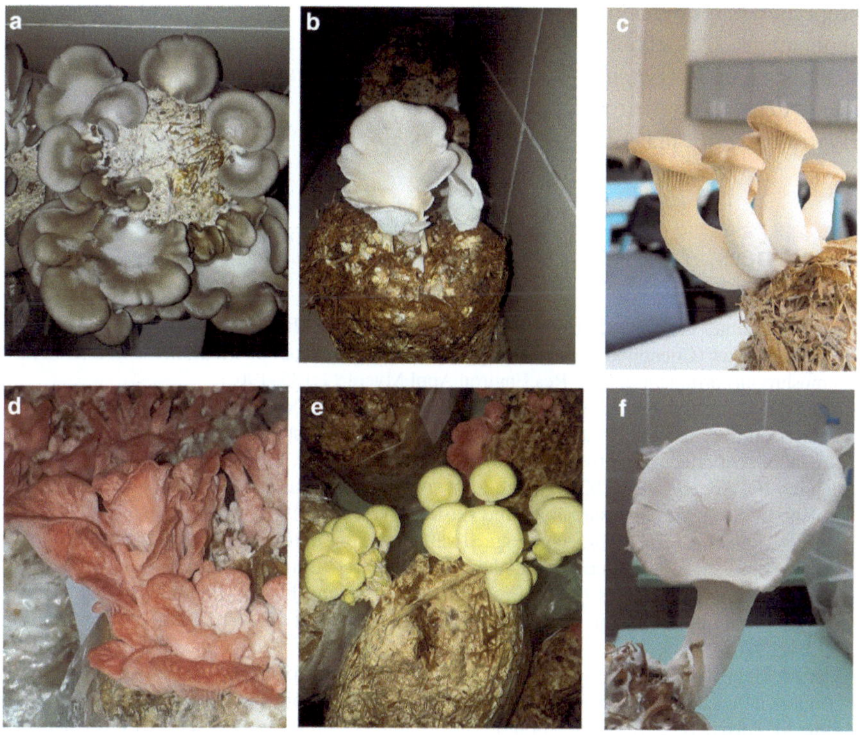

Fig. 5 *Pleurotus* species (**a**: *P. ostreatus*, **b**: *P. pulmonarius*, **c**: *P. eryngii*, **d**: *P. djamor*, **e**: *P. citrinopleatus*, **f**: *P. nebrodensis*)

3 Conclusion

This chapter provides fundamental insights into the cultivation of various *Pleurotus* species for human consumption and culinary purposes. Mushroom cultivation has become a significant component of the agricultural economy, supported by diverse actors ranging from amateurs to research institutions and industrial enterprises. Different *Pleurotus* species, reflecting species-specific and habitat-based diversity, have been cultivated worldwide, with culture collections made available internationally. Their strong enzymatic systems, simple cultivation methods, efficient utilization of agricultural wastes, rapid mycelial growth, early and abundant yields, year-round production capability, and nutritional and therapeutic properties highlight their global importance in terms of production, diversity, and market value. Comprehensive research has expanded the recognition and practical use of these species, transforming agricultural waste into a valuable resource. Furthermore, the identification of species, aroma profiles, flavor, nutritional content, and therapeutic properties has rendered *Pleurotus* indispensable for gourmets and gastronomic applications. Consequently, focusing on less-studied species with limited market share, such as *P. eryngii* and its varieties, *P. nebrodensis*, and *P. tuoliensis*, rather than extensively researched species like *P. ostreatus*, *P. pulmonarius*, *P. djamor*, *P. citrinopleatus*, may offer greater benefits.

References

Abed IA, Hamad HBA, Owaid MN, Hamdan NT, Lafi AS, Mutlaq HH (2021) Effect of using desert weeds (Chenopodiaceae) as supplements in substrates of *Pleurotus ostreatus* (oyster mushroom) production. Curr Res Environ Appl Mycol 11:185–196

Adebayo EA, Elkanah FA, Afolabi FJ, Ogundun OS, Alabi TF, Oduoye OT (2021) Molecular characterization of most cultivated *Pleurotus* species in sub–western region Nigeria with development of cost effective cultivation protocol on palm oil waste. Heliyon 7:e06215

Adejumo IO, Adebiyi OA (2020) Agricultural solid wastes: causes, efects, and efective management. In: Strategies of sustainable solid waste management. Intech Open, London, pp 139–145

Ajayi VA, Lateef A (2023) Biotechnological valorization of agrowastes for circular bioeconomy: melon seed shell, groundnut shell and groundnut peel. Clean Circ Bioecon 4:100039

Akcay C, Ceylan F, Arslan R (2023) Production of oyster mushroom (*Pleurotus ostreatus*) from some waste lignocellulosic materials and FTIR characterization of structural changes. Sci Rep 13:12897

Akter M, Halawani RF, Aloufi FA, Taleb MA, Akter S, Mahmood S (2022) Utilization of agro–industrial wastes for the production of quality oyster mushrooms. Sustainability 14:994

Akyüz M, Kırbağ S (2022) Cultivation of king eryngii (*Pleurotus eryngii* (DC. ex Fr.) Quel.) isolates on various local agro–residues. Biomass Convers Biorefinery 14:12197–12205

Alves LDS, Moreira BRDA, Viana RDS, Pardo-Gimenez A, Dias ES, Noble R, Zied DC (2021) Recycling spent mushroom substrate into fuel pellets for low emission bioenergy producing systems. J Clean Prod 313:127875

Amirullah NA, Abdullah E, Abidin NZ, Abdullah N, Manickam S (2023) Influence of extraction technologies on the therapeutic properties of *Pleurotus* spp.(oyster mushrooms) – a critical review. Food Bioscience 56:103352

Araújo NL, Avelino KV, Halabura MIW, Marim RA, Kassem ASS, Linde GA, do Valle, J.S. (2021) Use of green light to improve the production of lignocellulose–decay enzymes by *Pleurotus* spp. in liquid cultivation. Enzym Microb Technol 149:109860

Atila F (2017) Evaluation of suitability of various agro–wastes for productivity of *Pleurotus djamor*, *Pleurotus citrinopileatus* and *Pleurotus eryngii* mushrooms. J Exp Agric Int 17:1–11

Atila F, Çetin M (2024) Bioconversion of lavender oil extraction wastes through cultivation of *Pleurotus eryngii* var. *ferulae*: its effects on yield, nutritional content and antioxidant capacity of the mushroom. Biocatal Agric Biotechnol 58:103138

Bakratsas G, Polydera A, Katapodis P, Stamatis H (2021) Recent trends in submerged cultivation of mushrooms and their application as a source of nutraceuticals and food additives. Future Foods 4:100086

Bánfi R, Pohner Z, Kovács J, Luzics S, Nagy A, Dudás M, Vajna B (2015) Characterisation of the large–scale production process of oyster mushroom (*Pleurotus ostreatus*) with the analysis of succession and spatial heterogeneity of lignocellulolytic enzyme activities. Fungal Biol 119:1354–1363

Bellettini MB, Fiorda FA, Maieves HA, Teixeira GL, Ávila S, Hornung PS, Ribani RH (2019) Factors affecting mushroom *Pleurotus* spp. Saudi J Biol Sci 26:633–646

Blasi A, Verardi A, Lopresto CG, Siciliano S, Sangiorgio P (2023) Lignocellulosic agricultural waste valorization to obtain valuable products: an overview. Recycling 8:61

Bumanlag C, Kalaw S, Dulay R, Reyes R (2018) Optimum conditions for mycelial growth and basidiocarp production of *Pleurotus djamor* on corn–based media. Int J Biol Pharm Allied Sci 7:558–575

Caldas LA, Zied DC, Sartorelli P (2022) Dereplication of extracts from nutraceutical mushrooms *Pleurotus* using a molecular network approach. Food Chem 370:131019

Carrasco-González JA, Serna-Saldívar SO, Gutiérrez-Uribe JA (2017) Nutritional composition and nutraceutical properties of the *Pleurotus* fruiting bodies: potential use as food ingredient. J Food Compos Anal 58:69–81

Catalogue of Life Search (2020) Retrieved January 15, 2020, from http://www.catalogueoflife.org/col/search/all/key/pleurotus/fossil/1/match/1

Chai WY, Krishnan UG, Sabaratnam V, Tan JBL (2021) Assessment of coffee waste in formulation of substrate for oyster mushrooms *Pleurotus pulmonarius* and *Pleurotus floridanus*. Future Foods 4:100075

Chang ST (2006) The world mushroom industry: trends and technological development. Int J Med Mushrooms 8:297–314

Chilanti G, da Rosa LO, Poleto L, Branco CS, Camassola M, Fontana RC, Dillon AJ (2022) Effect of different substrates on *Pleurotus* spp. cultivation in Brazil–Ergothioneine and lovastatin. J Food Compos Anal 107:104367

Choi DB, Nam HG, Cha WS (2006) Studies on cultivation and biological activities of *Pleurotus nebrodensis* inzenga. Korean J Chem Eng 23:241–246

Correa RCG, Brugnari T, Bracht A, Peralta RM, Ferreira IC (2016) Biotechnological, nutritional, and therapeutic uses of *Pleurotus* spp. (oyster mushroom) related with its chemical composition: a review on the past decade findings. Trends Food Sci Technol 50:103–117

De A, Mridha D, Roychowdhury T, Bandyopadhyay B, Panja AS (2023) Substrate level optimization for better yield of oyster mushroom (*Pleurotus ostreatus*) production, using different ratio of rice straw and sugarcane bagasse. World J Microbiol Biotechnol 39:270

Dedousi M, Melanouri EM, Diamantopoulou P (2023) Carposome productivity of *Pleurotus ostreatus* and *Pleurotus eryngii* growing on agro–industrial residues enriched with nitrogen, calcium salts and oils. Carbon Resour Convers 6:150–165

Delmas J, Mamoun M (1983) Le *Pleurote* en corne d'abondance un champignon aujourd'hui cultivable en France. PHM–Revue Horticole 240:39–46

Deora A, Sharma SS, Kumari P, Dahima V, Kumar S, Rohith M (2021) Cultivation of kabul dhingri (*Pleurotus eryngii*) mushroom by standardizing protocols in subtropical zones of world. Sci Rep 11:14692

Devi PV, Islam J, Narzary P, Sharma D, Sultana F (2024) Bioactive compounds, nutraceutical values and its application in food product development of oyster mushroom. J Future Foods 4:335–342
Doroški A, Klaus A, Režek JA, Djekic I (2022) Food waste originated material as an alternative substrate used for the cultivation of oyster mushroom (*Pleurotus ostreatus*): a review. Sustainability 14:12509
Dudekula UT, Doriya K, Devarai SK (2020) A critical review on submerged production of mushroom and their bioactive metabolites. 3 Biotech 10:1–12
Durán-Aranguren DD, Meléndez-Melo JP, Covo-Ospina MC, Díaz-Rendón J, Reyes-Gutiérrez DN, Reina LC, Sierra R (2021) Biological pretreatment of fruit residues using the genus *Pleurotus*: a review. Bioresour Technol Rep 16:100849
Eliopoulos C, Markou G, Chorianopoulos N, Haroutounian SA, Arapoglou D (2022) Preliminary research concerning the enrichment of industrial hemp extract residues via solid state fermentation with Pleurotus ostreatus. Appl Sci 12:2376
El-Ramady H, Abdalla N, Fawzy Z, Badgar K, Llanaj X, Törős G, Prokisch J (2022) Green biotechnology of oyster mushroom (*Pleurotus ostreatus* L.): a sustainable strategy for myco–remediation and bio–fermentation. Sustainability 14:3667
Espinosa-Páez E, Hernández-Luna CE, Longoria-García S, Martínez-Silva PA, Ortiz-Rodríguez I, Villarreal-Vera MT, Cantú-Saldaña CM (2021) *Pleurotus ostreatus*: a potential concurrent biotransformation agent/ingredient on development of functional foods (Cookies). LWT Food Sci Technol 148:111727
Estrada AER, Pecchia J (2017) Cultivation of *Pleurotus ostreatus*. In: Edible and medicinal mushrooms: technology and applications. Wiley, pp 339–360
Gargano ML, Zervakis GI, Venturella G (2013) Cultivation and nutritional value of *Pleurotus nebrodensis*. In: Pleurotus nebrodensis, a very special mushrooms. Bentham Science Publishers, pp 99–20
Girmay Z, Gorems W, Birhanu G, Zewdie S (2016) Growth and yield performance of *Pleurotus ostreatus* (Jacq. Fr.) Kumm (oyster mushroom) on diferent substrates. AMB Express 6:1–7
Golak-Siwulska I, Kałużewicz A, Spiżewski T, Siwulski M, Sobieralski K (2018) Bioactive compounds and medicinal properties of oyster mushrooms (sp.). Folia Horticulturae 30:191–201
Grimm D, Wösten HAB (2018) Mushroom cultivation in the circular economy. Appl Microbiol Biotechnol 102:7795–7803
Grimm A, Eilertsen L, Chen F, Huang R, Atterhem L, Xiong S (2021) Cultivation of *Pleurotus ostreatus* mushroom on substrates made of cellulose fibre rejects: product quality and spent substrate fuel properties. Waste Biomass Valoriz 12:4331–4340
Guragain DP, Shrestha B, Bajracharya I (2024) A low–cost centralized IoT ecosystem for enhancing oyster mushroom cultivation. J Agric Food Res 15:100952
Hanaf FHM, Rezania S, Taib SM, Din MFM, Yamauchi M, Sakamoto M, Hara H, Park J, Ebrahimi SS (2018) Environmentally sustainable applications of agro–based spent mushroom substrate (SMS): an overview. J Mater Cycles Waste Manag 20:1383–1396
Huang C, Han X, Luo Q, Nie Y, Kang M, Chen Y, Chen Y (2023) Agro–based spent mushroom compost substrates improve soil properties and microbial diversity in greenhouse tomatoes. Agronomy 13:2291
İnci Ş, Kırbağ S, Akyüz M (2023) Growth period, yield, and nutrient contents of *Pleurotus citrinopileatus* Singer grown on some local agricultural wastes in Turkey. Biomass Convers Biorefinery 13:15029–15038
İnci Ş, Kirbağ S, Akyüz M (2024) Valorization of local agro–residues for the cultivation of *Pleurotus djamor* (Rumph. Ex Fr.) Boedijn and their effects on nutritional value. Biomass Convers Biorefinery. https://doi.org/10.1007/s13399-024-05515-3
Iossi MR, Cobos JDV, Alegria FJG, Zied DC (2018) *Pleurotus* spp. cultivation on *Brachiaria* sp. straw treatment with alkaline water. Braz J Microbiol 49:64–67

Jegadeesh R, Lakshmanan H, Kab-Yeul J, Sabaratnam V, Raaman N (2018) Cultivation of pink oyster mushroom *Pleurotus djamor* var. *roseus* on various agro–residues by low cost technique. J Mycopathol Res 56:213–220

Jeznabadi EK, Jafarpour M, Eghbalsaied S (2016) King oyster mushroom production using various sources of agricultural wastes in Iran. Int J Recycl Org Waste Agric 5:17–24

Karpagavalli S, Manisha R, Mageshwari S, Sowbharnika M (2024) Influence of growth substrates on bioactive compounds and yield of oyster mushroom (*Pleurotus florida*). Sci Hortic 329:112959

Kim YJ, Ha TM, Kim JH, Choi JY, Lee CY, Lee CJ, Lim GJ (2023) Characteristics and breeding of the new cultivar of *Pleurotus nebrodensis* 'Boram'. J Mushrooms 21:145–149

Kırbag S, Akyuz M (2008) Effect of various agro–residues on growing periods, yield and biological efficiency of *Pleurotus eryngii*. J Food Agric Environ 6:402–405

Koutrotsios G, Kalogeropoulos N, Kaliora AC, Zervakis GI (2018) Toward an increased functionality in oyster (*Pleurotus*) mushrooms produced on grape marc or olive mill wastes serving as sources of bioactive compounds. J Agric Food Chem 66:5971–5983

Koutrotsios G, Tagkouli D, Bekiaris G, Kaliora A, Tsiaka T, Tsiantas K, Zervakis GI (2022) Enhancing the nutritional and functional properties of *Pleurotus citrinopileatus* mushrooms through the exploitation of winery and olive mill wastes. Food Chem 370:131022

Krakowska A, Zięba P, Włodarczyk A, Kała K, Sułkowska-Ziaja K, Bernaś E, Muszyńska B (2020) Selected edible medicinal mushrooms from *Pleurotus* genus as an answer for human civilization diseases. Food Chem 327:127084

Kumar K (2020) Nutraceutical potential and processing aspects of oyster mushrooms (*Pleurotus* species). Curr Nutr Food Sci 16:3–14

Kumla J, Suwannarach N, Sujarit K, Penkhrue W, Kakumyan P, Jatuwong K, Lumyong S (2020) Cultivation of mushrooms and their lignocellulolytic enzyme production through the utilization of agro–industrial waste. Molecules 25:2811

Lee AML, Chin CFS, Seelan JSS, Chye FY, Lee HH, Rakib MRM (2023) Metabolites profiling of protein enriched oyster mushroom (*Pleurotus ostreatus* (Jacq.) P. Kumm.) grown on oivl palm empty fruit bunch substrate. LWT Food Sci Technol 181:114731

Leong CC, Ho WY, Yeap SK, Krishnen G, Chong ZX, Ho JS, Lim PT, Ten ST (2021) Assessment of phylogenetic, growth, and antioxidant capacity of *Pleurotus* spp. in Malaysia. J Food Process Preserv 45:e15554

Li C, Xu S (2022) Edible mushroom industry in China: current state and perspectives. Appl Microbiol Biotechnol 106:3949–3955

Li G, Wang Y, Yu D, Zhu P, Zhao G, Liu C, Zhao H (2022) Ligninolytic characteristics of *Pleurotus ostreatus* cultivated in cotton stalk media. Front Microbiol 13:1035040

Li X, Chen G, Li X, Yao F (2023) Three *Pleurotus* mushroom species cultivated in a mixed *Phragmites australis* substrate differ in nutrient utilization capacity. J Food Compos Anal 115:104924

Liu SR, Zhang WR, Kuang YB (2018) Production of stalk spawn of an edible mushroom (*Pleurotus ostreatus*) in liquid culture as a suitable substitute for stick spawn in mushroom cultivation. Sci Hortic 240:572–577

Llanaj X, Törős G, Hajdú P, Abdalla N, El-Ramady H, Kiss A, Prokisch J (2023) Biotechnological applications of mushrooms under the water–energy–food nexus: crucial aspects and prospects from farm to pharmacy. *Foods 12*:2671

Lu H, Lou H, Hu J, Liu Z, Chen Q (2020) Macrofungi: a review of cultivation strategies, bioactivity, and application of mushrooms. Compr Rev Food Sci Food Saf 19:2333–2356

Magdziak Z, Gąsecka M, Stuper-Szablewska K, Siwulski M, Budzyńska S, Jasińska A, Niedzielski P, Kalač P, Mleczek M (2021) A possibility to use selected crop post–extraction wastes to improve the composition of cultivated mushroom *Pleurotus citrinopileatus*. J Fungi 7:894

Mahari WA, Peng W, Nam WL, Yang H, Lee XY, Lee YK, Lam SS (2020) A review on valorization of oyster mushroom and waste generated in the mushroom cultivation industry. J Hazard Mater 400:123156

Mapayi TT, Oluduro AO, Omoboye OO, Taiwo AM (2021) Medicinal plant assisted cultivation of *Pleurotus forida* using diferent lignocellulosic waste substrates. Vegetos 34:485–494
Melanouri EM, Dedousi M, Diamantopoulou P (2022a) Cultivating *Pleurotus ostreatus* and *Pleurotus eryngii* mushroom strains on agro–industrial residues in solid–state fermentation. Part I: Screening for growth, endoglucanase, laccase and biomass production in the colonization phase. Carbon Resour Convers 5:61–70
Melanouri EM, Dedousi M, Diamantopoulou P (2022b) Cultivating *Pleurotus ostreatus* and *Pleurotus eryngii* mushroom strains on agro–industrial residues in solid–state fermentation. Part II: Effect on productivity and quality of carposomes. Carbon Resour Convers 5:52–60
Mujtaba M, Fraceto LF, Fazeli M, Mukherjee S, Savassa SM, de Medeiros GA, Vilaplana F (2023) Lignocellulosic biomass from agricultural waste to the circular economy: a review with focus on biofuels, biocomposites and bioplastics. J Clean Prod 402:136815
Munir N, Ramli ANM, Norsazali NFS, Bhuyar P (2023) Valorization of agro–industrial waste for the advancement of mushrooms and their production yield. Biomass Convers Biorefinery 14:26415–26426
Nam WL, Phang XY, Su MH, Liew RK, Ma NL, Rosli MHNB, Lam SS (2018) Production of bio–fertilizer from microwave vacuum pyrolysis of palm kernel shell for cultivation of oyster mushroom (*Pleurotus ostreatus*). Sci Total Environ 624:9–16
Niazi AR, Ghafoor A (2021) Diferent ways to exploit mushrooms: a review. All Life 14:450–460
Niego AG, Rapior S, Thongklang N, Raspé O, Jaidee W, Lumyong S, Hyde KD (2021) Macrofungi as a nutraceutical source: promising bioactive compounds and market value. J Fungi 7:397
Östbring K, Lager I, Chagas JCC, Ramin M, Ahlström C, Hultberg M (2023) Use of oyster mushrooms (*Pleurotus ostreatus*) for increased circularity and valorization of rapeseed residues. J Environ Manag 344:118742
Patel Y, Naraian R, Singh V (2012) Medicinal properties of *Pleurotus* species (oyster mushroom): a review. World J Fungal Plant Biol 3:1–12
Philippoussis A, Zervakis G, Diamantopoulou P (2001) Bioconversion of agricultural lignocellulosic wastes through the cultivation of the edible mushrooms *Agrocybe aegerita*, *Volvariella volvacea* and *Pleurotus* spp. World J Microbiol Biotechnol 17:191–200
Pilafidis S, Diamantopoulou P, Gkatzionis K, Sarris D (2022) Valorization of agro–industrial wastes and residues through the production of bioactive compounds by macrofungi in liquid state cultures: growing circular economy. Appl Sci 12:11426
Ragunathan R, Gurusamy R, Palaniswamy M, Swaminathan K (1996) Cultivation of *Pleurotus* spp. on various agro–residues. Food Chem 55:139–144
Raman J, Jang KY, Oh YL, Oh M, Im JH, Lakshmanan H, Sabaratnam V (2021) Cultivation and nutritional value of prominent *Pleurotus* spp.: an overview. Mycobiology 49:1–14
Ritota M, Manzi P (2019) *Pleurotus* spp. cultivation on diferent agri–food by–products: example of biotechnological application. Sustainability 11:5049
Royse DJ, Baars JJP, Tan Q (2017) Current overview of mushroom production in the world. In: Edible and medicinal mushrooms: technology and applications. Wiley, Hoboken, pp 5–13
Salmones D (2017) *Pleurotus djamor*, un hongo con potencial aplicación biotecnológica para el neotrópico. La Revista Mexicana de Micología 46:73–85
Sanchez C (2010) Cultivation of *Pleurotus ostreatus* and other edible mushrooms. Appl Microbiol Biotechnol 85:1321–1337
Sardar H, Ali MA, Anjum MA, Nawaz F, Hussain S, Naz S, Karimi SM (2017) Agro–industrial residues influence mineral elements accumulation and nutritional composition of king oyster mushroom (*Pleurotus eryngii*). Sci Hortic 225:327–334
Sekan AS, Myronycheva OS, Karlsson O, Gryganskyi AP, Blume Y (2019) Green potential of *Pleurotus* spp. in biotechnology. PeerJ Biochem Biophys Mol Biol 7:e6664
Sharma A, Sharma A, Tripathi A (2021) Biological activities of *Pleurotus* spp. polysaccharides: a review. J Food Biochem 45:e13748
Shirahigue LD, Ceccato-Antonini SR (2020) Agro–industrial wastes as sources of bioactive compounds for food and fermentation industries. Ciência Rural 50:e20190857

Shrikhandia SPP, Devi S, Sumbali G (2022) Lignocellulosic waste management through cultivation of certain commercially useful and medicinal mushrooms: recent scenario. In: Biology, cultivation and applications of mushrooms. Springer, Singapore, pp 497–534

Silva LA, Dulay RMR, Kalaw SP (2018) Mycelial growth of pink oyster mushroom (*Pleurotus djamor*) on banana sucrose gulaman and fruiting body production on banana–based substrate formulations. CLSU Int J Sci Technol 3:24–32

Soares TS, Haminiuk CWI, Maciel GM (2023) Fungal biorefinery for simultaneous production of enzymes and bioconversion of agro–industrial residues into renewable sugars and phenolic compounds. Biomass Convers Biorefinery 14:25995–26009

Subedi S, Kunwar N, Pandey KR, Joshi YR (2023) Performance of oyster mushroom (*Pleurotus ostreatus*) on paddy straw, water hyacinth and their combinations. Heliyon 9:e19051

Suwannarach N, Kumla J, Zhao Y, Kakumyan P (2022) Impact of cultivation substrate and microbial community on improving mushroom productivity: a review. Biology 11:569

Tesfay T, Godifey T, Mesfn R, Kalayu G (2020) Evaluation of waste paper for cultivation of oyster mushroom (*Pleurotus ostreatus*) with some added supplementary materials. AMB Express 10:1–8

Thakur MP (2020) Advances in mushroom production: Key to food, nutritional and employment security: a review. Indian Phytopathol 73:377–395

Törős G, El-Ramady H, Béni Á, Peles F, Gulyás G, Czeglédi L, Prokisch J (2024) *Pleurotus ostreatus* mushroom: a promising feed supplement in poultry farming. Agriculture 14:663

Tsujiyama SI, Ueno H (2013) Performance of wood–rotting fungi–based enzymes on enzymic saccharification of rice straw. J Sci Food Agric 93:2841–2848

Umor NA, Ismail S, Abdullah S, Huzaifah MHR, Huzir NM, Mahmood NAN, Zahrim AY (2021) Zero waste management of spent mushroom compost. J Mater Cycles Waste Manag 23:1726–1736

Vega A, De León JA, Miranda S, Reyes SM (2022) Agro–industrial waste improves the nutritional and antioxidant profle of *Pleurotus djamor*. Clean Waste Syst 2:100018

Wachira JW, Nguluu S, Kimatu J (2022) Differential growth and productivity of oyster mushroom (*Pleurotus pulmonarius*) on agro–waste substrates in semi–arid regions of Kenya. Int J Recycl Org Waste Agric 11:375–383

Wang Q, Meng L, Wang X, Zhao W, Shi X, Wang W, Wang L (2022) The yield, nutritional value, umami components and mineral contents of the first–flush and second–flush *Pleurotus pulmonarius* mushrooms grown on three forestry wastes. Food Chem 397:133714

Wichai T, Sooksai S, Noitang S, Sukaead W, Chanprapai P, Piapukiew J, Sawangkeaw R (2022) Cost reduction of gray oyster mushroom (*Pleurotus sajor-caju* (Fr.) Singer) production using lemon basil (*Ocimum citriodorum* Vis.) straw as a substrate. Waste Biomass Valoriz 13:1449–1458

Yaashikaa PR, Kumar PS, Varjani S (2022) Valorization of agroindustrial wastes for biorefinery process and circular bioeconomy: a critical review. Bioresour Technol 343:126126

Zadrazil F (1978) Cultivation of *Pleurotus*. In: The biology and cultivation of edible mushrooms. Academic, New York, pp 521–558

Zárate-Salazar JR, Santos MN, Caballero ENM, Martins OG, Herrera ÁAP (2020) Use of lignocellulosic corn and rice wastes as substrates for oyster mushroom (*Pleurotus ostreatus* Jacq.) cultivation. *SN Appl Sci* 2:1–10

Zervakis GI, Koutrotsios G (2017) Solid–state fermentation of plant residues and agro–industrial wastes for the production of medicinal mushrooms. In: Medicinal plants and fungi: recent advances in research and development. Springer, pp 365–396

Zervakis GI, Polemis E (2013) The genus *Pleurotus* (Fr.) P. Kumm. (Pleurotaceae) in Europe. In: *Pleurotus nebrodensis* a very special mushroom. Bentham Science Publishers, pp 31–56

Zervakis G, Venturella G (2002) Mushroom breeding and cultivation enhances ex situ conservation of Mediterranean *Pleurotus* taxa. In: Managing plant genetic diversity. CABI Publishing, pp 351–358

Zhang Z, Shah AM, Mohamed H, Tsiklauri N, Song Y (2021) Isolation and screening of microorganisms for the efective pretreatment of lignocellulosic agricultural wastes. Biomed Res Int 2021:5514745

Zhou Y, Li Z, Zhang H, Hu Q, Zou Y (2022) Potential uses of scallop shell powder as a substrate for the cultivation of king oyster mushroom (*Pleurotus eryngii*). Horticulturae 8:333

Zhou Y, Li Z, Xu C, Pan J, Zhang H, Hu Q, Zou Y (2023) Evaluation of corn stalk as a substrate to cultivate king oyster mushroom (*Pleurotus eryngii*). Horticulturae 9:319

Zied DC, Pardo JE, Noble R, Pardo-Giménez A (2024) Efficiency of mushrooms for food production–fundamental strategic decision–making. J Food Compos Anal 125:105734

Zou Y, Du F, Hu Q, Yuan X, Dai D, Zhu M (2020) Integration of *Pleurotus tuoliensis* cultivation and biogas production for utilization of lignocellulosic biomass as well as its benefit evaluation. Bioresour Technol 317:124042

Zurbano LY, Bellere AD, Savilla LC (2017) Mycelial growth, fruiting body production and proximate composition of *Pleurotus djamor* on diferent substrate. CLSU Int J Sci Technol 2:20–30

When Should I Use a Pasteurized or Sterilized Substrate for Mushroom Production?

Diego Cunha Zied and Arturo Pardo-Giménez

Abstract The success of mushroom production is related to several factors and production stages; among these, the preparation of substrate—responsible for the nutrition of the cultivated fungus—stands out. A quality substrate must have an ideal quantity of nutrients and be free from competing microorganisms and pests. All this information must be considered when choosing the sanitary treatment method to be adopted in commercial crops, as a nutritionally rich substrate needs to be protected by a beneficial thermophilic microbiota or be managed in sterile conditions, called axenic. Other factors are also related to the sanitary treatment of the substrate, such as operation and maintenance costs, the size of the cultivation area, the type of packaging and shelves used, the level of qualification of the workforce, the distance or difficulty in accessing spawn, etc. This book chapter will cover all the points that will clarify the best substrate production method, considering the positive and negative aspects of pasteurization and sterilization.

Keywords Mushroom substrate · Substrate sanitation · Pasteurization · Sterilization · Mushroom cultivation systems

1 Introduction

The success of mushroom production hinges on several critical factors, with the preparation of a high-quality substrate being paramount. As the primary source of nutrition for cultivated mushrooms, the substrate must undergo a sanitary treatment before spawning. This process has several key objectives: to eliminate pests and competing microorganisms, to promote selective mycelial growth, and to maximize

D. C. Zied (✉) · A. Pardo-Giménez
Research Group of Centro de Estudos em Cogumelos, Faculdade de Ciências Agrárias e Tecnólogicas, Universidade Estadual Paulista Julio de Mesquita Filho, Campus de Dracena, São Paulo, Brazil
e-mail: diego.zied@unesp.br

M. N. Owaid et al. (eds.), *Mushroom Production*, Interdisciplinary Biotechnological Advances, https://doi.org/10.1007/978-981-95-7793-4_13

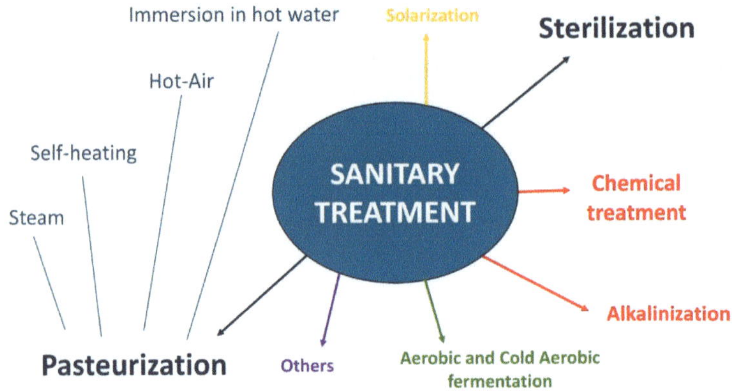

Fig. 1 Different sanitary treatments used in mushroom cultivation

the final yield of fruiting bodies, all while minimizing adverse changes to the substrate's chemical composition.

Fundamentally, substrate treatment strategies fall into one of two technical approaches. The first is pasteurization, which aims to create a *selective* substrate by reducing harmful organisms while preserving a beneficial thermophilic microbiota that protects the crop. The second is sterilization, which creates a completely sterile or *axenic* environment by eradicating all microbial life. Figure 1 presents the methods used for the sanitary treatment of the compost/substrate, considering pasteurization and sterilization as the most widely applied around the world in small-, medium-, and large-scale technological cultivation. On the other hand, chemical, alkalinization, and other treatments are used in nontechnological cultivation.

The choice between these and other alternative strategies and their associated methods is a complex decision, influenced by technical factors, such as the species to be cultivated, the type of base material and additives used, previous treatments in substrate preparation, substrate nutrient loss, contamination control, and substrate selectivity. It also depends on economic factors—including the capital investment necessary for infrastructure, production scale, energy demand, and labor costs—environmental factors associated with, for example, the use of chemicals, resulting emissions, or wastewater generation, as well as sanitary factors (Table 1).

This chapter provides a comprehensive overview of the various methods for substrate disinfection, from conventional thermal and chemical treatments to alternative approaches. Each technique is analyzed by examining its advantages and disadvantages in the context of these factors, offering practical examples derived from various research works to guide the selection of the most suitable procedure for different species and cultivation systems.

Table 1 Main factors concerning the selection criteria for different treatments

Factors	Points
Technical	Species to be cultivated
	Availability of base materials and additives
	Previous treatments in preparing the substrate
	Substrate nutrient loss
	Selectivity of substrate (chemical/biological)
	Contamination control
Economic	Capital investment necessary for infrastructure (construction and equipment)
	Production scale (small scale, industrial, research)
	Energy demand
	Water demand
	Labor costs
Environmental	Use of chemicals
	Emissions
	Generation of wastewater
Sanitary	Pesticide residues in fruitbodies (residual toxicity)
	Health of workers exposed to carcinogenic chemicals

2 Methods for Disinfecting Substrates for Edible Mushroom Production

2.1 *Steam Pasteurization*

Pasteurization is necessary to kill as many undesirable organisms as possible, such as flies, their eggs, larvae, mites, nematodes, and pathogenic fungi like green molds, bubbles, and cobweb molds (Buth 2017). Steam pasteurization is one of the most commonly used and effective substrate preparation techniques in mushroom cultivation worldwide, as it is a relatively inexpensive method compared to sterilization (Ghimire et al. 2021). This pasteurization method is a typical procedure for industrial-scale operations where large quantities of substrate must be prepared. The treatment of materials with steam is carried out inside tunnels, maintaining temperatures of 58–80 °C for 6–12 h (Macias et al. 2022). This technique requires a high initial investment but is applicable to both small-scale and industrial production. It provides high contaminant control with minimal nutrient loss, although the selectivity of the resulting substrate is low (Muñoz et al. 2014).

In the conventional two-phase composting process for *Agaricus* sp., self-heating pasteurization is the first step of Phase II in aerated bulk tunnels, although steam is used when Phase II is conducted in growing rooms. Silva et al. (2009) described a composting process (14 days), followed by a two-step steam pasteurization (55–65 °C; 15 h each step) of a substrate based on sugarcane bagasse, coast-cross straw, and wheat bran prepared for the production of *Agaricus subrufescens* (syn. *Agaricus brasiliensis*).

For the production of *Pleurotus* species, a substrate with chemical selectivity is not required, although biological selectivity is. In this case, any accompanying flora, if present, must be protective and not compete with *Pleurotus* spp. Steam pasteurization at 65 °C for 8 h is used for substrates based on wheat straw and other alternative materials—such as maize stems, maize cobs, pseudo-banana stems, and other recyclable wastes—in the production of *Pleurotus sajor-caju* (gray oyster mushroom) (Dhar 2017). This methodology can also be used to prepare substrates for growing *Pleurotus ostreatus* (black oyster mushroom), *P. ostreatus* var Florida (white oyster mushroom), *Pleurotus djamor* (pink oyster mushroom), and *Pleurotus citrinopileatus* (yellow oyster mushroom). *Calocybe indica* (milky mushroom) is also grown on wheat straw after pasteurization at 65 °C for 8 h in polyethylene bags. In the case of *Pleurotus eryngii* (king oyster mushroom), a wheat straw substrate pasteurized at 65 °C with steam does not support an effective spawn run (Dhar 2017). Steam pasteurization (80 °C, 1 h) yielded maximum mycelial growth and a higher yield compared to hot water treatment with boiling water (30 min) and chemical sterilization with formalin in three *Pleurotus* species (*P. ostreatus* var. Florida, *Pleurotus pulmonarius*, and *P. ostreatus*) on a cotton waste substrate (Ali et al. 2004, 2007).

In the production of *P. ostreatus*, substrates based on corn stubble and rice straw were immersed in tap water for 24 h and then drained for another 24 h. The wet substrate was subsequently bagged and disinfected with steam (60–70 °C) for 4 h. This method yielded the lowest biological efficiency (an average of 55.12%) compared to immersion in hot water and sterilization (Muñoz et al. 2014).

Volvariella volvacea is grown on paddy straw bundles that are steam pasteurized at 65 °C for 6 h, while a temperature of 62 °C is maintained for 2 h when cotton waste is used (Ahlawat and Tewari 2007; Dhar 2017). *Flammulina velutipes* (enoki mushroom) grows best on a sawdust and rice bran mixture after steam pasteurization of the substrate in polypropylene bags (Dhar 2017).

Pasteurization in enclosed chambers using injected steam is suited for larger operations in the production of *P. ostreatus* due to the high volumes of substrate to be processed. Temperatures are held between 60 and 100 °C for 1–12 h, with the occasional addition of fungicides (Rodriguez and Pecchia 2017). In general, the commercial procedure maintains temperatures in the substrate mass between 60 and 80 °C for a period that usually varies between 6 and 12 h (Muez and Pardo 2002).

Barley straw, both inoculated with *Paenibacillus polymyxa* and noninoculated (control), was fermented for 0, 3, or 5 days prior to heat treatment (steam pasteurization for 12 h at 60 °C) to produce the substrate for cultivating *P. pulmonarius*. Inoculation decreased the colonization ability of *Trichoderma viride* and favored mycelial growth, although mushroom yield was negatively affected (Llarena-Hernández et al. 2019).

A bulk substrate treatment was developed for the cultivation of shiitake mushrooms (*Lentinula edodes*) on a 1:1 mixture of cotton and wheat straws. The straw mixture was moistened to 70–75% humidity and pasteurized for 24 h at 68 °C in a bulk-pasteurization tunnel. Yields were 70–80%, measured as biological efficiency (Levanon et al. 1993).

Steam pasteurization (55–70 °C for 30 min) of paddy straw provided a biological efficiency of 12.75% in *V. volvacea* cultivation. A yield of 233 g kg^{-1} of wet substrate was also obtained with steam-pasteurized paddy straw (80 °C, 2 h) for *P. ostreatus* var. Florida (Saritha and Pandey 2010).

Composted substrates with basic formulations using *Eucalyptus* sawdust, either alone or mixed with sugarcane bagasse and subjected to different composting periods, were evaluated for *Lentinula edodes* (shiitake) cultivation. The best results, similar to those obtained with noncomposted substrates, were achieved with 6 days of composting, followed by pasteurization for 12 h at 80 °C, or 4 days of composting, followed by autoclaving for 1 h (Castro et al. 2024).

A temperature of 70–80 °C for up to 90 min can be used in the production of a rice straw-based substrate for milky white mushroom (*C. indica*) cultivation (Ahmed et al. 2020).

A variant of this pasteurization technique used to prepare substrates for growing oyster mushrooms is the so-called xerothermic process, in which dry substrates are treated with steam at 100 °C for 1 h in a small tunnel, after which cool water is added (Martínez-Carrera 1998).

2.2 *Self-Heating Pasteurization*

Self-heating pasteurization, which serves to reduce competitor microbiota, is the first step of Phase II in the composting process for *Agaricus* sp. cultivation, following the general criteria described by Fermor et al. (1985). The temperature of the compost rises due to its own activity and is controlled via the air temperature. In general, when a compost spontaneously reaches a high temperature during pasteurization, this is considered a positive factor (Gerrits 1988). Subsequent thermophilic conditioning provides the mushroom substrate with chemical and biological selectivity. In addition, ammonia is reduced to levels that are nontoxic to *Agaricus bisporus*. The warming of compost for pasteurization is normally achieved using steam when Phase II occurs in rooms, but steam is not necessary when the process takes place in tunnels. Temperatures between 56 and 60 °C for approximately 6–10 h are desirable for effective pasteurization in *A. bisporus* (Buth 2017). However, a temperature above 60 °C is undesirable, as it is harmful to beneficial microorganisms and should therefore be prevented (Gerrits 1988). As an example, a pasteurization period of 8–10 h at 60 °C has been proposed during Phase II of indoor composting of vine by-products to produce substrates for white button mushroom cultivation (Pardo et al. 2007). Similarly, temperatures between 57 and 59 °C for 10–24 h are proposed for compost production in *A. subrufescens* cultivation (Zied et al. 2017). Figure 2 shows high-tech and rustic pasteurization tunnels.

Several alternative methods have been developed to grow *A. bisporus*. These consist of preparing a noncomposted, sterilized, or pasteurized substrate at different temperatures to avoid the lengthy conventional composting procedure. A self-heating pasteurization of different substrates based on agricultural by-products

Fig. 2 (**a**) Insulated panel pasteurization tunnels with double doors (compost inlet and outlet). (**b**) Concrete pasteurization tunnels with only one door

combined with 2% hydrated lime ($Ca(OH)_2$) was carried out by allowing the temperature to rise to 60–65 °C in a 1 m^3 wooden crate. After 30 h, the substrate was removed, turned, and placed back in the crate. The process was completed after 45 h by aerating the substrate to decrease the temperature. The results were comparable to those obtained with the Phase II compost traditionally used for *A. bisporus* cultivation, particularly with supplemented substrates used to produce *A. bisporus* 'Portobello' mushrooms (Colmenares-Cruz et al. 2017).

Sánchez et al. (2011) studied an alternative to steam application by forming a substrate compost pile with sufficient size and moisture to generate the heat necessary for pasteurization. Pangola grass plus 2% lime ($Ca(OH)_2$) was adjusted to moisture contents between 55% and 70% and subjected to an incomplete composting process (48 h long). The optimal treatment was the one that provided enough heat to the substrate to reach at least 50 °C for no less than 10 h. The 65% moisture pile produced a time/temperature profile sufficient to kill or inhibit contaminant microorganisms. Biological efficiency values varied between 70.5% and 88.1% after two flushes of *Pleurotus ostreatus* on Pangola grass pasteurized by self-heating and supplemented with different mixtures of soybean, wheat bran, and ground cowpeas. Higher biological efficiency values (112%) had been previously obtained by Barrios et al. (2009) with the same substrate subjected to a 48-h composting process, which were similar to those obtained with steam pasteurization (102%) and higher than those from alkaline immersion (62%). In both studies, no significant differences were found between steam pasteurization and self-heating.

2.3 Sterilization

The general procedure consists of sterilization in an autoclave with steam at 121 °C for 1–4 h, depending on the container/packaging size. This method requires a high capital investment and is very expensive to operate due to its high energy demand.

Fig. 3 Industrial process that produces a mechanized substrate. On the right you can see three autoclaves in a rectangular shape, the middle one with the door open. These autoclaves also have double doors

However, there is scope for developing low-cost, energy-efficient pasteurization and sterilization methods using solar energy to produce hot water and steam (Gowda and Manvi 2019). This technique is applicable to both research and industrial production (Fig. 3). It provides very high contaminant control with minimal nutrient loss, although the resulting substrate is nonselective (Muñoz et al. 2014).

High-quality mushrooms of *Pleurotus eryngii* (king oyster mushroom) are obtained from substrates based on sawdust supplemented with wheat or rice bran and sterilized at 121 °C for 90 min. Wheat-straw-based substrates can also be used, but they typically yield less. The cultivation technology for *Auricularia polytricha* is similar for both wheat straw and sawdust substrates. *Hypsizygus tessellatus* (buna shimeji mushroom) is also grown on a sterilized sawdust substrate (Dhar 2017).

Sterilization promoted high biological efficiency (95.51%) in a corn stubble substrate for *P. ostreatus* cultivation. The substrate was kept immersed in tap water for 24 h and drained for 24 h. The wet substrate was then bagged and disinfected in an autoclave at 121 °C and 103 kPa for 1 h (Muñoz et al. 2014).

In the production of *P. ostreatus* in Brazil, sterilization by autoclaving at 121 °C for 1 h with local lignocellulosic residues was the most efficient disinfection method, providing higher mushroom productivity than immersion in hot water or 2% calcium hydroxide (Bentolila de Aguiar et al. 2021). On straw and sawdust-based substrates packed in polyethylene bags, autoclaving at 121 °C also provided good results in oyster mushroom production, although straw was more efficient than sawdust (Caral Dinesh et al. 2013).

Biological efficiency values of 94.91% were obtained with wheat straw hydrated to a final moisture content of 70% and sterilized for 2 h at 120 °C in *P. ostreatus* cultivation (Jaramillo and Albertó 2013).

For *P. ostreatus* var. Florida cultivation, a paddy straw substrate autoclaved at 121 °C for 20 min after soaking in cold water for 12 h provided a biological efficiency of 89.24% (Kalita 2015).

Rathod et al. (2023) also evaluated different methods for *P. ostreatus* var. Florida cultivation on 15 substrates based on bread grass, soybean straw, sawdust, rice straw, wheat straw, cotton straw, sugarcane bagasse, and various mixtures. Steam sterilization in an autoclave at 121 °C for 20 min after soaking in fresh water for 8 h provided biological efficiencies between 41.56% (sugarcane bagasse substrate) and 75.55% (a mixture of all materials in different proportions).

Three different types of sterilization methods were evaluated for the growth parameters and productivity of *P. ostreatus* cultivated on rice straw: chemical sterilization, steam sterilization, and hot-water sterilization. The oyster mushroom performed best under steam sterilization as it required the shortest time for pinhead formation, fruiting body formation, and cropping duration, and it produced the highest biological efficiency (101.38%) (Shrestha et al. 2021).

Autoclaving a mixture consisting of equal parts of maize straw, faba bean straw, wheat straw, and soy straw yielded a biological efficiency of 66.8% in the production of *Pleurotus ostreatus*. In terms of environmental impact, this sterilization method requires low water usage but high energy usage (Grimm et al. 2024).

A sterilization process at 121 °C for 2 h has been proposed by Zied et al. (2017) for the cultivation of *A. subrufescens*. A similar treatment can be used in the production of *C. indica* on substrates based on rice straw in polyethylene bags (Ahmed et al. 2020).

For *Ganoderma lucidum* (lingzhi) production, wood logs in plastic bags are sterilized under high-pressure steam at 121 °C for 1.5–2 h. Optionally, sterilization can also be carried out at atmospheric pressure in a sterilizer for 10–12 h after reaching 100 °C. Substrate formulations using heat-resistant polypropylene bags or glass bottles as containers, which are autoclaved at 121 °C for 1–2 h, can also be used. When sterilization is performed at atmospheric pressure at temperatures around 100 °C, the sterilization time of the substrates must also be increased to 10–12 h (Zhou 2017).

L. edodes (shiitake) can be produced on a sawdust substrate supplemented with rice or wheat bran and sterilized at 121 °C for 90 min (Dhar 2017). Four-day composting and 1-h autoclaving of substrates based on *Eucalyptus* sawdust and sugarcane bagasse gave results similar to those obtained for noncomposted substrates in shiitake cultivation (Castro et al. 2024).

Most mushroom production companies in Japan, China, Korea, and Thailand adopt bottle cultivation utilizing fully automated systems with specialized equipment. They mainly produce *F. velutipes*, *H. marmoreus*, and *P. eryngii*. Bottle and small bag systems are also used in the production of *Pholiota nameko*, *P. ostreatus*, *Grifola frondosa*, and *L. edodes*. Small-scale plastic bag cultivation does not require a large initial investment and is therefore the world's most popular method for cultivating specialty mushrooms. However, two companies in Japan produce *G. frondosa* in small polyethylene bags in facilities equipped with fully automated machinery. Base materials are mainly sawdust (cedar, pine), cottonseed hulls, and ground corncobs. During substrate preparation, the filled bottles and bags are loaded into large autoclaves. Bottle systems commonly use pressurized sterilization cycles of 50–60 min at 118–120 °C, while low-tech bag growers run sterilization cycles of 5–6 h at 99–100 °C (Yamanaka 2017).

Currently, several species of *Pleurotus*, *L. edodes*, *Sparassis crispa*, and *Tremella aura* are being cultivated in sterilized substrates in cylindrical form in China and South Korea (Fig. 4).

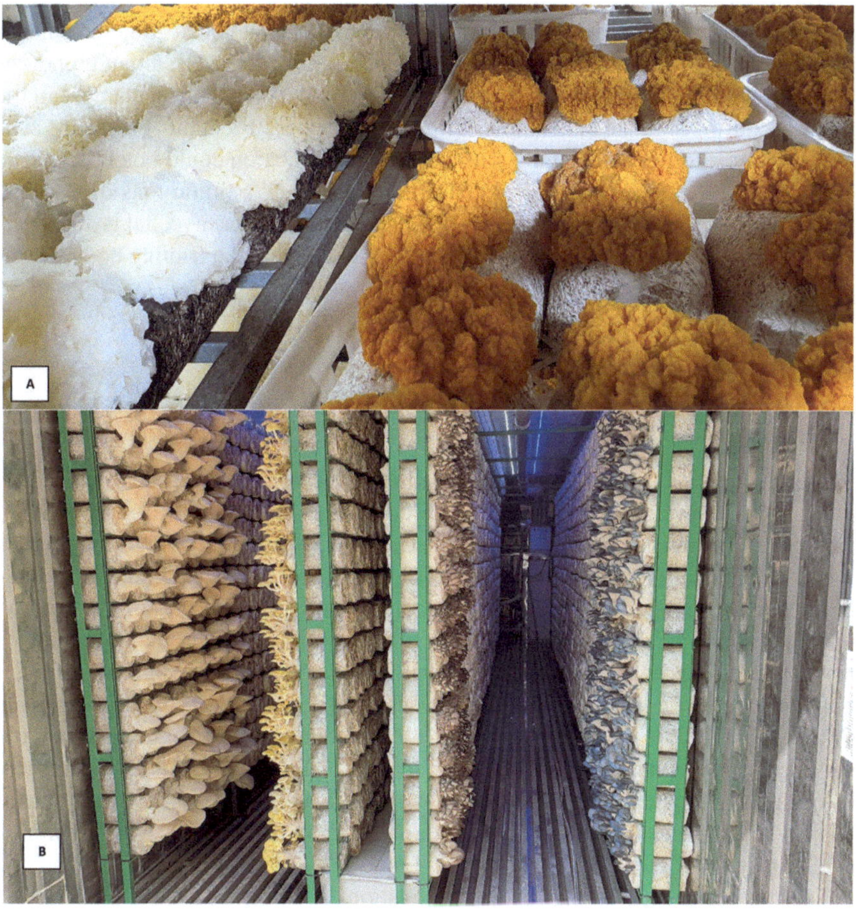

Fig. 4 (**a**) Sterilized sawdust-based substrate for the production of *Sparassis crispa* (left) and *Tremella aura* (right). (**b**) Vertical production of *Pleurotus nebrodensis*, *Pleurotus citrinopileatus*, *Pleurotus ostreatus*, and *Pleurotus* var. blue (shelves—from left to right)

2.4 Hot-Air Pasteurization

Grimm et al. (2024) evaluated the hot-air pasteurization of maize straw substrate in an oven for 3 h at temperatures of 75, 85, or 100 °C with two different soaking periods. The highest biological efficiency (116.0%) was recorded with a treatment at 85 °C and a 96-h soaking time during the cultivation of the gray oyster mushroom *P. ostreatus*. In terms of environmental impact, this system uses less energy and water compared to autoclaving and hot-water pasteurization.

Xiong et al. (2019) and Wei et al. (2020) compared the effects of hot-air oven pasteurization and autoclaving (121 °C) on birch-sawdust-based substrates used for shiitake mushroom cultivation. Treatment temperatures were set between 75 and 100 °C for 1.5–3 h. Hot-air pasteurization and autoclaving did not result in significant differences in the degradation of lignin and carbohydrates. Major compositional changes were associated with the cultivation process itself, although hot-air pasteurization had a similar effect as autoclaving on the biochemical changes in the substrate during shiitake cultivation. With regard to energy demand, the results suggest that substituting autoclaving with hot-air pasteurization could potentially save up to 9.9 TWh of energy in the global production of 7.5 million tons of shiitake. Substrate pasteurization by hot-air treatment resulted in faster mycelial growth, earlier fructification, and a higher or comparable yield of shiitake fruiting bodies compared to autoclaving with pressurized steam (Xiong et al. 2019).

2.5 Pasteurization by Immersion in Hot Water

Hot-water treatment is an inexpensive and easy method for substrate disinfection that is used in several parts of the world. It consists of preparing a hot-water bath in which the substrate is immersed. Once the treatment is complete, a draining process is carried out to eliminate excess water. Different temperatures and immersion times have been recommended by various researchers. The disadvantages of this method include its considerable energy and water consumption, as well as the generation of contaminated wastewater. Furthermore, the time/temperature ratio of pasteurization must be carefully considered, as it influences the degradation of substrate nutrients (Macias et al. 2022). In fact, yield may be reduced due to the loss of important nutrients during immersion in hot water (Ghimire et al. 2021). This method requires low investment, is applicable to small-scale production, and provides high contaminant control and a selective substrate, although nutrient loss is very high (Muñoz et al. 2014).

In small-scale operations for the cultivation of *P. ostreatus*, the substrate is immersed in water at 80–85 °C for 30–35 min (Rodriguez and Pecchia 2017). Muez and Pardo (2002) had previously indicated treatments at 70–80 °C for 15 min for straw-based substrates, while for other materials, they cited temperatures between 60 and 80 °C with treatment durations of 30–60 min. Atila (2016) compared four different treatments (scalding in hot water at 60, 80, and 100 °C for 1 h; chemical disinfection with 1% formaldehyde solution; and sterilization by autoclave at 121 °C for 90 min) in the cultivation of *P. ostreatus* using poplar sawdust (40%) and wheat straw (40%) as a base medium. Scalding in hot water at 80 °C resulted in the highest productivity in terms of yield and biological efficiency. During the cultivation of *P. ostreatus*, Bentolila de Aguiar et al. (2021) found high levels of contamination in substrates composed of different local lignocellulosic residues that were immersed in hot water at 80–90 °C for 90 min.

Hot-water immersion resulted in high biological efficiency values for *P. ostreatus* on a rice straw substrate (86.98%). The substrate was kept immersed in hot water (70–80 °C) for 1 h and then drained for 24 h (Muñoz et al. 2014).

Chopped wheat straw (3–5 cm long) immersed in hot water (80 °C) for 90 min and drained for 12 h to a final moisture content of 70% yielded a biological efficiency of 75.83% in *P. ostreatus* cultivation. These results were lower than those obtained with substrates subjected to steam sterilization (120 °C, 2 h) or chemical treatment with carbendazim (Jaramillo and Albertó 2013). Soaking of wheat straw substrate in hot water provided biological efficiency values between 34.0% and 119.0% for different *Pleurotus* species (Kerketta et al. 2019).

Pasteurization of different substrates in hot water at 70 °C for 60 min provided low biological efficiency values, between 10.00% and 60.71%, in *P. ostreatus* var. Florida cultivation. Higher values, between 16.11% and 67.69%, were obtained when substrates were immersed in boiling water for 60 min (Rathod et al. 2023).

Ejigu and Kebede (2015) evaluated the performance of different organic substrates and treatments in the cultivation of *P. sajor-caju*. The highest yield of oyster mushroom (395 g kg^{-1} of substrate) was recorded from shredded maize stalk pasteurized by soaking in hot water at 70 °C for 1 h. This same species provided low biological efficiency values (13.86%) on sugarcane bagasse substrates pasteurized in water at 80 °C for 2 h. This low-average biological efficiency could be partially attributed to losses resulting from contamination that occurred during the second and third flushes.

Hot-water pasteurization at 63–70 °C for 90 min applied to a substrate composed of equal parts of maize straw, faba bean straw, wheat straw, and soy straw yielded a biological efficiency of 75.7% in the production of *P. ostreatus*. However, in terms of environmental impact, this high-yielding pasteurization method is not very sustainable due to its extensive energy and water requirements (Grimm et al. 2024).

Soaking bundles of paddy straw in boiling water for 20–30 min is a common method used in the production of *V. volvacea* (Ahlawat and Tewari 2007).

For *P. ostreatus* var. Florida cultivation, a paddy straw substrate boiled at 85 °C for 30 min provided a biological efficiency of 88.62% (Kalita 2015), while a treatment at 80 °C for 60 min yielded 189.5 g kg^{-1} of wet substrate (Saritha and Pandcy 2010).

A chopped rice straw substrate, poured into a net bag, is treated with hot water at 60 °C in a drum for 60 min and then allowed to drain by hanging the bag for 20 h for the production of *C. indica*. This procedure is considered better than steam pasteurization and autoclaving for the cultivation of milky white mushroom (Ahmed et al. 2020).

Gowda et al. (2014) developed and evaluated a paddy straw pasteurizer for mushroom cultivation. Different temperature-time combinations and compression levels were applied. The resulting substrate was used to cultivate *Hypsizygus ulmarius*, *P. sajor-caju*, and *C. indica*, yielding the highest biological efficiencies (79.6%, 69.6%, and 93.8%, respectively) without any contamination when unsoaked straw was pasteurized at 80 °C for 2 h with maximum compression.

2.6 Chemical Treatment

This technique is proposed as a low-cost alternative to steam pasteurization, using chemicals with fungicidal action. Mushroom growers often prefer chemical sterilization to decontaminate the substrate because autoclaving, hot-water treatment, and steam pasteurization are expensive due to high equipment costs (Gowda and Manvi 2019). However, this procedure has certain limitations, as only a few chemicals are recommended for use in mushroom cultivation against competitive fungi, and it is necessary to choose effective fungicides with a low inhibitory effect on the desired mushroom species. Furthermore, this technique has several additional drawbacks, including the use of carcinogenic chemicals, which can affect the health of workers who inhale these fumes; environmental pollution; toxic chemical residues in mushrooms, which can cause chronic diseases; and the fact that the repeated use of chemicals may lead to the development of pesticide resistance among competitor molds (Gowda and Manvi 2019; Ghimire et al. 2021). These concerns have led to a general discouragement of this practice.

Initially, formaldehyde and carbendazim were used, alone or in combination, in the preparation of wheat-straw-based substrates for the production of *P. pulmonarius* (Muez and Pardo 2002).

Corn cobs shredded into 1–5 cm pieces and soaked in a carbendazim solution (0.01%) for 48 h have been evaluated for the production of *P. ostreatus*. This chemical treatment of cellulosic substrates can be a good alternative to pasteurization or sterilization, although fungicide accumulation has been detected in the mushroom fruiting bodies (Ficior et al. 2006).

A high biological efficiency value (106.93%) was obtained in the cultivation of the oyster mushroom *P. ostreatus* with a wheat straw substrate hydrated to a final moisture content of 70% with a commercial formula of carbendazim solution (0.5%) (Jaramillo and Albertó 2013).

In chemical sterilization with formaldehyde, 250 mL of formalin (40% formaldehyde) was added to 10 L of water, and the substrate was soaked in the chemical solution overnight. Finally, the substrate was aerated for a few minutes and then filled into polyethylene bags. Compared with other methods, this chemical disinfection technique required a longer period for the completion of the mycelial growth of *P. ostreatus* but yielded the highest biological efficiency, which was not significantly different from that of substrates disinfected by immersion in hot water at 80 °C (Atila 2016).

Lower concentrations of formaldehyde (10 mL per 10 L of water) have provided good results in the production of *P. ostreatus* on straw- and sawdust-based substrates, although these were inferior to the results obtained from substrates sterilized in an autoclave at 121 °C (Caral Dinesh et al. 2013).

Kalita (2015) evaluated two chemical treatments of paddy straw substrate in *P. ostreatus* var. Florida cultivation: soaking in a formaldehyde solution (50 mL/L water) for 12 h (biological efficiency 66.05%) and soaking in a Bavistin (carbendazim) solution (2 g/L water) for 12 h (biological efficiency 89.03%). Previously, Saritha and Pandey (2010) had recorded yields of 129.5 g/kg of wet substrate for this same species when paddy straw was soaked for 18 h in an aqueous solution containing 75 ppm Bavistin and 500 ppm formaldehyde. The same substrate and chemical treatment (Bavistin and formaldehyde, 18 h) provided a biological efficiency of 72% in *P. sajor-caju* cultivation, which was higher than that obtained with autoclaving (68%), hot water (67%), and ultraviolet (UV) radiation (61%) treatments (Oys et al. 2014).

Substrate sterilization using thiophanate-methyl in the cultivation of oyster mushroom (*P. ostreatus* var. Florida) was also evaluated. A wheat straw substrate was chemically treated with different solutions of thiophanate-methyl (30–60 ppm) + formalin (500 ppm) for 12 h. The straw was then sieved to drain off the excess water and air-dried to a 65% moisture content. The highest biological efficiencies were recorded in treatments containing 50 and 60 ppm of thiophanate-methyl (50.32% and 63.68%, respectively) compared with all other treatments. No residue of thiophanate-methyl or its metabolite carbendazim was detected at a 50 ppm concentration of thiophanate-methyl in any of the assays performed (Sharma et al. 2020).

Rathod et al. (2023) evaluated two fungicide treatments applied to different growing media for *P. ostreatus* var. Florida. One consisted of a solution containing Bavistin (50 ppm) and formalin (500 ppm), and the other consisted of a solution containing 500 ppm of commercial fungicide formulated with carbendazim (12%) and mancozeb (63%). The substrate bags were allowed to soak in these chemical solutions for 18 h. The biological efficiency values obtained ranged between 24.10% and 72.42% for the first treatment and between 21.11% and 71.01% for the second.

Paddy straw bundles were soaked in a solution of formalin (500 ppm) + carbendazim (75 ppm) for 12–18 h before spawning in the preparation of substrates for *V. volvacea* cultivation, yielding a biological efficiency of 14.12%. A combination of carbendazim and formaldehyde provided biological efficiency values between 45.2% and 123.0% for different *Pleurotus* species on a wheat straw substrate (Kerketta et al. 2019).

In experiments for the cultivation of white oyster mushroom (*P. ostreatus* var. Florida), Saritha and Pandey (2010) used different hydrogen peroxide (H_2O_2) and bleaching powder solutions for soaking chopped paddy straw substrate. The best results (125.5 g kg^{-1} wet substrate) were obtained with a 6% solution of H_2O_2, although these were lower than those obtained with treatments with hot water (189.5 g kg^{-1} wet substrate) and steam (233.0 g kg^{-1} wet substrate).

Chlorine dioxide (ClO_2) and sodium hypochlorite are the most commonly used chemical agents in the bleaching procedure for disinfecting mushroom substrates (Ghimire et al. 2021). Atila (2020) evaluated immersion treatment in different ClO_2 concentrations for substrates based on cottonseed hull and wheat straw for the production of *P. ostreatus*. The highest biological efficiency was obtained at a 4 mL L^{-1} concentration of ClO_2 and 30 min immersion time for wheat straw (89.7%) and at a 4 mL L^{-1} concentration of ClO_2 and 20 min immersion time for cottonseed hull substrate (116.2%).

2.7 *Alkalinization*

The alkaline immersion technique, which is inexpensive and nonthermal, is a simple chemical disinfection method (Gowda and Manvi 2019). In the alkalinization process, substrates are disinfected by soaking them for 12–48 h in an alkali solution, such as calcium oxide or calcium carbonate. Adjusting the substrate pH to an alkaline level has been found to deactivate harmful microbes present in the substrate, as it inhibits the growth of competitive fungi without severely lowering the growth of the desired fungus (Ghimire et al. 2021; Rathod et al. 2024). This procedure requires low energy usage but high water usage. It requires low investment, is applicable to small-scale production, and provides high contaminant control and a selective substrate, although nutrient loss is high (Muñoz et al. 2014).

Different substrates for cultivating *P. ostreatus* were prepared by soaking them in water containing 0.5% commercial lime (81.5–82% calcium hydroxide content) with no further thermal treatment. The best results for biological efficiency (126.0%) were provided by a grass substrate immersed in 0.5% lime for 48 h. Proper control of substrate moisture must be ensured to limit the proliferation of undesirable bacteria (Contreras et al. 2004). Soaking of Pangola grass straw substrate in water containing 2% lime for 48 h provided a biological efficiency of 62% in *P. ostreatus* cultivation (Barrios et al. 2009). Similarly, immersion of wheat straw substrate in water containing 2% lime provided biological efficiency values between 33.0% and 118.2% for different *Pleurotus* species (Kerketta et al. 2019).

Soaking bundles of paddy straw in a 2% calcium carbonate solution ($CaCO_3$) for 12 h before spawning has been employed in the preparation of substrates for *Volvariella volvacea* cultivation (Ahlawat and Tewari 2007). Similarly, used a 2% calcium oxide solution for 12–18 h, achieving a biological efficiency of 15.76%.

Immersion in 2% calcium hydroxide ($Ca(OH)_2$) for 24 h inhibited the colonization and growth of mushrooms on different substrates for the cultivation of *P. ostreatus* (Bentolila de Aguiar et al. 2021). In contrast, immersion in a solution of 5 g of calcium hydroxide per liter for 8 h applied to a substrate composed of equal parts of maize straw, faba bean straw, wheat straw, and soy straw provided a biological efficiency of 52.0% in the production of gray oyster mushroom (Grimm et al. 2024).

Muñoz et al. (2014) also achieved good results with *P. ostreatus* on a corn stubble substrate. In this case, the substrate was kept immersed in water with 5 g L^{-1} $Ca(OH)_2$ for 24 h and then drained for 24 h.

2.8 *Other Alternative Methods*

2.8.1 Aerobic Fermentation

This procedure, used to obtain substrates for growing *Pleurotus* spp., combines conventional pasteurization and thermophilic fermentation conditioning, applied in different variations. In this way, the substrate is provided with biological selectivity through the action of bacteria of the genus *Bacillus*. Pasteurization treatments vary, with temperatures ranging from 60 to 70 °C and durations between 2 and 24 h, while conditioning is in the range of 50–55 °C for 24–72 h (Muez and Pardo 2002).

2.8.2 Cold Anaerobic Fermentation

This low-cost process has been developed for the production of *Pleurotus* spp. from various agricultural and agro-industrial by-products. It is based on the principle that different materials, when immersed in water, undergo anaerobic fermentation through the action of naturally occurring lactic acid bacteria. Between temperatures of 12 and 30 °C, fermentation begins spontaneously. The metabolic action of these bacteria eliminates sugars, hindering the development of competing fungi while enhancing enzymatic action on the substrate. Under the most suitable temperatures, at 20–30 °C, the process typically takes 5–7 days (Muez and Pardo 2002).

2.8.3 Solarization

In this process, described by Saritha and Pandey (2010), wet paddy straw was spread on a black polyethylene sheet to a depth of 10 cm, and the top was covered with a transparent polyethylene sheet under direct sunlight. The next day, the transparent sheet was removed, and the substrate was dried to the desired moisture content. The maximum temperatures recorded during the process were 69.6 °C (on the surface), 55.9 °C (5 cm from the surface), and 40.3 °C (10 cm from the surface). A yield of 128.5 g kg^{-1} of wet substrate was obtained in *P. ostreatus* var. Florida cultivation.

2.8.4 UV Irradiation

Ultraviolet (UV) radiation exhibits germicidal properties in the UV-C region, with maximum microbial inactivation occurring in the 254 to 264 nm range. Several studies suggest that microorganisms are destroyed by UV-C light penetrating their cells, which causes deoxyribonucleic acid (DNA) damage (Macias et al. 2022). A paddy straw substrate exposed to UV light for 1 h and 30 min in *P. sajor-caju* cultivation provided a biological efficiency of 61% (Oys et al. 2014).

2.8.5 Microwaves

Internal friction between molecules due to the polarization effects of microwaves causes volumetric heating of the material. This thermal effect of microwave irradiation is the primary cause of microbial death and, consequently, the disinfection of substrates. Microwave preparation of the substrate consumes much less energy than traditional technologies (Macias et al. 2022).

2.8.6 Ozone-Based Sterilization

Sidik et al. (2015) introduced an ozone treatment system for sterilizing mushroom substrates. A wet substrate (composed of 93% sawdust, 5% rice bran, and 2% gypsum), filled into black polyvinyl chloride (PVC) bags, was injected with different ozone concentrations ranging from 50 to 500 ppm for 30 min. These processes were carried out to determine the optimum ozone concentration required to effectively kill all bacteria and fungi in the substrate within a specified time, which was verified by monitoring mycelial colonization after inoculation with oyster mushroom spawn. The results showed that substrates treated with an ozone concentration of 150 ppm or higher for 30 min exhibited normal mycelial growth. The authors concluded that using ozone injection as a treatment system for sterilizing mushroom substrates is effective, less time-consuming, and more cost-effective than current methods. Due to the shorter duration of this treatment, an estimated sixfold more substrate can be sterilized per day using the ozone sterilization process than with conventional techniques (Sidik et al. 2015; Rathod et al. 2024).

2.8.7 Cold Plasma Technology

Cold plasma is an ionized gas that can be produced at atmospheric pressure using a dielectric barrier discharge (DBD) system. Cold plasma technology is of great technical and commercial interest in the mushroom industry due to its ability to be applied under ambient conditions, speed up the cultivation period, increase treatment speed, and enable industrial applicability. The effectiveness of a dielectric barrier discharge cold plasma (DBD-CP) mixer system for the sterilization of agricultural residue substrates used in oyster mushroom production has been investigated. Cold plasma technology is a viable and effective technique for sterilizing crop residue used as a mushroom substrate, considerably reducing the duration of the process compared to conventional steaming techniques. Nevertheless, more research is required to maximize processing conditions and promote mushroom development (Rathod et al. 2024).

References

Ahlawat OP, Tewari RP (2007) Cultivation technology of paddy straw mushroom (Volvariella volvacea). National Research Centre for Mushroom (ICAR)

Ahmed F, Rahman J, Islam N, Ullah J, Sarker NC (2020) Effect of substrate sterilization and spawning method on the yield and yield attributes of milky white mushroom (*Calocybe indica*). Univ J Plant Sci 8(2):17–21

Ali MA, Hussain S, Nawaz R, Ahsan A, Siddiq M (2004) Effect of pasteurization techniques on mycelial growth of oyster mushroom, *Pleurotus* spp. J Agric Res 42(2):201–207

Ali MA, Mehmood MI, Nawad R, Hanif MA, Wasim R (2007) Influence of substrate pasteurization methods on the yield of oyster mushroom (*Pleurotus* species). Pak J Agric Sci 44(2):300–303

Atila F (2016) Effect of different substrate disinfection methods on the production of *Pleurotus ostreatus*. J Agric Stud 4(4):52–64

Atila F (2020) Chlorine dioxide as an alternative disinfectant for disinfection of oyster mushroom growing media. J Hortic Sci Biotechnol 95(1):121–127

Barrios BM, Moreno L, Sánchez JE (2009) Composteo en cajones de madera como pretratamiento del sustrato para cultivar *Pleurotus ostreatus* [Composting in wooden boxes as substrate pretreatment for cultivating *Pleurotus ostreatus*]. Rev Mex Mic 29:51–59

Bentolila de Aguiar LV, Sales-Campos C, Gouvêa PR, Vianez BF, Dias ES, Chevreuil LR (2021) Substrate disinfection methods on the production and nutritional composition of a wild oyster mushroom from the Amazon. Ciênc Agrotec 45:e010321

Buth J (2017) Compost as a food base for *Agaricus bisporus*. In: Edible and medicinal mushrooms: technology and applications. John Wiley & Sons Ltd., Chichester, pp 129–147

Caral Dinesh R, Vinay P, Manasa P, Vinothkumar D, Ramesh Babu NG (2013) Comparative study of oyster mushroom (*Pleurotus ostreatus*) cultivation by physical and chemical method of sterilization using two different substrates. Int J Sci Eng Res 4(9):898–902

Castro CP, Abreu CG, Junqueira de Moraes TS, Zied DC, Siqueira FG, Dias ES (2024) Shiitake mushroom cultivation in composted substrate: is it possible? Cienc Agrotec 48:e013224

Colmenares-Cruz S, Sánchez JE, Valle-Mora J (2017) *Agaricus bisporus* production on substrates pasteurized by self-heating. AMB Express 7:135

Contreras EP, Sokolov M, Mejía G, Sánchez JE (2004) Soaking of substrate in alkaline water as a pretreatment for the cultivation of *Pleurotus ostreatus*. J Hortic Sci Biotechnol 79(2):234–240

Dhar DL (2017) Mushroom farm design and technology of cultivation. In: Edible and medicinal mushrooms: technology and applications. John Wiley & Sons Ltd., Chichester, pp 271–308

Ejigu D, Kebede T (2015) Effect of different organic substrates and their pasteurization methods on growth performance, yield and nutritional values of oyster mushroom (*Pleurotus ostreatus*) for small scale cultivation at Arsi University, Ethiopia. Glob J Sci Front Res C Biol Sci 15(6):v1

Fermor TR, Randle PE, Smith JF (1985) Compost as a substrate and its preparation. In: The biology and technology of the cultivated mushroom. John Wiley & Sons Ltd, Chichester, UK. pp 81–109

Ficior D, Indrea D, Apahidean AS, Apahidean M, Pop R, Moldovan Z, Maniutiu D, Ganea R, Paven I (2006) Importance of substrate disinfection on oyster mushroom (*Pleurotus* sp.) culture. Not Bot Horti Agrobot Cluj Napoca 34(1):48–53

Gerrits JPG (1988) Nutrition and compost. In: The cultivation of mushrooms. Mushroom Experimental Station Horst, Horst, The Netherlands. pp 29–72

Ghimire A, Pandey KR, Joshi YR, Subedi S (2021) Major fungal contaminants of mushrooms and their management. Int J Appl Sci Biotechnol 9(2):80–93

Gowda NAN, Manvi D (2019) Agro-residues disinfection methods for mushroom cultivation. Agric Rev 40(2):93–103

Gowda NAN, Kumaran GS, Pandey M (2014) Performance evaluation of paddy straw pasteurizer for mushroom cultivation. Agric Mech Asia Afr Lat Am 45(3):28–36

Grimm D, Sonntag E, Rahmann G (2024) Evaluation of different pasteurization and sterilization methods for oyster mushroom substrates. J Microbiol Biotechnol Food Sci 13(5):e10428

Jaramillo S, Albertó E (2013) Heat treatment of wheat straw by immersion in hot water decreases mushroom yield in *Pleurotus ostreatus*. Rev Iberoam Micol 30(2):125–129
Kalita MK (2015) Impact of various sterilization methods on growth and yield of oyster mushroom (*Pleurotus florida*). Int J Agric Sci 11(1):104–107
Kerketta V, Shukla CS, Singh HK, Kerketta A, Singh V, Tamin (2019) Effect of substrate treatment methods on yield of *Pleurotus* spp. Int J Curr Microbiol App Sci 8(12):1792–1799
Levanon D, Rothschild N, Danai O, Masaphy S (1993) Bulk treatment of substrate for the cultivation of Shiitake mushrooms (*Lentinus edodes*) on straw. Bioresour Technol 45:63–64
Llarena-Hernández RC, Alonso-López A, Hernández-Rosas F, López CJ, Murguía J, Savoie J-M, Mata G (2019) Aerobic fermentation prior to pasteurization produces a selective substrate for cultivation of the mushroom *Pleurotus pulmonarius*. Biotechnol Agron Soc Environ 23(3):165–173
Macias AA, Crespo LM, Bordons A, Rodríguez-Porrata B (2022) Pasteurisation of agricultural substrates for edible mushrooms production. J Microbiol Biotechnol Food Sci 12(1):e5729
Martínez-Carrera (1998) Oyster mushrooms. In: McGraw-Hill yearbook of science & technology 1999. McGraw-Hill Book Co., New York, pp 242–245
Muez MA, Pardo J (2002) La preparación del sustrato. In: La biología y el cultivo de Pleurotus spp. El Colegio de la Frontera Sur, pp 157–186
Muñoz EN, Zárate-Salazar JR, Silva SA, Martins OG, Santos MN, Herrera AAP (2014) Evaluation of agronomic waste disinfection methods for oyster mushroom (*Pleurotus ostreatus* Jacq. P. Kumm) cultivation. Proc Indian Natl Sci Acad 91:184–193
Oys S, Yadav S, Mishra R, Singh R (2014) Effect of substrate disinfection on the biological efficiency of *Pleurotus sajor-caju* (Fr.) Singer. Plant Arch 14(1):205–209
Pardo A, Perona MA, Pardo J (2007) Indoor composting of vine by-products to produce substrates for mushroom cultivation. Span J Agric Res 5(3):417–424
Rathod MG, Dhotare JM, Kamble GT, Dhawale PI, Dhuldhaj UP, Chakke AB, Puri JB, Pathak AB (2023) Influence of various substrate treatment methods on yield and biological efficiency of *Pleurotus florida* mushroom. World J Adv Eng Technol Sci 08(1):216–224
Rathod A, Prajapati B, Solanki H (2024) Impact of different substrate sterilization methods in *Pleurotus* sp. mushroom cultivation: a comprehensive review. Int Res J Mod Eng Technol Sci 6(4):831–843
Rodriguez AE, Pecchia J (2017) Cultivation of *Pleurotus ostreatus*. In: Edible and medicinal mushrooms: technology and applications. John Wiley & Sons Ltd., Chichester, pp 339–360
Sánchez JE, Moreno L, Andrade-Gallegos R (2011) Pasteurization of substrate for growing *Pleurotus ostreatus* by selfheating. In: Mushroom biology and mushroom products: proceedings of the 7th international conference on mushroom biology and mushroom products. French National Institute of Agronomical Research (INRA), pp 398–405
Saritha B, Pandey M (2010) Evaluation of alternate substrate pasteurization techniques for culinary-medicinal white oyster mushroom, *Pleurotus ostreatus* var. *florida* (Agarcomycetideae) cultivation. Int J Med Mushrooms 12(3):309–316
Sharma VP, Kumar A, Kumar S, Barh A, Kamal S (2020) Substrate sterilization with thiophanate-methyl and its biodegradation to carbendazim in oyster mushroom (*Pleurotus ostreatus* var. florida). Environ Sci Pollut Res 27:899–906
Shrestha S, Bhattarai S, Shrestha RK, Shrestha J (2021) Effect of different substrate sterilization methods on performance of oyster mushroom (*Pleurotus ostreatus*). J Agric Sci 32:127–132
Sidik MAB, Buntat Z, Che Razali M, Buntat Y, Nawawi Z, Jambak MI, Smith IR (2015) A new method to sterilise mushroom substrate for oyster mushroom cultivation. In: Handbook on emerging trends in scientific research, vol 4. Pak Publishing Group, Faisalabad, Pakistan. pp 45–51
Silva CF, Azevedo RS, Braga C, da Silva R, Dias ES, Schwan RF (2009) Microbial diversity in a bagasse-based compost prepared for the production of *Agaricus brasiliensis*. Braz J Microbiol 40(3):590–600

Wei M, Xiong S, Chen F, Geladi P, Eilertsen L, Myronycheva O, Lestander TA, Thyrel M (2020) Energy smart hot-air pasteurisation as effective as energy intense autoclaving for fungal preprocessing of lignocellulose feedstock for bioethanol fuel production. Renew Energy 155:237–247

Xiong S, Martín C, Eilertsena L, Wei M, Myronycheva O, Larsson SH, Lestander TA, Atterhem L, Jönsson LJ (2019) Energy-efficient substrate pasteurisation for combined production of shiitake mushroom (*Lentinula edodes*) and bioetanol. Bioresour Technol 274:65–72

Yamanaka K (2017) Cultivation of mushrooms in plastic bottles and small bags. In: Edible and medicinal mushrooms: technology and applications. John Wiley & Sons Ltd., Chichester, pp 309–338

Zhou X-W (2017) Cultivation of *Ganoderma lucidum*. In: Edible and medicinal mushrooms: technology and applications. John Wiley & Sons Ltd., Chichester, pp 385–413

Zied DC, Pardo JE, Dias ES, Pardo-Giménez A (2017) Characteristics, production, and marketing of the Sun Mushroom: the new medicinal cultivated mushroom. In: Edible and medicinal mushrooms: technology and applications. John Wiley & Sons Ltd., Chichester, pp 361–384

Culinary-Valued Basidiomycetous Mushrooms

Muhammad Akram

Abstract Both ecological systems and human gastronomy depend on basidiomycetous fungi, which include a broad range of macrofungi, such as puffballs, bracket fungi, and mushrooms. This study focuses on a few chosen basidiomycetous specimens that have good organoleptic qualities, including taste, scent, texture, and appearance. The specimens were identified by morphological and DNA analysis after being collected from different ecological zones. A group of professionals conducted a sensory analysis, focusing on characteristics that are critical to both culinary and commercial value. The findings demonstrated that species such as *Pleurotus ostreatus*, *Lentinula edodes*, and *Agaricus bisporus* consistently obtained high scores on organoleptic evaluations, despite notable regional variations in preference. The findings open the door for a broader incorporation of less well-known but organoleptically rich basidiomycetes into food systems, with potential applications in sustainable agriculture, culinary innovation, and the creation of functional foods.

Keywords Pleurotus · Lentinula · Basidiomycetes · Edible mushrooms · Mushroom farming · Culinary fungi · Sensory evaluation · Fungal biodiversity · and Food with functional properties

1 Introduction

For centuries, people have consumed mushrooms as part of their diet because of their health benefits, medicinal value, and taste. Mushrooms are found in many different species within the fungal kingdom. Various types of mushrooms are grown for both food and medicine. Edible mushrooms are rich in nutrients like protein, fiber, vitamins (including vitamins D and B), minerals (like potassium, zinc, and selenium), and antioxidants (Smith et al. 2023). They support several bodily

M. Akram (✉)
Department of Eastern Medicine, Government College University, Faisalabad, Pakistan

M. N. Owaid et al. (eds.), *Mushroom Production*, Interdisciplinary Biotechnological Advances, https://doi.org/10.1007/978-981-95-7793-4_14

functions, including boosting the immune system, improving mental health, and promoting heart health. Mushrooms are often included in meals because they are low in fat and can be cooked without adding extra oil (Lee and Kim 2022). Beyond their role in food, mushrooms are also valuable for health-related purposes. They have been tested for their ability to reduce inflammation, fight viruses, help the body resist infections, and boost immunity (Patel et al. 2024). Some of the most studied mushrooms include *Lentinula edodes* (shiitake), *Ganoderma lucidum* (reishi), and *Grifola frondosa* (maitake). The connection between polysaccharides found in mushrooms and their health benefits has become very important. This is due to a lot of research on the medicinal properties of mushrooms. These studies show great potential for developing treatments for diseases like cancer, diabetes, and heart disease (Wang et al. 2023). The role of polysaccharides in mushroom functions affects both global health and food safety. Over the past few decades, there has been a growing interest in cultivating mushrooms, which has led to more efficient and sustainable production methods. The industrial production of mushrooms, like *Pleurotus ostreatus* (button mushroom) and *Agaricus bisporus* (oyster mushroom), plays a significant role in the global market. Now, mushrooms are widely used around the world, and their production is an important part of the agricultural economy in many countries. It is important that mushroom farming remains sustainable, as it has a smaller environmental impact compared to other forms of agriculture. Mushrooms are a sustainable food source because they grow using minimal space, energy, and water (Johnson et al. 2022). This sustainability is especially important given the challenges of resource scarcity, climate change, and the need for food systems that are environmentally friendly. Mushrooms are nutritious, sustainable, and healthy. More research and attention are needed to fully utilize their biological activities, cultivation techniques, and ecological benefits in order to make them a key part of renewable food systems (Garcia and Singh 2023).

2 The Role of Mycology in Human Nutrition and Gastronomy

In addition to eating mushrooms for their improved nutrition, people also enjoy them for their great taste. This makes mushrooms an important part of a healthy diet. Mushrooms contain a lot of fiber and protein, and even though they have some fat and calories, they are rich in essential vitamins and minerals. They contain several B vitamins, including riboflavin (B2), niacin (B3), and pantothenic acid (B5), which help with energy metabolism, cellular processes, and maintaining healthy skin, hair, and nerves (Rodriguez and Lopez 2021). They also provide important minerals, like potassium, phosphorus, selenium, and copper. Potassium helps control blood pressure, while copper is essential for making red blood cells and supporting the immune system (Kim et al. 2019). When mushrooms are exposed to UV light, they can produce vitamin D, which helps the body absorb calcium and supports bone health (Singh and Chatterjee 2020). The fiber in mushrooms, especially beta-glucan, helps boost the immune system, lowers cholesterol, and may reduce the risk of diseases like heart disease and type 2 diabetes (Nguyen et al. 2018). Mushrooms are also rich in antioxidants, which help fight diseases linked to aging by reducing oxidative stress and neutralizing harmful free radicals like ergothioneine and glutathione (Chen et al. 2022). Around the world, fungi are used in many different diets. They have a distinct umami taste, often described as a rich, savory flavor that enhances other foods because of their glutamic acid content. Mushrooms can be added to salads, stews, soups, and sauces and are often cooked fresh, dried, or powdered using methods like sautéing, grilling, roasting, or steaming. The densely cooked oyster mushroom (*Pleurotus ostreatus*) is known for its delicate texture and mild, sweet flavor, making it a great side dish for rice, risotto, and vegetable meals (Martinez 2020). Certain mushrooms, such as *Lentinula edodes* (shiitake) and *Grifola frondosa* (maitake), are especially valued in Asian cuisine for their unique flavors and health benefits. The strong umami flavor of shiitake mushrooms can complement other intensely flavored ingredients, like soy sauce in soups or stews. Maitake mushrooms are often used in Chinese and Japanese cooking because they can absorb flavors and help thicken soups. Mushrooms have also been found to increase levels of polysaccharides, which support immunity and offer other health benefits (Zhao 2021). They are also used as a meat substitute in vegetarian dishes, especially when prepared with spices. For a meat-like texture, mushrooms are often grilled or roasted, especially large portobello mushrooms, which make a great alternative to beef burgers. Delicious vegetarian dishes that require a meaty texture include sausages, burgers, and even mushroom sandwiches. These include shiitake and chanterelle mushrooms (Ali et al. 2022). Dried mushrooms like porcini and morel are highly prized in gourmet cooking and are often used in pasta, risotto, and sauces. Mushrooms continue to be a staple in kitchens today, contributing nutrition, taste, and versatility. They are part of a healthy, sustainable food system, offering affordable nutrition, along with the taste and health benefits of expensive gourmet dishes.

3 Useful Mycological Substances for Nutrition and Health

Edible mushrooms are very nutritious and have been regarded as more than just food; they were once considered to have medicinal qualities. The reason mushrooms are so beneficial is that they are low in fat and calories but rich in bioactive compounds like vitamins, phenolic acids, terpenoids, and polysaccharides. Eating mushrooms regularly has been linked to better immune and heart health and a lower risk of chronic diseases like diabetes, cancer, and neurodegenerative disorders. Research shows that mushrooms have therapeutic properties that help maintain health and prevent various diseases (Fernandez et al. 2020). Many bioactive molecules in mushrooms, such as reishi, shiitake, and maitake, help boost the immune system. Beta-glucan, a type of polysaccharide, helps protect the body from infections and cancer by stimulating immune cells like macrophages, dendritic cells, and natural killer (NK) cells. Because mushrooms naturally support immune function, they contain many nutrients that can be used to improve immunity against infections (Olsen and Smith 2018). Mushrooms like oyster and button mushrooms are low in fat and cholesterol and are linked to reduced blood pressure. They also contain compounds, such as ergosterol, that help control cholesterol and lipid levels (Lee et al. 2017). Mushrooms are high in cholesterol-lowering fiber and antioxidants, like ergothioneine and selenium, which help fight oxidative stress and free radicals, both of which are linked to heart disease (Martinez and Alvarez 2021). Eating mushrooms can help reduce the risk of heart attacks, strokes, and other cardiovascular issues by improving blood flow and preventing plaque buildup in arteries (Gonzalez et al. 2023). Certain bioactive compounds in mushrooms, such as those in turkey tail, shiitake, and maitake, have antitumor effects. Studies show that lentinan and polysaccharides from these mushrooms can help regulate the immune system, slow cancer growth, and promote cancer cell death (Sakamoto et al. 2022). Shiitake mushrooms, in particular, can help prevent liver, colon, and breast cancers by enhancing the immune system's ability to detect and destroy cancer cells. In the end, edible mushrooms are more than just a tasty and nutritious food—they also act as a powerful medicine. The combined effects of their bioactive compounds offer wide-ranging health benefits through immune support, heart health, cancer prevention, blood sugar regulation, and the protection of the nervous system.

4 Edible Fungi's Bioactive Potential: Antimicrobial and Antioxidant Qualities

In addition to their possible health benefits, fungi are also widely recognized for their strong antibacterial and antioxidant properties. Some of the active compounds found in mushrooms, such as proteins, terpenoids, phenols, and polysaccharides, are responsible for many of their therapeutic effects. These mushrooms help fight infections, balance oxidative stress, and promote overall health (Garcia et al. 2021). Here, we will explore the antibacterial activity of edible mushrooms.

Research has shown that some fungi can stop the growth of bacteria, fungi, and viruses. *Ganoderma lucidum* (reishi), *Trametes versicolor* (turkey tail), and *Lentinula edodes* (shiitake) are known for their strong antibacterial activity due to compounds like terpenoids, polysaccharides, and lectins. These substances stop the growth of harmful organisms by disrupting the membranes of diseased cells. *Ganoderma lucidum* has been found to have antibacterial and antiviral effects against bacteria such as *Staphylococcus aureus* and *Escherichia coli* (Martinez and Lopez 2019). Mushrooms can help in the prevention and treatment of infections, especially for people with weakened immune systems. According to Singh et al. (2020), they have antibacterial properties and can be used in medicine to control or treat infections, especially in immunocompromised individuals. Foods made from mushrooms have strong antioxidant activity, along with antibacterial properties, which help in neutralizing free radicals and reducing oxidative stress in the body. The removal of free radicals can help prevent chronic diseases like diabetes, heart disease, and neurodegenerative conditions such as Alzheimer's. Mushrooms are a good source of antioxidants, including flavonoids, phenolic acids, and ergothioneine. Ergothioneine is a unique antioxidant found in mushrooms, with high activity against free radicals. It has been reported that antioxidant activity helps reduce cell damage and the onset of chronic diseases, inflammation, and aging (Lopez and Ramirez 2022). These components help protect the body from oxidative damage and inflammation. Consuming shiitake mushrooms daily has been shown to significantly lower oxidative stress markers, reducing the loss of cell integrity and the risk of chronic diseases. Another mushroom that shows strong antioxidant and anti-inflammatory activity is *Grifola frondosa* (maitake). Ergothioneine, present in high concentrations in maitake mushrooms, protects the physical defenses and tissues from oxidative stress (Kim and Park 2018). Oyster mushrooms (*Pleurotus ostreatus*) also exhibit high antioxidant activity due to their rich content of flavonoids, phenolic acids, and vitamin C. These mushrooms help neutralize free radicals effectively. In addition to their antioxidant potential, oyster mushrooms can also help stimulate the body's antioxidant defense system by increasing the production of certain enzymes. They also have the ability to boost the immune system and provide therapeutic benefits against infectious and other diseases. Based on the current knowledge of food mushrooms and their antibacterial and antioxidant content, they are considered functional foods with specific health benefits.

5 Potential Immunomodulation and Sustainable Edible Fungi Cultivation

Food mushrooms are valued for their taste and health benefits, but they are especially known for their strong immune-boosting effects. These mushrooms contain various active compounds, like proteins, terpenoids, phenolic acids, and polysaccharides, which help support the immune system. Beta-glucans, found in many mushrooms, are of particular interest because they act as immune modulators. Beta-glucan can boost both the body's innate and adaptive immune responses. Studies show that beta-glucan works by activating key immune cells, such as dendritic cells, macrophages, and natural killer (NK) cells. It also helps regulate cytokine production and has immune-enhancing properties. Some common mushrooms with strong immune benefits include *Pleurotus ostreatus* (oyster mushroom), *Ganoderma lucidum* (reishi), Lentinula edodes (shiitake), and *Grifola frondosa* (maitake). Shiitake mushrooms are well known for their ability to support the immune system, and they contain antioxidants like ergothioneine and selenium, which help reduce oxidative stress. Reishi mushrooms (*Ganoderma lucidum*) are known for their effects on the immune system, as well as their anti-inflammatory and anticancer properties. They are rich in triterpenes and polysaccharides, which activate immune responses. The regular consumption of reishi can improve NK cell activity and increase interferon release, which is important for the immune system. Maitake mushrooms (*Grifola frondosa*) are nutrient-dense and help the immune system due to their beta-glucan content. These fungi support both natural and adaptive immunity, which provides protection against infections and cancer. As the global population grows and environmental challenges increase, sustainable agriculture becomes more important than ever. Traditional farming methods use a lot of water, land, and chemicals, leading to deforestation, water shortages, and greenhouse gas emissions. Food mushrooms are one example of products that support both ecological sustainability and human health. Edible mushrooms offer sustainable alternatives for meeting global food demands while addressing long-term challenges like natural resource depletion, climate change, and food insecurity. Mushroom cultivation requires fewer environmental resources and less capital compared to traditional farming methods, making it a promising solution to many of these issues.

6 Fungal Species Selective Breeding and Genome Editing

Genetic changes in mushroom cultures are a major way to improve biotechnology used in growing rice and edible mushrooms. Genetic modification is a dependable method because fungi have unstable genetics and keep changing over time. This makes it hard to use old breeding techniques for mushrooms. Scientists use recombinant deoxyribonucleic acid (DNA) technology to create fungi with better traits, like higher yields, faster growth, and stronger resistance to pests and diseases. For example, oyster mushroom (*Pleurotus ostreatus*) has been genetically improved to better handle environmental stress, such as temperature changes and lack of nutrients. Genetically modified organism (GMO) button mushrooms (*Agaricus bisporus*) are more nutritious because their antioxidant properties have been enhanced, particularly with substances like polyphenols and selenium (Singh et al. 2023). Genetic changes also improve mushrooms in terms of taste, texture, and shelf life, which is great for both farmers and consumers. Using too much pesticide is bad for human health and the environment, and traditional chemicals do not meet the long-term needs of sustainable agriculture. In some cases, molecular biotechnology has helped transfer disease resistance genes into mushroom breeding lines. For example, *Agaricus bisporus* has been made resistant to *Verticillium fungicola*, a deadly pathogen, by incorporating antifungal genes. Using chemical pesticides is not the way forward, and focusing on biotechnology helps make food safer while protecting the environment (Kim and Park 2021). Additionally, ribonucleic acid interference (RNAi) technology has been used to lower the risk of disease without killing the mushrooms, as it targets the genes that cause fungal infections (Rodriguez and Martinez 2022). Bioreactors give mushrooms the best growing conditions by controlling factors like moisture, temperature, light, and nutrient content. This method is especially useful for making medicinal mushrooms, though there is still a need to standardize their quality and effectiveness. Bioreactor cultures can grow all year round and in much larger amounts than traditional farming methods. For example,

Ganoderma lucidum, also known as reishi mushroom, is highly valued for its possible use in medicine. It is especially useful for its anti-inflammatory and immune-boosting effects, and growing it in bioreactors helps produce more of the active compounds, like triterpenoids and polysaccharides (Cheng et al. 2024). The fast-growing nutrition and functional food industry can benefit from large-scale, cost-effective production methods.

7 Enzymatic Biosynthesis and Biochemical Fermentation Processes

Fermentation techniques used in biotechnology can boost the medical and nutritional benefits of mushrooms. Some mushrooms can be fermented, which releases bioactive molecules that are not naturally found in them. These molecules are active peptides that can increase the effectiveness of antioxidant enzymes, secondary metabolites, and other beneficial substances in the fungi. For example, fermenting oyster mushrooms like *Pleurotus ostreatus* with probiotic microbes such as lactic acid bacteria can improve the availability of nutrients like vitamins, proteins, and amino acids, making them easier for the body to absorb (Zhang et al. 2022). Through enzyme technology, mushrooms can now be enriched with proteins, amino acids, polysaccharides, beta-glucans, and other bioactive compounds that enhance their nutritional value and immunomodulatory effects (Chen et al. 2021). These mushrooms, enhanced with enzymes, are widely used in traditional medicines, dietary supplements, and functional foods (Kim et al. 2023; Silva et al. 2020).

8 An Approach to Waste Management Using Recovery Circulation Management

In addition, biotechnology contributes to the promotion of bacteria in environmentally friendly activities. Previously, mushrooms were discovered to reduce lignocellulose substrates, primarily agricultural waste. Thanks to biotechnology support, researchers were able to develop funds to improve the potential of mushrooms in the decomposition of organic waste, such as wood chips, plastics, and agricultural residues. This activity on mushroom farms is not only environmentally friendly but also contributes to the treatment of agricultural waste. For example, we examined the degradation of cellulose, lignin, and other organic farm wastes. Because this fungus is likely to lead to biological degradation, biotechnology has been explored for the utilization of such agricultural waste in compost or biofuel production. These practices promote the circular economy and reduce the ecological destruction caused by waste decay (Zhang et al. 2022).

9 Conclusion

Mushrooms offer not just good nutrition but also beneficial antibacterial, antioxidant, and immune-boosting properties. Advances in agricultural biotechnology have led to the development of green and waste-free farming methods. Growing mushrooms can help address global food security and health concerns because they are rich in nutrients and have medical benefits. As more research is being done, mushrooms are becoming part of more food systems and agricultural practices, which makes them beneficial for both the environment and public health.

References

Ali R, Hussain M, Khan S (2022) Mushrooms as meat substitutes: nutritional and culinary perspectives. J Food Sci 57(3):145–160

Chen X, Li J, Zhao L (2021) Bioactive peptides and enzyme technology in fermented mushrooms: a review. Food Res Int 140:109866

Chen L, Zhao X, Wang J (2022) Antioxidant compounds in edible mushrooms and their role in aging-related diseases. Nutr Biochem 40(6):225–239

Cheng L, Zhao Q, Li W (2024) Bioreactor cultivation of Ganoderma lucidum: optimizing bioactive compound production for medicinal use. J Appl Mycol 30(1):45–58

Fernandez A, Martinez P, Ruiz L (2020) Health benefits and therapeutic potential of edible mushrooms. J Med Food 23(6):545–559

Garcia M, Singh R (2023) Sustainability and ecological functions of edible mushrooms. J Environ Agric 45(2):112–126

Garcia MA, Torres SR, Hernandez JL (2021) Bioactive compounds and therapeutic potential of edible mushrooms. J Nutr Biochem 87:108536

Gonzalez R, Torres M, Vargas A (2023) Cardiovascular protective properties of dietary mushrooms. Heart Circ J 12(3):98–110

Johnson A, Brown T, Lee C (2022) Energy and water efficiency in mushroom cultivation. Sustain Food Syst 37(1):55–70

Kim JH, Park S (2018) Ergothioneine content in Maitake mushrooms and its antioxidant properties. Food Sci Technol Int 24(7):569–576

Kim J, Park H (2021) Biotechnology advances in Agaricus bisporus resistance against fungal diseases. Appl Environ Microbiol 87(19):e01567–e01521

Kim H, Park J, Lee S (2019) Mineral content and health benefits of edible mushrooms. J Nutr Biochem 31(4):320–330

Kim D, Park S, Lee H (2023) Probiotic fermentation of edible mushrooms: effects on antioxidant enzymes and immunomodulatory activity. Appl Microbiol Biotechnol 107(3):1231–1242

Lee H, Kim J (2022) Nutritional benefits of edible mushrooms in human diet. Nutr Health 40(3):198–210

Lee S, Park J, Kim H (2017) Ergosterol and its role in cholesterol metabolism: a mushroom perspective. J Nutr Biochem 38:32–41

Lopez F, Ramirez C (2022) Role of antioxidants in preventing oxidative stress-related diseases: focus on mushrooms. Antioxid Redox Signal 36(3):245–258

Martinez D (2020) Culinary uses and flavor profiles of Pleurotus ostreatus. Int J Gastron 15(2):88–99

Martinez J, Alvarez C (2021) Antioxidant capacity of edible mushrooms: focus on ergothioneine and selenium. Food Res Int 147:110–121

Martinez D, Lopez E (2019) Antibacterial and antiviral effects of Ganoderma lucidum extracts. Int J Med Mushrooms 21(6):553–562
Nguyen T, Tran P, Hoang L (2018) Beta-glucans from mushrooms and their immunomodulatory effects. J Funct Foods 42:109–118
Olsen M, Smith T (2018) Immunomodulatory effects of beta-glucans from edible mushrooms. Int J Immunol 42(4):300–315
Patel S, Verma R, Gupta P (2024) Pharmacological activities of medicinal mushrooms. J Med Mycol 12(1):33–47
Rodriguez M, Lopez F (2021) Vitamin B complex in edible mushrooms: metabolism and health benefits. Food Chem 129(1):55–64
Rodriguez M, Martinez F (2022) RNA interference applications in controlling fungal pathogens of edible mushrooms. Mol Plant Pathol 23(4):608–620
Sakamoto Y, Tanaka H, Fujimoto K (2022) Mushroom polysaccharides in cancer therapy: Immunoregulation and apoptosis induction. Oncol Rep 47(5):1231–1242
Silva T, Mendes F, Gomes A (2020) Fermentation in functional food development: case study on mushrooms as dietary supplements. Crit Rev Food Sci Nutr 60(11):1828–1840
Singh N, Chatterjee S (2020) UV light-induced vitamin D synthesis in mushrooms and bone health implications. Nutrients 12(9):2721
Singh A, Verma R, Kumar S (2020) Immunomodulatory effects of medicinal mushrooms in infection control. J Clin Mycol 45(2):120–130
Singh R, Gupta P, Verma S (2023) Genetic improvement of Pleurotus ostreatus for stress tolerance and nutritional quality. Front Microbiol 14:1123–1135
Smith J, Roberts L, Wang X (2023) Antioxidant properties of edible mushrooms. Food Sci Technol 59(4):302–315
Wang Y, Chen Z, Liu Q (2023) Polysaccharides from mushrooms and their biomedical potential. Int J Biochem 58(2):77–93
Zhang L, Sun H, Wang Y (2022) Enhancement of nutritional and medicinal properties of Pleurotus ostreatus by lactic acid bacteria fermentation. J Food Sci Technol 59(8):3140–3150
Zhao Y (2021) Polysaccharides from shiitake and Maitake mushrooms: immunological and medicinal properties. Mycol Res 25(3):145–157

MIX
Papier aus verantwortungsvollen Quellen
Paper from responsible sources
FSC® C105338

If you have any concerns about our products,
you can contact us on
ProductSafety@springernature.com

In case Publisher is established outside the EU,
the EU authorized representative is:
Springer Nature Customer Service Center GmbH
Europaplatz 3, 69115 Heidelberg, Germany

Printed by Libri Plureos GmbH
in Hamburg, Germany